Science Education for Diversity

Cultural Studies of Science Education
Volume 8

Series Editors
KENNETH TOBIN, *City University of New York, USA*
CATHERINE MILNE, *New York University, USA*
CHRISTINA SIRY, *University of Luxembourg, Walferdange, Luxembourg*

The series is unique in focusing on the publication of scholarly works that employ social and cultural perspectives as foundations for research and other scholarly activities in the three fields implied in its title: science education, education, and social studies of science.

The aim of the series is to establish bridges to related fields, such as those concerned with the social studies of science, public understanding of science, science/technology and human values, or science and literacy. *Cultural Studies of Science Education*, the book series explicitly aims at establishing such bridges and at building new communities at the interface of currently distinct discourses. In this way, the current almost exclusive focus on science education on school learning would be expanded becoming instead a focus on science education as a cultural, cross-age, cross-class, and cross-disciplinary phenomenon.

The book series is conceived as a parallel to the journal *Cultural Studies of Science Education*, opening up avenues for publishing works that do not fit into the limited amount of space and topics that can be covered within the same text.

For further volumes:
http://www.springer.com/series/8286

Nasser Mansour • Rupert Wegerif

Editors

Science Education for Diversity

Theory and Practice

 Springer

Editors
Nasser Mansour
Graduate School of Education
University of Exeter
Devon, United Kingdom

Rupert Wegerif
University of Exeter
Devon, United Kingdom

ISSN 1879-7229   ISSN 1879-7237 (electronic)
ISBN 978-94-007-4562-9   ISBN 978-94-007-4563-6 (eBook)
DOI 10.1007/978-94-007-4563-6
Springer Dordrecht Heidelberg New York London

Library of Congress Control Number: 2013941042

Printed on acid-free paper

Springer is part of Springer Science+Business Media (www.springer.com)

Contents

Why Science Education for Diversity?

Nasser Mansour and Rupert Wegerif

Introduction

There are obvious ethical and political reasons for being concerned about the apparently unequal access to science education of students depending on their gender or cultural background or even their social identity. However, recently interest in how cultural and gender differences interact with science education has been driven by a pragmatic concern from many governments that the take up of science subjects is declining. There is now something of a consensus from governments around the world that effective science education is vital to economic success in the emerging knowledge age. It is also widely accepted that knowledge of science and scientific ways of thinking is essential to participation in democratic decision-making. However, in the developed world at least, we see a decreasing engagement of young people with science subjects at school and university. Some research suggests that issues of cultural diversity lie behind this looming crisis. The diversity at issue here is not simply the traditional diversity of gender and ethnicity, although that plays a part, but a new diversity of cultural identity positions flourishing amongst what might be called the facebook generation (Sjøberg and Schreiner 2005). Where young people use education to help construct an identity rather than only to help prepare for a career, their relationship to science education changes. The implication of this suggestion is that the authoritarian one-size fits all model of the traditional science curriculum is no longer appropriate in a context of increasing diversity wedded to increasing globalisation.

In this introductory chapter we offer a brief exploration of some of the tensions and dilemmas raised by the issue of diversity in science education. These point to a research agenda. This research agenda, we argue, needs to be global in order to

N. Mansour (✉) • R. Wegerif
Graduate School of Education, University of Exeter, St Luke's Campus, Heavitree Road,
EX1 2LU Exeter, Devon, UK
e-mail: n.mansour@exeter.ac.uk; r.b.wegerif@exeter.ac.uk

explore the interaction between science education and culture in different contexts. Finally we describe the sections and chapters of this book in terms of how they address this research agenda, the partial answers that they provide and the further research that they point to.

What Do We Mean by Diversity?

Sheets (2005) unpacks diversity by assigning the phenomenon a dual perspective which consists of predetermined and reversible characteristics. 'Diversity', she defines, 'refers to dissimilarities in traits, qualities, characteristics, beliefs, values, and mannerisms present in self and others'. It is displayed through (a) predetermined factors such as race, ethnicity, gender, age, ability, national origin, and sexual orientation; and (b) changeable features, such as 'citizenship, worldviews, language, schooling, religious beliefs, marital, parental, and socioeconomic status, and work experience' (p. 14). One can dispute whether some of the components Sheets places in (a) and (b) have universal acceptance, but in general the way in which diversity is used in the literature of science education reflects the dual nature and potential ambiguity between pre-determined features like skin colour and changeable features such as worldview that Sheers points out. In general the studies in this book, with two exceptions focussing on gender difference, suggest that the emphasis has shifted towards Sheets category (b) differences that stem from worldviews and identity issues. In this context Lemke (2001) argues that diversity and its needs are not matters of exceptionality and exotic and radical difference. Diversity in some degree is the condition of every community. However, our individual ways of living and making meaning are different according not only to which communities we have lived in, but also to which roles we chose or were assigned to by others.

Social and Historical Context

The current social and historical context of research on science education and diversity is summarised well in a 2008 paper on Science Education in Europe reporting findings that emerged from a Nuffield hosted set of two seminars involving leading science educators from nine European countries (Osborne and Dillon 2008). This report found that while science education across the developed world has many differences, there is a common trend in the decline of student attitudes to science. Data from the ROSE project (Sjøberg and Schreiner 2005) shows that students' response to the statement 'I like school science better than other subjects' is more likely to be negative the more developed the country is (see Fig. 1). Indeed, there is a 0.92 negative correlation between responses to this question and the UN Index of Human Development.

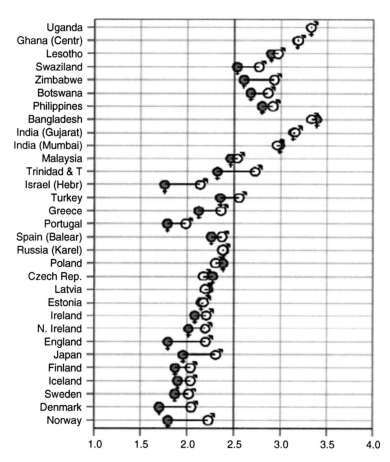

Fig. 1 Data from the ROSE study showing students responses to the question 'I like school science better than most other school subjects' (*1* – strongly disagree, *4* – strongly agree; *dark symbols* – female/*light* – male)

Most European countries have seen a reduction in the numbers of students choosing to pursue the study of physical sciences, engineering and mathematics at university with even more marked falls from 1993 in the number seeking to do a PhD in these areas in all European countries (OECD 2006).

Although the negative correlation with the level of a country's development makes these decreases in student's attitudes to school science appear as if almost inevitable, the many European science education specialists involved in the Nuffield study did not think this was the case but pointed instead to a mismatch between the experiences and expectations of young people in Europe and the current curricula and pedagogy of science education. In fact there is no shortage of interest in science but the topics that are of contemporary interest; global warming, the ethics of animal experimentation, cosmology, space exploration, medical advances and many more, seldom appear within school science.

While there is a great deal of variety in the structure of the school science curricula in Europe, with the teaching of different sciences at different times in different countries, nonetheless the Nuffield report concludes that almost all science curricula in Europe and beyond are essentially similar. That this is a more general finding is confirmed in some of the research of projects (e.g. BouJaoude's survey of science teaching in the Arab world, 2003; and Mahajan and Chunawala account of science education on health issues in India, 1999). The general pattern of science curricula is to start with teaching basic concepts that are returned to later in more depth. However, as a result of this curriculum, the experience of the students can be of a series of separate ideas lacking relevance to their concerns or any real-world context. The reason for what they are learning is seldom apparent to students. Assessment appears to focus too much on memory and recall of facts. There are few links made to the real life science issues that often dominate the news and touch the everyday reality of students. And finally there is an over-reliance on a pedagogy of transmission and copying (BouJaoude 2003; Osborne and Dillon 2008).

The negative correlation between interest in science education and development is not because science becomes less relevant as countries develop, in fact the opposite is true, but it might be because the lifestyles and choices and cultural diversity of young people increases in a way that, for many, becomes incompatible with their experience of school science.

The relationship between young people and science education is complexly mediated by culture. Helen Haste (2004) conducted a survey of the values and beliefs that 704, 11–21-year-old individuals held about science and technology and found four distinct groups, 'Greens' interested in environmental issues but with a specific agenda, 'Techno-investors' enthusiastic about the potential of science, the 'Science oriented' keen on science as a way of thinking and the 'Alienated from science' who were mostly young and female.

Similar cultural groupings are found in analysis of the ROSE data (Sjoberg 2005) which found five distinct groups of students with different values. These included a mainly male group fascinated by technology, and, mainly female group who just wanted to work with others and develop themselves as people.

We would agree with Osborne and Dillon's conclusion from the evidence that science curricula have failed to respond to the changing needs of children and young people and that what is needed is 'a new vision of why an education in science matters that is widely shared by teachers, schools and society'.

Cultural Diversity and Science Education

There is a body of literature that deals with students' orientation to science and pursuing scientific careers in relation to science education (for reviews see Osborne 2003, 2007). This literature generally points to single factors that influence this orientation, such as cultural background, religious beliefs, and gender. Some of these studies, too, point to the inherent culturally biased nature of science education as a reason why many young people from ethnic minorities decide to opt out of science education and hence scientific careers (reviewed by Hodson 1993). However, such studies do not clarify in detail the mechanisms at work that influence students' orientations towards science and scientific career choices. This more detailed work on the mechanisms of attraction to science and exclusion from science is the contribution made by some of the studies in this book. (See chapters in this volume by Xiufeng Liu; Lynn Dierking; Sibel Erduran and Siu Ling Wong.)

Sociocultural and Dialogic Perspectives

Most of the studies in this book take a sociocultural perspective to science and science education. From a sociocultural perspective, science is not just about content (i.e., making sense of the world), it is also about context (i.e., relationships with people and with oneself, and relationships between events) and is therefore heavily loaded with cultural connotations related to values, beliefs and emotions, involving both the social and individual aspects of the learner. Considering the context of both science and learners will lead to a much better way of empowering students in a culturally sensitive approach to science teaching which is to explicitly address the historical, philosophical, and sociocultural dimensions of science. Instead of avoiding epistemological distinctions between different ways of knowing, we should teach students about the cultural background, epistemological assumptions, and methodological procedures of developing scientific concepts, theories, and models (El-Hani and Mortimer 2007).

Not only is it important to understand the underlying ideas children have while teaching them in a science classroom – it is also important to pay attention to culture and context, and explore where ideas and perceptions come from (Aikenhead 2001; Barton and Osborne 2001; Lee 2001; Roth and Barton 2004). From this sociocultural perspective of science education, scientific knowledge can be seen as a meaning system in which scientific words have meaning not in themselves but in relation to social settings in science as a whole (Halliday and Martin 1993; Lemke 1990, 1994). Such a meaning system is not built through step by step logic but rather is a particular way of viewing and dividing up the world, based on particular beliefs, values and practices. This contrasts with the way science is often taught in schools as though it is a set of facts and definitions which present stand-alone, self-evident truths, with meaning not being dependent on or relative to understanding a much broader system of ideas, beliefs, values and social practices (Hanrahan 2002).

In the sociocultural view, what matters to learning and doing science is primarily the socially learned cultural traditions of what kinds of discourses and representations are useful and how to use them, far more than whatever brain mechanisms may be active while we are doing so (Lemke 2001). Kuhn's (1962) and other scholars view the modern sciences as embedded in particular social, historical, political, and cultural contexts (Harding 1994; Keller 1992). Kuhn (1962) used detailed, historical examples to portray science as a social activity, as a set of practices and a body of knowledge created by a community of people called scientists.

A growing number of educators have argued that such sociocultural perspectives must inform descriptions of science if teachers are to interest and engage students from underrepresented ethnic groups usually positioned on the margins of the science classroom (Stanley and Brickhouse 1994, 2001; Bianchini and Solomon 2002; Mansour 2013). Sociocultural perspectives include the social-interactional, the organizational, and the sociological; the social-developmental, the biographical, and the historical; the linguistic, the semiotic, and the cultural. For many researchers they also include the political, the legal, and the economic, either separately or as implicit in one of the others (Lemke 2001, p. 297).

A dialogic perspective on science education is emerging within the broad socio-cultural approach suggested by Kuhn and outlined in more detail by Lemke for the context of science education. Longino (1993) for example extended Kuhn's notion of science as a communal activity to argue that there are always multiple voices and diverse perspectives involved in dialogues that decide on the legitimacy of scientific claims. Science moves forward by critical dialogue, Longino explained, and this requires the presence and expression of alternative points of view. This view of science as very much a social rather than an individual product highlights the need for a diverse membership of scientific communities. Interestingly it also suggests that it is quite useful to have multiple, sometimes incompatible, theories, each responding to different local community standards. The influence of this dialogic turn can be seen in the chapters by Roth and by Wegerif et al.

Voice is a key dialogic concept that arises in a number of studies. For students with little experiences in academic communities, the struggle to develop an effective voice though which to 'speak' the discourse, whether in writing or in class, can be long and difficult. Yet, until they do, their grades suffer, since their progress can only be registered through speaking the discourse. Support in establishing voice is a vital component of courses for students from diverse backgrounds. (For further discussion of voice see, e.g. Clark and Ivanic. 1997). Lemke (2001) raises the intellectual and social consequences of ignoring people's identities and voices when working on conceptual change models in science education. He wrote:

> An apparent assumption of conceptual change perspectives in science education is that people can simply change their views on one topic or in one scientific domain, without the need to change anything else about their lives or their identities. This modularism runs contrary to the experience of sociocultural research. Let me give a simple but telling example: the evolutionist creationist controversy. To adopt an evolutionist view of human origins is not, for a creationist, just a matter of changing your mind about the facts, or about what constitutes an economical and rational explanation of the facts. It would mean changing a core element of your identity as a Bible-believing (fundamentalist) Christian. It would mean breaking an essential bond with your community (and with your god). It could lead to social ostracism and the ruin of your business or job prospects. It could complicate your family life or your marriage chances (p. 301)

Lemke also explained the impracticality of a conceptual change model in a classroom that aims simply to change students' mind from A to B. From a sociocultural perspective, Lemke argues that the practical reality is that we are dependent on one another for our survival, and all cultures reflect this fact by making the viability of beliefs contingent on their consequences for the community. This is no different in fact within the scientific research community than it is anywhere else. It is another falsification of science to pretend to students that anyone can or should live by extreme rationalist principles. It is often unrealistic even to pretend that classrooms themselves are closed communities, which are free to change their collective minds. Students and teachers need to understand how science and science education are always a part of larger communities and their cultures, including the sense in which they take sides in social and cultural conflicts that extend far beyond the classroom.

Tensions and Dilemmas

One tension that emerges in the research is that between cultural groupings and individual differences. Grotzer (1996) argues that while paying attention to culture is important, students need to be treated first as individuals who are influenced by the contributions of their culture, before treating them as part of a larger stereotyped cultural group.

Another version of this same tension can be seen in the contrast between the study of culturally situated indigenous sciences and the study of the cultural diversity of individuals. Carter (2004) claims that the literature on cultural diversity in science displays a number of related tendencies that seem to draw together into two main positions: one focused on the identities/subjectivities of those learning science, that is, the culturally and linguistically diverse students themselves, and the second, on considerations of science as culturally located, Western and non-Western knowledge, frequently identified as multicultural approaches to science.

The socio-cultural turn in science education raises the question of how we understand science, whether we accept its ideology of decontextualised knowledge or locate knowledge in the context of cultural practices and interests. This turns out to be a key question for science education for diversity since it is the authoritarian voice of science as decontextualised truth that many of our authors claim is alienating students from different backgrounds.

Another issue that is raised by applying dialogic theory is how we conceptualise diversity. Previously most research on diversity in education has focused on externally definable groups that are visibly different, but the dialogic concept of voice suggests a more fluid definition of cultural difference as more like a distinct voice in a dialogue.

Finally all the studies in this volume raise, implicitly or explicitly, the question of who science education is for. All students require enough scientific literacy to cope in the modern world but most will not be actively engaged in science as a career. On the other hand, science education in schools needs to engage and inspire the minority of students who will become career scientists. Is science education for the minority of students who may wish to become scientists or for all the students who need to understand science? These two different aims imply two different ways of teaching science that are not easy to unite into one science curriculum.

Argument and Structure of This Book

By exploring the cultural factors that influence engagement and non-engagement in science education around the world, the chapters collected in this book shed light on these various tensions and dilemmas. What emerges is that the issue of cultural diversity may be central to the reason why science education often fails to connect and to the ways in which science education could respond more effectively to the current challenge.

A theme that emerges strongly through these varied studies is the way in which science education around the world is experienced as authoritarian and unfriendly to diversity. Many chapters argue for the need to reform science education. The direction that emerges is one of reform that places openness to diversity at the heart of science education. This approach goes beyond trying to respond to specific issues of diversity such as addressing gender differences or the needs of particular ethnic groups but suggests that a new understanding of diversity is required in science education, diversity as a way of thinking about science and about education into science, which does not define specific diversities in advance of practice but embraces openness, responsiveness and responsibility in the nature of the practice of science education itself. Here the application of socio-cultural theory informed by the insights of Bakhtin's dialogism helps to frame an emerging new approach to developing a science education curriculum for diversity which could respond to the challenges of the emerging global knowledge society.

This book explores the issue of science education for diversity from a range of perspectives. All the chapters take a broadly socio-cultural theoretical approach to investigate the contexts of science education. The scope is international drawing on experiences from all the major continents.

Diversity is our main theme and several chapters discuss how we can conceptualise diversity. As well as the more traditional approach to diversity focussing on externally evident differences of gender and ethnicity, we explore the more subtle impact of the diversity of constructions of personal identities and the ways in which these identities interact with existing science education. As with Nasser Mansour's study of the impact of personal religion on science teacher's practice in the classroom the approach taken is socio-cultural in assuming that beliefs which impact on actions are mediated by cultural constructs. The chapters of this book all explore how different cultural constructs influence both the teaching of science and the learning of science in a variety of contexts.

The book is arranged in four parts. Part I, 'Science Education Reform for Diversity' introduces new approaches to science education that challenge existing thinking about science education and diversity and seek to define new ways forward.

Wegerif, Postlethwaite, Skinner, Mansour, Morgan and Hetherington argue that a dialogic understanding of the nature of science should lead to a dialogic approach to science education, which is more open to engagement with different cultural voices. It proposes a new approach to understanding diversity as 'voices' within a dialogue defined from within rather than as based on features defined from outside. In the next chapter Liu makes a similar sort of argument about scientific literacy. Scientific literacy is not a definable set of skills and competencies because it cannot be defined from outside of the issues that lead to interest in science. Liu proposes that we expand the notion of scientific literacy into the larger and more relevant concept of scientific engagement and instead of teaching scientific literacy in the abstract we promote engagement in the real issues that concern students. Roth explores the subtle impact of the diversity of constructions of personal identities across different contexts and the ways in which these identities interact with existing science education. The point that emerges from his ten ethnographic studies is that the impact of cultural diversity is not a simple causal notion but requires an analysis of the whole of life. In a way all of the chapters in Part I are arguing for the importance of taking cultural context more seriously into account through engaging with participants' points of views rather than equating understanding with categories imposed from outside. Van Eijck adds to this theme with a brief introduction to cultural studies of science education and their relevance for the teaching and learning of science in general and diversity issues in particular. His chapter reflects on the current state of this research field in Europe and the reasons why cultural studies have not had more impact on science education research before now.

The chapters in Part II, 'From Learning to Pedagogy', explore diverse experiences of learning science within interrelated historical, cultural, institutional, and communicative contexts. As with Part I, the approach is broadly sociocultural, describing learners embedded within and constituted by a matrix of social relationships and processes. Parker and Krockover review the research literature for learning science in informal settings, and offer some valuable examples and conclude with the impact of science education in informal learning in relation to cultural and diversity issues. Taylor, Taylor and Chow focus on science education for sustainability, drawing on results of two case studies that show how ethical dilemma story teaching can be made compelling for students. They argue that this approach supports students in finding their *voice* as critically aware decision-makers and prospective leaders in the Knowledge Age. Max, Ziegler, and Kracheel presents research in early childhood classrooms in Luxembourg, a European country with a complex multilingual situation. They shed light on the discourse practices of 6- to 12-year-old children and examine the co-construction of the children's growing understanding of science in collaborative inquiries. Arguing from a context-sensitive perspective, this chapter approaches the learning of science as an interactional achievement in situ, one that encompasses the enactment of science as shared discourse and therefore as a cultural accomplishment. Taber offers an analysis of the relationship between science education and religion arguing that science education should explicitly represent the diversity of views within the scientific community on whether, and if so how, science and religion are related and should make a clear distinction between metaphysical commitments and frameworks for understanding that are part of scientific discourse. Erduran and Wong offer case studies in the UK and Hong Kong that have implications for policy by showing the effective implementation of scientific literacy in school science. In this chapter, they draw on classroom-based research projects such as the *Mind the Gap* and *S-TEAM* Projects in England and the *Learning Science* series of research and teacher development projects in Hong Kong in order to highlight the different senses of diversity found in play within science education.

Part III consists of studies of science teacher education programs looking at the dimension of cultural diversity. Each of these studies has implications for the nature and goals of science education in general. Mansour reports research on the sociocultural contexts in which Egyptian science teachers are embedded and the ways in which these contexts help in understanding Egyptian science teachers' pedagogical beliefs and practices. His chapter presents an empirical case study exploring the interplay between teachers' personal religious beliefs (as a case for their cultural beliefs) and their pedagogical beliefs and practices. Thompson and Tippins looked at the needs for preparing pre-service science teachers to teach climate change and argue through this that we need a new more interdisciplinary and holistic approach to teacher education. This chapter raises the daunting challenges facing science teachers and provides a useful perspective on what science teachers need to know in preparation for teaching in their twenty-first-century classrooms. Ramos and Espinet offer a detailed study of a classroom integrating science and language education and find evidence that multilingualism in science classrooms can be a rich context for science learning. They focus on ways in which students expand their agency in the use of multimodal resources as a way of overcoming the difficulties derived from the need to construct a scientific explanation of natural phenomena using English as a foreign language.

Part IV ends the book with various examples of how we can respond to cultural issues (e.g. religion, gender and language) in science classroom. Dierking explores the growing potential of 'free-choice' informal science education outside of school contexts offering evidence that this can be effective and arguing that it should be part of any reforms of science education that are intended to address issues of diversity. Scantlebury, Hussénius, Andersson and Gullberg provide a feminist critique of science education, using examples from an ongoing feminist research project in teacher education of how gender theory and feminist perspectives could help to generate new knowledge about gender and science education. Reiss explores the vexed and topical issue of when and how science educators need to deal with question of religious belief in their classroom practice. He examines two possible circumstances where one might wish matters of religion to be included within the teaching of science: when teaching about the nature of science and when teaching about evolution. Billingsley continues the theme raised by Reiss, drawing on studies of young people's thinking about science and religion which find that social constraints in science classrooms can mask what pupils think from teachers in a way that is not helpful to their engagement in science education. She argues that action is required to open spaces in which these issues can be raised. Finally, BouJaoude and Gholam offer an overview of the status of science education in the Arab states with a focus on the status of women and the socio-cultural factors that constrain their ability to go beyond a certain stage in development and role in society. They explore factors that have been shown to influence student achievement in general, and girls' achievement more specifically. They conclude that, despite many efforts to promote women in science, science and technology careers in Arab states are still male-dominated. The lack of attention to socio-cultural factors influencing the choices that females make is proposed as one possible reason why interventions to change this situation have had limited impact.

References

Bianchini, J., & Solomon, E. (2003). Constructing views of science tied to issues of equity and diversity: A study of beginning science teachers. *Journal of Research in Science Teaching, 40*, 53–76.

BouJaoude, S. (2003). *Achievement levels of new school project students in grades 1 to 5 in Arabic, math, and science*. Project funded by the Education Development Center Inc., Newton, MA, USA.

Carter, L. (2004). Thinking differently about cultural diversity: Using postcolonial theory to (re)read science education. *Science Education, 88*(6), 819–836.

Clark, R., & Ivanic, R. (1997). *The politics of writing*. London: Routledge.

El-Hani, C.-N., & Mortimer, E. F. (2007). Multicultural education, pragmatism, and the goals of science teaching. *Cultural Studies of Science Education, 2*, 657–702.

Grotzer, T. A. (1996). *Math/Science matters: Issues that impact equitable opportunities for all math and science Learners*. Cambridge: Harvard Project on Schooling and Children, Exxon Education Foundation. Essay #1: Teaching to Diversity: Math and Science Learning for All Children.

Halliday, M. A. K., & Martin, J. R. (1993). *Writing science: Literacy and discursive power*. London: Falmer Press.

Hanrahan, M. (2002, July 11–14). *Learning science: Sociocultural dimensions of intellectual engagement*. Paper prepared for the ASERA Conference, Townsville, Queensland, Australia.

Harding, S. (1994). Is science multicultural? *Configurations, 2*, 301–330.

Haste, H. (2004). *Science in my future: A study of values and beliefs in relation to science and technology amongst 11–21 year olds*. London: Nestle Social Research Programme.

Hodson, D. (1993). In search of a rationale for multicultural science education. *Science Education, 77*, 685–711.

Kuhn, T. S. (1962). *The structure of scientific revolutions*. Chicago: University of Chicago Press.

Lemke, J. L. (1990). *Talking science: Language, learning, and values*. Norwood: Ablex.

Lemke, J. L. (1994). *The missing context in science education: Science*. Paper presented at American Educational Research Association annual meeting, Atlanta, GA, April 1992. Arlington, VA (ERIC Document Reproduction Service No. ED 363511).

Lemke, J. L. (2001). Articulating communities: Sociocultural perspectives on science education. *Journal of Research in Science Teaching, 38*(3), 296–316.

Longino, H. (1993). Subject, power and knowledge: Description and prescription in feminist philosophies of science. In L. Alcoff & E. Potter (Eds.), *Feminist epistemologies* (pp. 101–120). New York: Routledge.

Mahajan, B. S., & Chunawala, S. (1999). Indian secondary students' understanding of different aspects of health. *International Journal of Science Education, 21*(11), 1155–1168.

Mansour, N. (2013). Modelling the sociocultural contexts of science education: The Teachers' perspective. *Research in Science Education, 43,* 347–369. doi:10.1007/s11165-011-9269-7.

OECD. (2006). *Evolution of student interest in science and technology studies policy report.* Paris: OECD.

Osborne, J. (2003). Attitudes towards science: A review of the literature and its implications. *International Journal of Science Education, 25,* 1049–1079.

Osborne, J. (2007, May 28–29). Engaging young people with science: Thoughts about future direction of science education. In C. Linder, L. Östman, & P. Wickman (Eds.), *Promoting scientific literacy: Science education research in transaction* (pp. 105–112). Proceedings of Linnaeus Tercentenary Symposium held at Uppsala University, Uppsala, Sweden.

Osborne, J., & Dillon, J. (2008). *Science education in Europe: Critical reflections.* London: Nuffield Foundation.

Sheets, R. H. (2005). *Diversity pedagogy: Examining the role of culture in the teaching-learning process.* Boston: Allyn and Bacon.

Sjøberg, S., & Schreiner, C. (2005). How do learners in different cultures relate to science and technology? *Asia-Pacific Forum on Science Learning and Teaching, 6*(2), 1–17.

Stanley, W. B., & Brickhouse, N. W. (1994). Multiculturalism, universalism, and science education. *Science Education, 87,* 387–398.

Stanley, W. B., & Brickhouse, N. W. (2001). Teaching sciences: The multicultural question revisited. *Science Education, 85,* 35–49.

Part I
Science Education Reform for Diversity

This part introduces approaches to science education that seek to reform thinking and learning about diversity and value new possibilities in changes and differences which might have a pivotal role in shaping the future.

Dialogic Science Education for Diversity

Rupert Wegerif, Keith Postlethwaite, Nigel Skinner, Nasser Mansour, Alun Morgan, and Lindsay Hetherington

Introduction: The Science Education for Diversity Project

The current crisis in science education was discussed in the Introduction to this book (Mansour and Wegerif 2013). In Europe there has been a decline in the number of young people who are interested in pursuing science topics for further education. This has led to concern from the European Commission (EC), expressed in several reports such as 'Europe needs more scientists' (EC 2004) that there will not be enough scientists to make the discoveries leading to new products on which it is assumed the knowledge economy will depend. But there is also concern that citizens with insufficient knowledge of science will not be able to participate rationally in democratic decision-making about the increasing number of controversial issues that require some scientific literacy, including issues associated with nuclear power, global warming and vaccination. The evidence suggests that young people in the 'facebook generation' choose school topics to support their developing sense of personal identity and in this context most find science education unappealing or not as appealing as other subjects (Sjøberg and Schreiner 2007; Osborne and Dillon 2008). This challenging situation has led to considerable investment in research on how to teach science in a more engaging way. One such project, the one million Euro 'science education for diversity' (SED), takes the innovative approach of seeking to learn from the experience of countries where science education remains a highly popular choice amongst young people.

The SED project focuses on issues of diversity, including cultural diversity and gender diversity, and seeks to explore how science education interacts with diverse populations and how it could be redesigned to respond better to the challenges that diversity raises. The partners include science education researchers

R. Wegerif • K. Postlethwaite • N. Skinner, B.Sc., Ph.D., PGCE • N. Mansour (✉) •
A. Morgan • L. Hetherington
Graduate School of Education, University of Exeter, Exeter, UK
e-mail: n.mansour@ex.ac.uk

N. Mansour and R. Wegerif (eds.), *Science Education for Diversity: Theory and Practice*,
Cultural Studies of Science Education 8, DOI 10.1007/978-94-007-4563-6_1,
© Springer Science+Business Media Dordrecht 2013

in Malaysia, India, Lebanon, Turkey, the Netherlands and the UK (see the list in Acknowledgements below).

In each country, we conducted a literature review of the science curriculum and initiatives that addressed issues of diversity in science education (SED WP2 2011). We then worked with ten schools in each country, five primary (focusing on ages 10–12) and five secondary (focusing on ages 12–15), thus addressing the point raised by previous research that the key window for engagement or disengagement with science appears to be from age 10 to 14 (Osborne and Dillon 2008). Questionnaires were developed addressing attitudes to science and understandings of science. Within each primary school 50–100 students aged between 10 and 12 completed questionnaires, and within each secondary school around 200 students aged between 12 and 15 completed questionnaires. The project team in each country selected schools that represented the mix of locations including urban, rural and suburban communities and, where relevant, the main religions of the country. We focused on state schools accessible to students of all backgrounds and looked for mixed gender schools or a balance of boy-only schools with girl-only schools. Where possible we sought out schools that have cultural diversity and students from a diversity of socio-economic backgrounds.

We selected four schools in each partner country for further study, and within each of these schools, a number of students were interviewed and recorded participating in focus groups. All the science teachers in each of these four schools were also interviewed (where possible) alongside other key member of staff involved in deciding on the science curriculum such as the head teacher.

The data from the questionnaires and the interviews were then analysed and synthesised as far as possible in a discussion between all the partners to produce a framework for the design of science education that could address the issue of diversity. In this chapter, we do not present the findings of our survey in detail. Some of the analysis is continuing and will form the basis of other publications. Instead we focus on one aspect of our research, the understanding of the nature of science, and how the initial analysis of the questionnaires and workshops with teachers helped us to reconceptualise science education for diversity.

Thinking about the nature of science in relation to science education is the focus of the first part of this chapter. The second half of this chapter describes and reports on the development of a framework for the design of effective science education for diversity. The development of this framework will form the basis for the next stage of the project, which focuses on designing, implementing and evaluating interventions based on the framework.

What Is Science?

Despite the relative uniformity of science education traditions in all six partner countries, questionnaires and interviews revealed that students' understanding of the nature of science differed greatly between the countries. We explored the question of what is science with students by asking them to identify what sort of things should be called science. Answers to this question revealed that in India, Malaysia, Lebanon and Turkey, practical aspects of science (including farming and building a bridge) were more likely to be included in a definition of science than in the Netherlands and the UK (SED WP3 2011). This finding is consistent with research that suggests that the understanding of science in Japan and in Korea also includes these more practical aspects in a way that is less common in the west (Kang et al. 2005; Kawasaki 2004).

In the Netherlands social science was mostly included within the concept of science, but this was not the case for most students in the UK. This difference may be due to the different normal usage of the Dutch term 'Wetenschap' which is used as a translation for the English term science.

Interviews with individual children in the UK revealed some quite narrow images of science, which were, as one might expect, closely connected to their attitudes towards science and their attitudes towards careers in science. One image conjured in several interviews in the UK was of a man in a white coat in a lab mixing chemicals or inventing things. There was often some reference to large machines and/or electronics when the young people interviewed were asked why some subjects were science and others not. The narrow images of science thrown up in the interviews suggest that, in order to address the issue of how to improve science education so that it better responds to the needs of diverse audiences, it is necessary to raise once again the controversial issue of what is understood by the word 'science'.

Alters (1997) surveyed the members of the US Philosophy of Science Association and found 11 distinct positions on the nature of science. He concluded that there is no shared ground to serve as a basis for teaching the nature of science in science education. However, this claim was immediately disputed (Smith et al. 1997). Another survey using the Delphi method, and including not only philosophers but also leading scientists, science educators and communicators about science, found considerably more consensus (Osborne et al. 2003). Using a measure of 66 % agreement as representative of consensus, Osborne et al. found consensus on nine themes to be taught as part of the nature of science. According to a review of a number of existing standards in science education for the nature of science within the USA, Canada, England and Wales and Australia, most of these themes are already addressed by science education curricula (McComas and Olson 1998). The eight areas of overlap between the two reviews were, using the more normative formulation of McComas and Olson:

1. Scientific knowledge is tentative.
2. Science relies on empirical evidence.
3. Scientists require replicability and truthful reporting.
4. Science is an attempt to explain phenomena.
5. Scientists are creative.
6. Science is part of social tradition.
7. Science has played an important role in technology.
8. Scientific ideas have been affected by their social and historical milieu.

However, the one area where Osborne et al. found a discrepancy between existing standards and their Delphi review of the experts is a crucial one. This is the theme they labelled 'Diversity of Scientific Thinking' which refers to the growing consensus within philosophy of science that there is no single 'scientific method' but many methods appropriate for different areas and different problems. Osborne et al. illustrate the importance of this theme by pointing out two areas where the UK science curriculum fails to include reference to important methods within the different sciences. Firstly, the distinction between historical reconstruction and empirical testing: as Rudolph (2000) points out, historical reconstructions such as the phylogenetic history of various species or records of climate change might use models to help them at times, but they are not really about testing models but more essentially about establishing correct chronologies. To give a well known example that perhaps illustrates Rudolph's point: finding out exactly when *Tyrannosaurus rex* became extinct is of scientific interest even if this extinction cannot be predicted by a model because it was due to contingent factors. Secondly, Osborne et al. claim the correlational methods common to many media reports of science and basic to medical science are absent from the curriculum, perhaps, they speculate, because school science in the UK focuses only on the three large natural sciences: biology, chemistry and physics.

This issue of the unity or diversity of scientific methods is obviously relevant to science education, but it has been even more central to debates in the philosophy of science. To help explicate this issue, the Exeter research team organised a seminar with Professor John Dupré, a leading philosopher of science working at the University of Exeter where he heads up the Egenis Centre for Research in Genomics in Society. The argument that follows is influenced by John Dupré's work (e.g. Dupré 1993, 2001). There is now near consensus in the philosophy of science community (certainly considerably more than 66 % agreement!) that there is no single scientific method but a variety of methods and practices used for different purposes in different contexts. While there is no easy or simple way to demarcate science from nonscience, established sciences often share a number of epistemological criteria which some claim can be used to distinguish them from nonsciences. Examples of these epistemological criteria are:

- The use of empirical evidence
- Consistency with known facts and theories
- Elegance and simplicity
- The power to generate useful implications
- Testability, i.e. that they could be proved wrong by the right observation

However, it is possible to (a) find established and respected areas of science that violate each of these criteria and (b) find areas of knowledge not normally called science that meet each of them. For example, taxonomies in biology and elsewhere are based to some extent on empirical evidence and can be very useful, but they are not testable and no single taxonomy is likely to be consistent with all relevant known facts and theories (Dupré 1993, p. 18). Much work in theoretical physics has little relation to empirical observations, for example, the Penrose-Hawkins singularity theorems relating to black holes have already been celebrated as a breakthrough in physics yet have no supporting empirical evidence to our knowledge.

Many more such examples of accepted and effective methods of knowledge generation that do not fit any unified account of 'scientific method' could be found if we were to consider the full range of practices found in all the established natural sciences from astronomy through zoology. Therefore empirical evidence from the sociology of science, i.e. looking at what scientists actually do, undermines the claim that there is a single 'scientific method'. Despite these examples many science educators still hold to the view that science can be distinguished from non-science through epistemological criteria. A Google search on 'scientific method', for example, finds over 7 million hits, and first 100 hits thrown up consist mainly of unself-critical accounts of exactly what the scientific method is and how to teach it, often accompanied by a flow-chart diagram.

Of course, some methods are better than others for answering particular types of questions in particular areas. The experimental method, which involves building a model and testing its predicted consequences on changes made in one variable while holding all other variables the same, has proven particularly effective in many contexts within the physical sciences. However, as we have noted looking at taxonomies and historical reconstructions, even this very vague account of scientific method is not universally applicable. What counts as a model and what counts as a valid observation are very much subject to debate in different areas of inquiry. If, in reality, there is no foolproof method, we can apply to find the truth, then in every case we need to resort to dialogue within communities to justify ourselves, which means that we need to be creative and flexible and open to alternative perspectives.

The implication of arguments from Richard Rorty (1991) and Jurgen Habermas (1984) is that communicative virtues, such as honesty, trust, relying on persuasion rather than force and respect for the opinions of others, led to more effective knowledge construction in some areas of inquiry. In the process, these virtues become institutionalised in cultural practices, such as the transparent publication of all methods, meetings where all have the right to challenge views and a blind peer review procedure to avoid the influence of status on the criticism of ideas. In this way, the social ground rules of scientific institutions and scientific communities have been, to some extent, designed to encourage criticisms and the considerations of alternative views and to try and prevent the imposition of views through manipulative or coercive means (Habermas 1984; Rorty 1991). If this reconstruction of the logic at work in the history of the development of science is true, then it seems that the success of some sciences in generating consensus behind their claims to knowledge may be more to do with the quality of their dialogues than with the power of any specific methods they advocate.

The significance of these arguments against there being any unique scientific method is not to undermine science, but it is to return science to the larger human dialogue within which we try collectively to make sense of our situation. Some methods seem to work to solve some problems for some periods of time, but they are always open to question and that questioning returns them to dialogue.

The court that decides if a method is plausible or not is the relevant community in dialogue together. The decision about whether or not a new method is scientific cannot, by definition, be made according to any pre-existing rigorous scientific method but requires the reaching of consensus within a community. The success of argumentation in achieving consensus implies the need for communicative virtues such as intellectual integrity and respect for the views of others, within a community, as well as institutional procedures for reaching and maintaining consensus.

Monologic, Dialogic and Diversity

The issue of how we treat the variety of methods in science points to a larger issue. For Ayer and other logical positivists, the essential distinction was between 'sense' and 'nonsense'. By 'sense' Ayer meant claims that could in theory be grounded on empirical observations and/or logic, and by 'nonsense' he meant everything else. For Popper the distinction was between science, which produced claims that could be falsified, and pseudoscience, which could never be tested. These attempts to draw a boundary around science have failed to convince (Gillies 1998). The more fundamental distinction that these attempts reveal impacts on how science education deals with diversity. This is the distinction between monologic and dialogic.

Science and scientists have a long tradition of aspiring to monologic. This, as the name suggests, is the ideal of the single voice, the one true perspective outside of any dialogue. The dialogic alternative that Bakhtin (1984) articulated is that truth is not found in a single utterance but always in a dialogue. Different positions held together in a dialogue do not take away from the truth; they enable truth: not truth as a proposition but what Bakhtin refers to as 'polyphonic truth', truth in action which is found through and across a number of different voices (Bakhtin 1984). Bakhtin was not referring to the truism that there can be many different but compatible perspectives on the same object but to the more radical idea that meaning takes place as an event only in the gap opened up by different perspectives in dialogue. Facts are, he pointed out, answers to questions, and those questions are forged within dialogues (Bakhtin 1986, p. 114). Bakhtin defined dialogue as shared inquiry in which answers give rise to further questions (Bakhtin 1986, p. 168). Since our dialogues develop and change over time, our questions also change and so the facts we find in response to those questions change or even dissolve as the dialogue moves on.

In practice, science is dialogic but its monologic image, which is often a self-image, remains hard to shift and is reinforced by much science education. The intransigence of this monologic image is significant for addressing the issue of diversity within science education. Where there is a diversity of views, a dialogic approach to education suggests the need for engagement and the need for a greater focus on the quality of dialogue. A diversity of perspectives gives meaning and is an opportunity to teach science as shared inquiry and to explore not only the alternative voices but also how, if at all, consensus can be built in answer to some specific questions. Bakhtin relates monologic and dialogic to the difference between an authoritative voice and a persuasive voice. The authoritative voice remains outside of me and orders me to do something in a way that forces me to accept or reject it without engaging with it, whereas the words of the persuasive voice enter into the realm of my own words and change them from within (Bakhtin 1981, p. 343). This distinction gets to the heart of the approach needed to engage young people in science in the context of cultural diversity (see Roth 2009 for similar arguments).

How Do We Conceptualise Diversity?

The evidence we gathered from questionnaires and interviews in the science education for diversity (SED) project confirms the findings of the earlier Relevance of Science Education (ROSE) project that young people in more developed countries have less interest in pursuing science as a career than those in developing countries (Sjøberg and Schreiner 2007). The key problem is expressed well by Osborne and Dillon:

> ... one of the issues behind the decrease in those opting to study (science) is the diversity of life-styles, religions and youth cultures, not all of which are appealed to by the somewhat limited approach to science education that dominates throughout Europe. (Osborne and Dillon 2008)

Provisional analysis of the interview data further supports the claim made by Sjøberg and Schreiner (2005) that identity formation is an important factor behind this relative lack of interest in a career in science. In more economically developed 'Western' countries, Sjoberg and Schreiner claim, young people are expected to construct their own identities rather than having these ascribed to them by their parents and the culture around them. This analysis fits with some sociological accounts of the continuum between more traditional and post-traditional or 'modern' societies (e.g. Giddens 2000). In this context the image young people have of science and of being a scientist does not always fit with their own identity project. Our data show young people in the UK and in Holland were less interested in science as a school subject than children in Malaysia, India, Lebanon and Turkey. The data offers some suggestions from interview analysis that this relative lack of interest might be linked to a narrow image of science and of the life of a scientist, images which did not always fit with their image of themselves and of what they wanted to be in the future (SED WP3 2011).

This issue of the 'image' of science in relation to the identity formation of young people questions the relevance of some approaches to diversity in education. In many guides for practitioners, diversity is defined in terms of gender, ethnicity and ability. While all three factors are significant, their impact is mediated by the identity-formation projects of young people, and these identity projects lead to other potential groupings. For example, the group of those who do not identify with science because of what they see as the negative ecological impact of science and technology is not defined by gender, ethnicity and ability alone, although these factors have relevance, but there is a more direct reflection of issues of adolescent identity-formation project linked to a particular lifestyle and a particular youth culture.

The literature review of science curricula and innovations in the partner countries of the project indicates that most science education initiatives designed to respond to the issue of cultural diversity in the last 10 years have been based on external categories such as membership of a particular ethnic group (SED WP2 2011). We would not reject this approach and would want to judge the impact of each intervention on the evidence; however, there is an obvious potential danger of imposing an identity on students that they themselves might not find empowering. Nanda (1997), for example, claims that the greatest advocates of indigenisation all have secure transnational cultural identities and children in Western schools. Carter (2004) sums up some recent criticisms of traditional approaches to educational diversity based on cultural comparison; thus,

> Comparison is seen to compartmentalize difference within continually reasserting borders, paradoxically putting a break on those processes of intercultural understanding multicul-turalism seeks to promote. Further, it does not take account of the newly emergent mixed, hybrid, and diverse identities consequent to intensified globalization and diaspora.

In interpreting our data in the light of the literature (e.g. Aikenhead and Lewis 2001; Lee 2001; Barton and Tobin 2001, 2002), we have found that the Bakhtinian notion of 'voice' is more useful than more objective and externally visible categorisations previously used in the classification of cultures. Our use of 'voice' here to indicate a lived perspective on the world that is both cultural and individual is articulated by Hermans in his article 'The dialogical self: towards a theory of personal and cultural positioning' (2001). Drawing on both Bakhtin and William James, Hermans argues that both selves and cultures are made up of 'a multiplicity of positions amongst which dialogical relations can be established'. In other words individuals form their sense of themselves by taking up positions that they first find outside themselves in the culture: cultures are in turn formed by the way in which individuals take up, mix and transform positions. The value of this theoretical perspective is that it breaks down fixed views of cultures and cultural differences of the kind criticised by Carter and others in favour of an understanding of cultural differences as fluid. 'Voices', in this sense, are cultural rather than purely individual and are in dialogue with each other. So, for example, the voice of techno-scepticism tends to be articulated in relationship to the voice of techno-enthusiasm, and while this 'voice' only exists in the utterances of individuals who identify with it while they are speaking, it has a cultural rather than a purely personal existence.

Because voices are internal to cultural dialogues and define themselves in relation to each other, the number of possible voices is not limited or determined by any external or objective features. However, in practice, a relatively small number of clear cultural voices emerged in our study and emerge in every similar study. For example, our team member Helen Haste (2004) conducted a survey of the values and beliefs that 704, eleven- to twenty-one-year-old individuals held about science and technology and found four distinct groups defined by their identifications, which we would now call cultural voices, 'greens' interested in environmental issues but with a specific agenda, 'techno-investors' enthusiastic about the potential of science, the 'science-oriented' keen on science as a way of thinking and the 'alienated from science'. Similar groupings were found in analysis of the ROSE data (Sjøberg and Schreiner 2005). These included a mainly male group fascinated by technology and mainly female group who just wanted to work with others and develop themselves as people. So far the analysis of our interview data suggests that similar identity-based concerns and cultural voices mediate the interest in pursuing science at school.

Understanding diversity in terms of a range of cultural voices has significance for pedagogy. Sjøberg and Schreiner write:

> When young people make their educational choice, they have a range of options. Young people wish to develop their abilities and their identities, and they want a future that they find important and meaningful. Only by being aware of the values and priorities of the young generation can we have a hope to show them that S&T studies may open up meaningful jobs in their lives. (Sjøberg and Schreiner 2007)

The implication for this is that what is required to engage young people in science is a more dialogic approach that is responsive to their interests and concerns. This is a challenging proposal for science education. Our questionnaire responses from teachers indicated that a majority in all countries claimed that they responded to cultural diversity by treating all students the same way. This implies that there may be a gulf in attitude to overcome if we are to adopt a dialogic approach because a dialogic pedagogy takes the opposite approach to diversity. Meanings in dialogues with different voices are never 'all the same' because they are always co-constructed between voices. A dialogic pedagogy in science education implies engaging with the diverse voices of students in such a way that these different voices are respected and enter into the joint construction of scientific knowledge.

In summary, diversity does not only refer to obvious and easily counted differences such as gender, ethnicity or religious tradition; it also refers to the many differences that there are between students due to their different attitudes and identifications. We refer to these as cultural voices. As we mentioned before, while sometimes these 'voices' coincide with obvious religious, ethnic and gender divisions, sometimes they do not. We want to address all the forms of diversity that might impact on how young people respond to science education. This is why the guidelines we developed for the design of educational activities and outlined below stress the need to be responsive to the different concerns, interests and experiences of all students. We call this approach to science education 'dialogic' in part because it is about engaging students in a dialogue in such a way that each feels able to express himself or herself secure in the knowledge that his or her voice will be listened to with respect. It follows from the dialogic view of the nature of science outlined earlier because this view of science argues that all particular science discourses are part of the general dialogue of humanity. This understanding of science puts the emphasis less on the content of what has been found out in the past and more on the process of shared inquiry and dialogic argumentation that leads to shared knowledge. This emphasis fits with another definition of dialogic education which is that dialogic education is education *for* dialogue as well as through dialogue (Wegerif 2007). In this case dialogic science education is education for participation in the dialogues that carry science and in the democratic decision-making dialogues that are informed by science.

Developing a Framework for Science Education for Diversity

How Do We Make Science Education More Relevant?

As already mentioned, we found considerable disengagement from science education between primary and secondary students, especially in the Netherlands and the UK and especially amongst girls. A wide range of factors has been proposed in the literature as contributing to this student disengagement from science. Amongst these are that teachers lack confidence because of their limited knowledge of science, the perception that science is 'dry' and abstract, and the lack of continuity across stages of schooling (Braund 2009; Diack 2009).

Our interview data suggest that one reason for this disengagement might be a sense that science often seems disconnected from the concerns of students and from other real-world concerns and motivating interests. From a dialogic theory of education, perspective motivation comes from participation. It follows that the more disconnected a dialogue is from (a) the internal dialogue of the student, (b) the local social face-to-face dialogues the student participates in and (c) the larger world-historical dialogues carried by communications media and the Internet, then the more likely it is to be experienced as boring and irrelevant.

To re-establish a relationship between school science and the interests of students, we should emphasise science content that is socially relevant (including science for development), science topics that are high-profile cutting edge science and science topics that impinge on students' everyday lives. Topics most likely to engage students' interest and commitment to in-depth study are topics where they feel they have the opportunity to shape their own learning about science topics that make a social impact and that they see as relevant to them as 'global citizens' or that empower them to make a difference in some way in their own local environment.

Will Inquiry-Based Science Education (IBSE) Help?

In Europe there is a tendency to see IBSE as the solution to the current crisis in science education. The recent Rocard et al. report, *Science Education Now* (2007), argues that IBSE

> has proved its efficacy at both primary and secondary levels in increasing children's and students' interest and attainments levels while at the same time stimulating teacher motivation. IBSE is effective with all kinds of students from the weakest to the most able

and is fully compatible with the ambition of excellence. Moreover IBSE is beneficial to promoting girls' interest and participation in science activities. Finally, IBSE and traditional deductive approaches are not mutually exclusive and they should be combined in any science classroom to accommodate different mindsets and age-group preferences.

However, the notion of IBSE encompasses a wide range of definitions and interpretations. A key idea is that students can 'inquire' by exploring existing information in science in ways that may be led by a teacher or by the students themselves, and by building on or contesting that knowledge, again through investigations led by a teacher or by the student (Minner et al. 2010; CILASS 2008). Since IBSE focuses on problem finding, data finding, argumentation and solution finding, it is clearly relevant to the teaching of socially relevant science in a way that draws young people into science as an open-ended process of shared inquiry.

A recent review of research on IBSE confirms to some extent Rocards' claims but points out that the evidence in favour of IBSE is not quite as overwhelming as this report implies:

> The evidence of effects of inquiry-based instruction from this synthesis is not overwhelmingly positive, but there is a clear and consistent trend indicating that instruction within the investigation cycle (i.e., generating questions, designing experiments, collecting data, drawing conclusion, and communicating findings), which has some emphasis on student active thinking or responsibility for learning, has been associated with improved student content learning, especially learning scientific concepts. This overall finding indicates that having students actively think about and participate in the investigation process increases their science conceptual learning. (Minner et al. 2010, p. 493)

The conclusion of this survey is that 'some emphasis on student active thinking and responsibility for learning' has a positive effect on understanding scientific concepts. In particular this survey did not find any advantage in what they call 'saturation' by inquiry learning. This would suggest that IBSE should be seen as one amongst a range of pedagogical approaches but not necessarily the only one to be used.

The evidence with regard to what is effective in IBSE is compatible with the broadly dialogic view expressed in the introduction that teachers need to listen to and respond to the voices of students taking up ideas from students and building on them, thereby allowing students to participate in a shared construction of knowledge. This dialogic approach to IBSE rejects the opposition between teacher-led and student-led science education. On the one hand teacher explanations only make sense to students once they have struggled for themselves with the problem for which the explanation offered by the teacher provides an answer, and so it follows that effective teacher transmission of conceptual understanding in science needs to be combined with active student engagement. On the other hand student inquiry is often naïve and requires the guidance of an expert learner if it is to result in conceptual learning rather than frustration (Rogoff 1994, p. 209; Brown 1992, p. 169). Polman and Pea (2001) try to isolate the key moves in what they refer to as the transformative dialogue between students and teachers that is required for IBSE to be effective:

The dialogue sequence we identified for achieving transformative communication is as follows:

1. Students make a move in the research process with certain intentions, guided, as well as limited by, their current knowledge.
2. The teacher does not expect the students' move, given a sense of their competencies, but understands how the move, if pursued, can have additional implications in the research process that the students may not have intended.
3. The teacher reinterprets the student move, and together students and teacher reach mutual insights about the students' research project through questions, suggestions and/or reference to artifacts.
4. The meaning of the original action is transformed, and learning takes place in the students' zone of proximal development, as the teachers' interpretation and reappraisal (i.e., appropriation) of the students' move is taken up by the student.

The reason this dialogue sequence is transformative is that it allows initial student actions and ideas to be incorporated into later teacher-influenced actions which push students' development and learning, while maintaining intersubjectivity between teacher and students. (Polman and Pea 2001, p. 227)

The literature suggests that IBSE offers one way to engage young people in a way that allows them to express their own voices and find themselves recognised and valued within the construction of scientific knowledge. However, this is not a simple or easy solution, since, as Polman and Pea bring out, to be effective it requires contingently responsive and creative teaching.

Explicitly Dialogic Pedagogy

Although the overall approach to science education that we are proposing is dialogic, there is a useful distinction to make between this theoretical framework for understanding science education and dialogic teaching and learning as a specific pedagogical technique. Mortimer and Scott (2003) help to clarify this relationship with their account of two dimensions in classroom talk in science education: authoritative versus dialogic on one axis and interactive versus non-interactive on the other. The four styles of talk identified by these dimensions are all argued to be valuable at times in science education, with choices dependent on the teacher's intentions at different stages of a lesson.

Dialogic pedagogy teaches students how to engage in dialogue for learning together as well as teaching content matter through dialogue (Wegerif 2007) and implies that all members of the class have a voice and that they expect to respect, listen to, discuss and develop a range of views including partly formed, tentative points of view. Such pedagogy provides one means of respecting the range of cultural explanations and the whole set of students' alternative frameworks, including misconceptions, held by members of the group. Through dialogue the conceptual foundations of the topic can be strengthened, its social significance explored and the opportunities for action considered. There are various specific techniques that can help teach for dialogue. Philosophy for Children offers a tried and tested programme for coaching effective dialogue for conceptual understanding, and this has been applied to science education (Sprod 2011). Similarly the Thinking Together approach which relies on coaching the use of 'exploratory talk' has proved effective in improving the quality of dialogue in science classrooms (Mercer et al. 2004; Webb and Treagust 2006). These and other explicit approaches to promoting and supporting dialogue in classrooms promote ground rules of debate that make it possible to tackle controversial issues of interest to students.

Connecting to Real Science

Both in the discussion of socially relevant science and in the exploration of high-profile science, it may be useful to make links with practicing scientists and people who use science in their careers: the industrial research scientist, the university lecturer, the high street optician or the local health worker. Either through face-to-face meetings, or visits to the workplace, or through electronic communication, such scientists could be expert witnesses whose role is to provide insight into the science involved, or they could be fully involved as contributors to or observers of the students' debates. A dialogic approach to science education means not teaching science in the abstract but drawing young people into the real dialogues and practices of science in action. Evidence from practice suggests that contact with real science could address the very narrow image that many young people have of science (an example is the success of the STEM Alliance programme at William and Mary University, http://stem.wm.edu/).

Mastery Learning Combined with Dialogic Science Pedagogy

Inquiry approaches are good for developing the process skills of science as shared inquiry but are not always the best way to provide the deep understanding of concepts from the tradition of science that is required to participate fully in the ongoing dialogue. As Oakeshott has argued, induction into the tradition of a dialogue should not be understood as a limitation to freedom of action but an essential empowerment giving the freedom to act as a participant in dialogue (Oakeshott 1989). In this context a modified version of mastery learning is suggested.

Master learning focuses on concepts and is teacher led. The teacher determines the core objectives to be learnt and plans whole-class teaching to cover those core objectives taking account of the prior learning of the class. This whole-class phase can make use of all the teaching tactics mentioned above: for example, approaches relevant to dialogic teaching and learning, to teaching controversial issues and to inquiry-based science education. There is then a formative assessment of the students' understanding. This might be based on observation of the students, on inspection of their written work, on a test or on any other methods of assessment.

The remaining time available for the topic (about half the total time) is then spent in different ways by different students. In this enrichment/remedial phase, those who have attained the core objectives work on enrichment and extension tasks (which may go just beyond the core or may, for the most able, be very challenging). Those who have not attained the core objectives revisit those parts of the topic that they have not yet mastered. They engage with the work in more individualised ways in order to address their remaining problems. The topic ends with a summative assessment.

One way to integrate the strengths of mastery learning into the kind of dialogic education required to address issues of diversity is to emphasise engagement with the voices of students. An example of this might be to work with students and the curriculum to identify a topic that is of interest to the students and to start with dialogue about this topic in order to find out more about the goals relevant to the class and about what the students need to find out. At the end of the topic, further dialogue could relate the concepts taught to the lives and concerns of the students (promoting further learning and serving as a summative assessment phase – in terms of which goals were met and how).

Teaching the Nature of Science

As we have already related, our interviews revealed that most of the young people we spoke to in the UK have a narrow image of science, sometimes linking this to men working in labs, authoritative knowledge and to particular technologies involving electronics and large machines. We propose two responses to the problem of a limited image of science. One is to include more contact with real working scientists. Young people who had such contact and knowledge often had a much more positive image of science and scientists than those who did not. The other is to explicitly reflect on and teach the 'nature of science' in a way that can lead young people to focus more on the process of science as flexible open-minded inquiry creatively and collectively seeking to answer real questions and solve real problems.

Teaching Thinking in Science and Through Science

Engaging with the beliefs and concerns of students in different contexts implies a focus more on the processes of science than on teaching specific facts or concepts. This coincides with a broader interest that science education should place value and emphasis on the processes of shared inquiry and argumentation that enable students to understand science as a way of knowing (Millar and Osborne 1998; Driver et al. 2000; Millar 2006). Several researchers have argued that science education needs to focus more on how evidence is used to construct explanations through examining the data and warrants that form the basis of belief in scientific ideas and theories, as well as exploring the criteria used to evaluate evidence (Osborne et al. 2004). Knowing how to read and understand such arguments is an important part of scientific literacy. However, research shows that reasoning and argumentation is not found in classrooms unless it is explicitly taught (Mercer et al. 2004; Lemke 1990; Mortimer and Scott 2003). Instead of being presented as a body of contestable findings that are part of an ongoing process of inquiry, science is often taught as a body of facts (Lemke 1990). Research suggests that argumentative discourse is important for engaging students in science education and that it can be taught (Mercer et al. 2004; Osborne 2007; Osborne et al. 2004; Zohar and Nemet 2002; Millar 2006; von Aufschnaiter et al. 2008).

The Role of ICT

In some of our partner countries, it is common to have separate schools for boys and for girls. In every country different faith groups and ethnic groups are often educated in different schools. Education for diversity that only operates within classrooms is therefore not enough to address diversity issues. Collaborative inquiry projects that bring together students from different backgrounds and different countries are one way to go beyond these school-based boundaries. For the dialogic science education for diversity proposed, it is important not only to relate to local dialogues but also to engage as a participant, in however modest a capacity, in science understood as a long-term global dialogue bringing together many voices from different backgrounds. The Internet offers a medium that facilitates this vision of science as a dialogue of diverse voices.

One study that inspired us was conducted in Exeter from 1997 onwards. Schools in Norway, Sweden, Denmark, Holland, the UK, France, Germany and Spain collaborated using the Internet to monitor the population of selected species of butterflies as indicators of climate change over a 6-year period (Seddon et al. 2008). European environmental experts worked with the teachers. One of these, for example, was Constanti Stefanescu who had co-authored an important study in nature entitled 'Poleward shifts in geographical ranges of butterfly species associated with regional warming' (Parmesan et al. 1999). In this way the collaboration engaged school students in real science that was cutting edge and in the news. Evidence gathered and analysed by students indicated a northerly shift in butterfly flight. Data collected in the collaboration was made available through a website, for partner schools to use and interpret, which allowed them to consider the implications of the appearance of butterfly species for climate change. Feedback from the teachers suggested that being able to see the work of other students, seeing their own work on the web and knowing that they were involved in real science was motivating for the students. Although this project was not designed to address diversity issues, it followed many of the principles that we suggest should be followed in education to address diversity. Through conducting a real science inquiry collaboratively mediated by the web, this project motivated and engaged students from schools with very different cultures from the Arctic Circle in the north to the Mediterranean coast in the south.

In Exeter we are currently working with science teachers in two local schools, one secondary and one primary, to implement an intervention based on the framework for the design of science education for diversity outlined above. These two schools plan to collaborate via the Internet, teaming up the group of mostly female 16–18-year-olds doing non-compulsory advanced level (A level) science courses in the secondary school with the 10- and 11-year-old primary school students. The idea is to use video links and text messages to allow the primary pupils to ask the advanced level students questions and get advice on their projects. This use of ICT for communication across schools will address an issue we noted in our initial survey: a marked fall-off in interest in science from primary to secondary that particularly impacts on girls.

The Need for Guided Collaborative Critical Reflection on Action

The discussion above provides no single straightforward specification of the ways in which science education might change to encompass more fully the diversity of students. It does not facilitate the construction of a detailed protocol for action that could be investigated in well-structured experimental designs. We argue that, instead, it provides a framework, different elements of which may be of greater relevance or particularly desirable to particular groups in constructing their own approaches to diversity.

Teachers working collaboratively to do action research supported by university partners and guided by the framework might prove an effective way to deepen professional development in relation to science education for diversity (Haggarty and Postlethwaite 2003).

Though each action research investigation is context specific and small scale, the methodology is a demanding one. In the SED project this process of collaborative research benefits hugely from the close partnership that exists between the university partners and the teachers in each country and from the sharing of ideas (interim as well as final) across the international partnership and beyond. We hope that our example of this kind of collaboration might serve as a useful model for the transformation of science education in response to the growing challenge of diversity.

Summary and Conclusion

This chapter began by pointing out a connection between theories as to the nature of science and how science education responds to diversity. A monologic theory of science focusing on finding correct and unique knowledge through a correct and unique method seems to lie behind and inform an approach to science education which does not respond to or engage with the many different world views of students. We argued that claims that there is a single scientific method have been exaggerated and that in fact there are many sciences with many methods all of which have to be justified ultimately by the same dialogic processes of argumentation as are found in other areas of human life. This led us to a dialogic vision of science and a dialogic vision of science education as being about drawing students into those ongoing scientific dialogues through which shared knowledge is constructed and human understanding is increased. The metaphor of dialogue applied to understanding science puts the emphasis on those virtues, skills and procedures which enable communities to reach consensus, which include intellectual integrity, listening with respect to alternative views, being open about procedures and the use of empirical evidence in combination with arguments in order to justify claims. Our argument is that a dialogic approach to science education which emphasises and promotes the virtues, skills and procedures required for the construction of understanding in the context of multiple voices would be the best way to engage with the increasing diversity of cultural voices.

Our initial research findings in the science education for diversity research project, combined with literature review, have led us to propose a number of principles for the design of science education that addresses the challenge of increasing cultural diversity:

1. The overall pedagogical approach should be *dialogic*: this means teaching 'for dialogue' as well as through dialogue and implies:

   (a) Being responsive to and engaging with multiple voices and perspectives
   (b) Teaching dialogic argumentation in science including the use of evidence and effective ways of 'talking science'

2. Science education needs to be *relevant to students* in some or all of the following ways:

   (a) Using science content that is related to events in the media
   (b) Using science content from the everyday world of students
   (c) Addressing controversial issues of interest to students
   (d) Involving real life work in science and technology

3. To engage with diversity the pedagogy should incorporate *reflection on knowledge and different ways of knowing*, including reflection and discussion on the nature of science.

4. We recommended the use of two approaches to pedagogy: *guided collaborative inquiry-based science education and dialogic mastery learning*. Guided collaborative inquiry-based learning implies student-led inquiry combined with guidance towards scientific concepts. Dialogic mastery learning implies planning teaching of key concepts in a way that is responsive to student's interests and understandings.

5. *Design-based research* in the form of *guided collaborative critical reflection on action* is proposed since any framework of principles should be continuously revisited by teachers working collaboratively to test, refine and revise it.

The dialogic focus of this framework reflects a vision of science education as drawing learners into participation into science understood as a form of shared inquiry. This means engaging them with inherited concepts and traditions of sciences in order to empower them to act in the future. Although this dialogic approach to science education is a response to the challenge of engaging and motivating the full range of cultural voices found in science classrooms, it also reflects a vision of science as a living dialogue, open to and engaged with the larger dialogue of humanity. Increasingly this living dialogue of science is carried by the Internet which is why drawing children to participate in scientific dialogues on the Internet also has an important potential role to play.

Acknowledgements This chapter draws on research funded by the EC Framework 7 programme, specifically the science education for diversity project. The team collecting and analysing the data included in addition to the authors:
Professor Helen Haste (Harvard) and Dr. Andrew Dean (The University of Exeter, UK); Professor Saouma BouJaoude, Dr. Rola Khishfe, Dr. Dian Sarieddine, Dr. Sahar Alameh and Dr. Nesreen Ghaddar (American University of Beirut, Lebanon); Professor Huseyin Bag and Dr. Ayse Savran Gencer (Pamukkale University, Turkey); Assistant Professor Michiel van Eijck and Dr. Ralf Griethuijsen (Eindhoven University of Technology); Dr. Ng Swee Chin and Dr. Oo Pou San (Tunku Abdul Rahman College, Malaysia); and Dr. Sugra Chunawala, Dr. Chitra Natarajan and Dr. Beena Choksi (Tata Institute of Fundamental Research, India).

References

Aikenhead, G. S., & Lewis, B. F. (2001). Introduction: Shifting perspectives from universalism to cross-culturalism. *Science Education, 85*, 3–5.
Alters, B. J. (1997). Whose nature of science? *Journal of Research in Science Teaching, 34*, 39–55.
Bakhtin, M. M. (1981). Discourse in the novel. In M. M. Bakhtin (Ed.), *The dialogic imagination: Four essays*. Austin: University of Texas Press.
Bakhtin, M. M. (1986). *Speech genres and other late essays*. Austin: University of Texas Press.
Bakhtin, M. M. (1984). *Problems of Dostoevsky's poetics*. Minneapolis: University of Minnesota Press.
Braund, M. (2009). Progression and continuity in learning science at transfer from primary and secondary school. *Perspectives on Education, 1*, 5–21. Available on: www.wellcome.ac.uk/perspectives. Accessed 23 Apr 2012.
Brown, A. L. (1992). Design experiments: Theoretical and methodological challenges in creating complex interventions in classroom settings. *The Journal of the Learning Sciences, 2*(2), 141–178.
Barton, A. C. (2000). Crafting multicultural science education with preservice teachers through service learning. *Journal of Curriculum Studies, 32*(6), 797–820.
Barton, A. C., & Tobin, K. (2001). Urban science education. *Journal of Research in Science Teaching, 38*(8), 843–846.
Barton, A. C., & Tobin, K. (2002). Learning about transformative research through other's stories: What does it mean to involve "Others" in science education reform? *Journal of Research in Science Teaching, 39*(2), 110–113.
Carter, L. (2004). Thinking differently about cultural diversity: Using postcolonial theory to (Re)read science education. *Science Education, 88*, 819–836.

CILASS. (2008). Inquiry-based learning: A conceptual framework. Available at: http://www.shef. ac.uk/cilass/resources. Last accessed 22 Sept 2011.

Diack A. (2009). A smoother path: managing the challenge of school transfer. *Perspectives on Education* (Primary Secondary Transfer in Science), *2*, 39–52.

Driver, R., Newton, P., & Osborne, J. (2000). Establishing the norms of scientific argumentation in classrooms. *Science Education, 84*(3), 287–312.

Dupré, J. (1993). *The disorder of things: Metaphysical foundations of the disunity of science.* Cambridge, MA: Harvard University Press.

Dupré, J. (2001). *Human nature and the limits of science.* Oxford: Oxford University Press.

European Commission. (2004). *Europe needs more scientists: Report by the high level group on increasing human Resources for science and technology.* Brussels: European Commission.

Giddens, A. (2000). *Runaway world.* London: Routledge.

Gillies, D. (1998). The Duhem thesis and the Quine Thesis. In M. Curd, J. A. Cover, et al. (Eds.), *Philosophy of science: The central issues* (pp. 302–319). New York: Norton. Hanson, Norwood Russell.

Habermas, J. (1984). *The theory of communicative action* (Vol. 1). Cambridge: Polity Press.

Haggarty, L., & Postlethwaite, K. (2003). Action research: A strategy for teacher change and school development? *Oxford Review of Education, 29*(4), 423–448.

Haste, H. (2004). *Science in my future: A study of values and beliefs in relation to science and technology amongst 11–21 year olds.* London: Nestle Social Research Programme.

Hermans, H. J. M. (2001). The dialogical self: Toward a theory of personal and cultural positioning. *Culture & Psychology, 7,* 243–281.

Kang, S., Scharmann, L. C., & Noh, T. (2005). Examining students' views on the nature of science: Results from Korean 6th, 8th, and 10th graders. *Science Education, 89,* 314–334.

Kawasaki, K. (2004). The concepts of science in Japanese and Western education. *Science Education, 5*(1), 1–20.

Lee, O. (2001). Culture and language in science education: What do we know and what do we need to know? *Journal of Research in Science Teaching, 38*(5), 499–501.

Lemke, J. L. (1990). *Talking science: Language, learning and values.* Norwood: Alex.

Mansour, N., & Wegerif, R. (2013). Science education for diversity: Theory and practices. Springer Science.

McComas, W. F., & Olson, J. K. (1998). The nature of science in international standards documents. In W. F. McComas (Ed.), *The nature of science in science education: Rationales and strategies* (pp. 41–52). Dordrecht: Kluwer Academic Publisher.

Mercer, N., Dawes, L., Wegerif, R., & Sams, C. (2004). Reasoning as a scientist: Ways of helping children to use language to learn science. *British Educational Research Journal, 30*(3), 359–377.

Millar, R. (2006). Twenty first century science: Insights from the design and implementation of a scientific literacy approach in school science. *International Journal of Science Education, 28*(13), 1499–1521.

Millar, R., & Osborne, J. F. (Eds.). (1998). *Beyond 2000: Science education for the future.* London: King's College London.

Minner, D., Levy, A., & Century, J. (2010). Inquiry-based science instruction – What is it and does it matter? Results from a research synthesis years 1984 to 2002. *Journal of Research in Science Teaching, 47*(4), 474–496.

Mortimer, E. F., & Scott, P. H. (2003). *Meaning making in secondary science classrooms.* Maidenhead: Open University Press.

Nanda, M. (1997). The science wars in India. *Dissent, 44*(1), 79–80.

Oakeshott, M. (1989). In T. Fuller (Ed.), *The voice of liberal learning: Michael Oakeshott on education.* New Haven/London: Yale University Press.

Osborne, J. (2007). Engaging young people with science: Thoughts about future direction of science education (pp. 105–112). In C. Linder, L. Östman & P. Wickman (Eds.) Promoting scientific literacy: Science education research in transaction. Proceedings of Linnaeus Tercentenary Symposium held at Uppsala University, Uppsala, May 28–29, 2007.

Osborne, J., & Dillon, J. (2008). *Science education in Europe: Critical reflections.* London: Nuffield Foundation.

Osborne, J., Collins, S., Ratcliffe, M., Millar, R., & Duschl, R. (2003a). What "ideas-about-science" should be taught in school science? A Delphi study of the expert community. *Journal of Research in Science Teaching, 40*(7), 692–720.

Osborne, J., Erduran, S., & Simon, S. (2004). Enhancing the quality of argument in school science. *Journal of Research in Science Teaching, 41*(10), 994–1020.

Osborne, J., Simon, S., & Collins, S. (2003b). Attitudes towards science: A review of the literature and its implications. *International Journal of Science Education, 25*(9), 1049–1079.

Parmesan, C., Ryrholm, N., Stefanescu, C., et al. (1999). Poleward shifts in geographical ranges of butterfly species associated with regional warming. *Nature, 399,* 579–583.

Polman, J. L., & Pea, R. D. (2001). Transformative communication as a cultural tool for guiding inquiry science. *Science Education, 85,* 223–238.

Rocard, M., et al. (2007). *Science education now: A renewed pedagogy for the future of Europe.* Luxembourg: Office for Official Publications of the European Communities.

Rogoff, B. (1994). Developing understanding of the idea of communities of learners. *Mind, Culture, and Activity, 1*(4), 209–229.

Rorty, R. (1991). *Objectivity, relativism, and truth: Philosophical papers* (Vol. 1). Cambridge: Cambridge University Press.

Roth, M. (2009). *Dialogism: A Bakhtinian perspective on science language and learning.* Rotterdam: Sense Publishers.

Rudolph, J. L. (2000). Reconsidering the "nature of science" as a curriculum component. *Journal of Curriculum Studies, 32*(3), 403–419.

SED WP2 (2011) SEDWP2D1: Documentary Analysis Synthesis Report. Available on: http://www.marchmont.ac.uk/Documents/Projects/sed/wp2_final_report.pdf. Accessed 21 Feb 2012.

SED WP3 (2011) SEDWP3D1: Survey Research Synthesis Report. Available on: http://www.marchmont.ac.uk/Documents/Projects/sed/201106_wp3.pdf. Accessed 21 Feb 2012.

Seddon, K., Skinner, N. C., & Postlethwaite, K. C. (2008). Creating a model to examine motivation for sustained engagement in online communities. *Education and Information Technologies, 13*(1), 17–34.

Sjøberg, S., & Schreiner, C. (2005). How do learners in different cultures relate to science and technology? *Asia-Pacific Forum on Science Learning and Teaching, 6*(2), 63–71.

Sjøberg, S., & Schreiner, C. (2007). Perceptions and images of science and science education. In M. Claessens (Ed.), *Communicating European research 2005.* Heidelberg: Springer.

Smith, M. U., Lederman, N. G., Bell, R. L., McComas, W. F., & Clough, M. P. (1997). How great is the disagreement about the nature of science: A response to alters. *Journal of Research in Science Teaching, 34,* 1101–1103.

Sprod, T. (2011). *Discussions in science: Promoting conceptual understanding in the middle school years.* Melbourne: Australian Council Educational Research (ACER).

von Aufschnaiter, C., Erduran, S., Osborne, J., & Simon, S. (2008). Arguing to learn and learning to argue: Case studies of how students' argumentation relates to their scientific knowledge. *Journal of Research in Science Teaching, 45,* 101–131.

Webb, P., & Treagust, D. (2006). Using exploratory talk to enhance problem-solving and reasoning skills in grade-7 science classrooms. *Research in Science Education, 36*(4), 381–401.

Wegerif, R. (2007). *Dialogic education and technology: Expanding the space of learning.* New York: Springer.

Zohar, A., & Nemet, F. (2002). Fostering students' knowledge and argumentation skills through dilemmas in human genetics. *Journal of Research in Science Teaching, 39*(1), 35–62.

Expanding Notions of Scientific Literacy: A Reconceptualization of Aims of Science Education in the Knowledge Society

Xiufeng Liu

Since early 1990s, science education reform documents in the USA and many other countries have been promoting scientific literacy (SL) as the aim of science education (e.g., American Association for Advancement of Science [AAAS] 1989; Council of Ministers of Education of Canada [CMEC] 1997; National Research Council [NRC] 1996). It is expected that "the scientifically literate person is one who is aware that science, mathematics, and technology are interdependent human enterprises with strengths and limitations, who understands key concepts and principles of science, who is familiar with the natural world and recognizes both its diversity and unity, and who uses scientific knowledge and scientific ways of thinking for individual and social purposes" (AAAS 1989, p. xvii). Similarly, the Canadian *Common Framework of Science Learning Outcomes* states that "all Canadian students, regardless of gender or cultural background, will have an opportunity to develop scientific literacy" (CMEC 1997, p. 4). The just released conceptual framework for science education standards in the USA (NRC 2011) expects that "by the end of 12th grade, students should have gained sufficient knowledge of the practices, crosscutting concepts, and core ideas of science and engineering to engage in public discussions on science-related issues, to be critical consumers of scientific information related to their everyday lives, and to continue to learn about science throughout their lives" (p. 6).

Although SL has been accepted as a common aim for science education, a universally accepted definition of SL remains unavailable. In order to help conceptualize the diversity of definitions of SL, Roberts (2007) proposes to use two competing visions to form a continuum of SL, Vision I and Vision II SL. Vision I SL emphasizes science as a distinct discipline, i.e., looking inward toward domains of science from scientists' perspectives that include propositional knowledge, procedural knowledge, metacognition, and disposition. Vision II SL

X. Liu (✉)
Department of Learning and Instruction, State University of New York, Buffalo, NY, USA
e-mail: xliu5@buffalo.edu

N. Mansour and R. Wegerif (eds.), *Science Education for Diversity: Theory and Practice*, Cultural Studies of Science Education 8, DOI 10.1007/978-94-007-4563-6_2, © Springer Science+Business Media Dordrecht 2013

emphasizes the context of science and its relation to technology, society, and environment. As Roberts (2011) points out, Vision I and Vision II are pointers instead of pigeonholes; specific science curriculum policy images (e.g., content standards, curriculum guides, instructional materials) most likely attend to both visions with relatively different emphasis on each of the visions. This differential emphasis on Vision I and Vision II is exemplified in two international assessment programs, i.e., TIMSS (Trends in International Mathematics and Science Study) and PISA (Programme for International Student Assessment). The TIMSS SL is school science curriculum oriented because all items are textbook questions without contexts, thus primarily Vision I, while the PISA SL is societal needs oriented because all items are framed within specific social and personal contexts, thus primarily Vision II. As one more example, Miller's operationalization of civic SL is in agreement with Vision I (Miller 1983, 1987, 1998), because it focuses on a core set of basic science understandings that are considered an intellectual foundation for reading and understanding contemporary issues and would remain relevant over a long period of time. Assessment is conducted by about 12 multiple-choice or true-false items (e.g., Lasers work by focusing sound – true or false?). Although civic SL would focus on Vision II, Miller's measurement of it is primarily about Vision I.

Despite the diversity in definitions of SL, there are common assumptions among them. Expanding from Layton et al. (1993), Liu (2009) summarizes three such common assumptions; they are: (a) deficit elimination, (b) commodity acquisition, and (c) one-way transport. In terms of the deficit elimination assumption, current definitions of SL assume that students and the general public lack SL, thus need to correct this deficiency (Bauer et al. 2007). This deficit assumption ignores the fact that students and the general public do have a wide range of informal knowledge and experiences about natural and life phenomena. Research has shown that the general public's knowledge of and attitude toward science is both context dependent and selective to their perceived uses and values (Bauer et al. 2007; Layton et al. 1993; Wynne 1995), and there is a difference between "inarticulate science" and "practical science" in everyday life (Layton et al. 1993).

In terms of the commodity acquisition assumption, current definitions of SL assume that if a person has achieved certain outcomes, then the person has obtained SL and will keep this status forever, because it is believed that there is a threshold between being scientifically literate and scientifically illiterate. This notion of SL simply ignores the fact that science is constantly changing and that even an expert in one field of science may be ignorant in other science fields thus in need of knowing more. Learning science is indeed a lifelong process instead of a one-time effort.

Finally, the one-way transport assumption assumes that SL is achieved through activities conducted by the knowledgeable to the less knowledgeable. This assumption gives scientists and the scientific community an unquestionable status by assuming that they can be trusted in deciding what scientific knowledge is beneficial to the society and how it may be conveyed. What scientists value may not necessarily be what the general public values. The one-way transport assumption of SL ignores the active role of learners by considering SL as being extrinsic to individuals, i.e., tools for economic development and national security (Laetsch 1987).

Garrison and Lawwill (1992) consider imposing SL on students as being immoral. They state that "chaining science and science education to the goal of maximizing the economic production function ... is immoral...because it treats students as means to the pecuniary ends of others" (p. 343). Critics have claimed that current definitions of SL serve to maintain the dominance of special interest groups such as the elites and technocrats, i.e., those with political and economic power, while excluding others – particularly minorities (Apple 1992; Osborne and Calabrese-Barton 2000). Furthermore, scientists may not necessarily be well equipped with skills to communicate science to the public (Bauer and Jensen 2011; Suleski and Ibaraki 2010).

Despite that efforts to achieve SL have been ongoing for over 50 decades, the success of these efforts has been limited (Liu 2009). The purpose of this chapter is to present a reconceptualization of SL as the aim of science education. I will first evaluate the inadequacy of current notions of SL from a historical perspective called the "two cultures" and from current recognition of grand challenges in the twenty-first century; I will then propose an expanded notion of SL based on science engagement and support it by social cultural learning theories. Finally, I will describe examples of science engagement to demonstrate the viability of this broadened notion of SL.

The "Two Cultures" and the Need for a Broader Notion of Scientific Literacy

More than 50 years ago, C. P. Snow expressed a concern about the divide between the scientific community and literary community, i.e., the "two cultures" (Snow 1959). In fact, his concerns about the two cultures were far beyond the above-mentioned two intellectual groups. The two cultures also referred to much larger issues confronting humanities of the twentieth century: the gaps between the rich and the poor, between the west and the east, between government officials and scientists, between basic science and technology, and the list goes on (Snow 1959, 1960, 1963). One fundamental belief underlying all gaps between the "two cultures" is that science is transforming the society in all the ways, and without understanding science by the rest of the society, humankind may be heading toward danger if not disaster.

Snow's concern remains just as applicable today, if not more so. According to a recent report by the Pew Research Center for the People and the Press (2009), although a majority of Americans (84 %) has high regards for science and scientists, a majority of scientists (85 %) identified as a major problem that the public does not know very much about science. Seventy-six percent of scientists also considered that the media does not distinguish between well-founded findings and those that are not. The gap between scientists and the public on more complex issues is even more striking (see Table 1).

Table 1 Differences between the public and scientists (Pew Research Center for the People and the Press 2009)

Issues	Public (%)	Scientists (%)
Think that humans and other living things have evolved due to natural processes	32	87
Think that the earth is becoming warmer because of human activities	49	84
Favor use of animals in scientific research	52	93
Favor building more nuclear power plants	51	70
Agree that all parents should be required to vaccinate their children	69	82

The divide between the scientific community and the general public also exhibits itself in the disconnect between school science and the society. It has been well documented that students gradually lose interest in science as they progress from elementary to middle and high school (Koballa and Glynn 2007; Osborne et al. 2003), and students who are interested in pursuing careers in science and engineering are getting small in number (NRC 2007). As a result, a group of prominent international science educators issued the following collective statement of concerns (Linder et al. 2011):

> Citizens' lives are increasingly influenced by science and technology at both the personal and societal levels. Yet the manner and nature of these influences are still largely unaddressed in school science. Few students complete a schooling in science that has addressed the many ways their lives are now influenced by science and technology. Such influences are deeply human in nature and include the production of the food we eat, its distribution, and its nutritional quality, our uses of transportation, how we communicate, the conditions and tools of our work environments, our health and how illness is treated, and the quality of our air and water.
>
> Science education is not contributing as it could to understanding and addressing such global issues as feeding the World's Population, Ensuring Adequate Suppliers of Water, Climate Change, and Eradication of Disease in which we all have a responsibility to play a role. Students are not made aware of how the solution of any of these will require applications of science and technology, along with appropriate and committed social, economic, and political action. As long as their school science is not equipping them to be scientifically literate citizens about these issues and the role that science and technology must play, there is little hope that these great issues will be given the political priority and the public support or rejection that they may need. (pp. 2–3)

There have been calls for actions to address the above divides. One of such calls is science communication by scientists. American writer Chris Mooney in his best-selling book entitled *Unscientific America: How Scientific Illiteracy Threatens Our Futures* (Mooney and Kirshenbaum 2009) states that

> scientists know what advances are under way and debate them regularly at their conferences, but they're talking far too much among themselves and far too little to everyone else. This isn't a gap the president or his administration can bridge, and certainly not alone. We need the experts themselves to launch new initiatives to bring these topics into spotlight, before it's too late to have a serious dialogue about them. (p. 10)

Mooney refers to an ever-widening divide between the scientific community and the general public, using such examples as the public debate and policy on global warming, the never-ending battle between evolution and creation/intelligent design, public displeasure with the excommunication of the Pluto from the solar system, prevalent unscientific information in Hollywood movies and the entertainment industries in general, and misinformation about science through social media such as blogs. As Mooney points out,

> there's another side to the scientific literacy tradition, one that goes beyond the standard emphasis on factual or theoretical scientific knowledge to stress a third aspect: citizens' awareness of the importance of science to politics, policy, and our collective future. This dimension has often fallen by the wayside in debates about scientific illiteracy, and yet we believe it is easily the most important. (p. 18)

Addressing the divide between the scientific society and the general public and between school science and the society requires a notion of SL beyond Vision I and Vision II, because understanding of science as well as its relations with technology and society by individuals is not enough; active participation and dialogue among all citizens on complex issues are needed. As Mooney and Kirshenbaum (2009) passionately argue, we need more scientifically literate people who can effectively communicate science to the general public and relate science to the public policy. The need to bridge science and public policy is also recognized by the National Academy of Engineering [NAE] (2008). After a wide consultation, NAE identified 14 grand challenges for the twenty-first century, such as engineering better medicines, prevent nuclear terror, and providing access to clean water, that require participation of not only the scientific community but also the rest of society in public policy, business, law, ethics, human behavior, and so on. As Fensham (2011) states, science and technology used to set the agenda for the society; now it is the society that sets agenda for science and technology.

Specifically, in today's knowledge society in which information and ideas flow freely and the society becomes more and more diverse in values and approaches to knowledge as a result of the "flat" world (Freeman 2006), diverse views exist on a wide range of topics, from the historical evolution and creation debate to such more recent issues as global warming, vaccination against diseases, genetically modified organism, and to emerging issues associated with advances in biotechnology, nanotechnology, and neuroscience. Aikenhead, Orpwood, and Fensham (2011), in citing Gilbert (2005), list the following implications of a knowledge society for education:

1. Knowledge is about acting and doing to produce new things; rather than being only an accumulation of established information, and
2. What one does with knowledge is paramount, not how much knowledge one possesses.

   Thus, it is highly valuable for

3. Knowing how to learn, knowing how to keep learning, and knowing when one needs to know more;
4. Knowing how to learn with others;
5. Using knowledge as a resource for resolving problems;
6. Acquiring important competences (skills) in the use of knowledge. (p. 30)

Accordingly, Aikenhead et al. (2011) argue that a vision of scientific literacy with a focus on *Science-Technology-knowing-in-Action*, or *ST-knowing-in-action*, is needed for a knowledge society. *ST-knowing-in-action* values both expert and citizen expertise in science-technology employment and on the capacity of its citizens to deal with ST-related situations in their everyday lives. Given that current ST-related situations are often complex involving social, cultural, political, and environmental issues, citizens with *ST-knowing-in-action* in a knowledge society must be capable of participating in dialogues with others of different views. Avoiding such complex issues completely or totally disregarding other views than their own will not be helpful for the society. Polarized views on all the issues will inevitably arise. What we need is a scientifically literate populace who can appreciate and value the diverse views and engage with each other and with the policy-makers to promote more balanced approaches to address the issues. Thus, what we need in today's knowledge society is more than getting scientists involved in science communication; we need every citizen to get engaged in science and technology-related issues. An expanded notion of SL by going beyond Visions I and II is necessary.

Scientific Literacy Reconceptualized

In accordance with Aikenhead et al. (2011)'s call for a new vision of SL and based on the *Science-Technology-knowing-in-Action*, it is proposed that science engagement (SE) be incorporated into the conceptual framework of scientific literacy. SE may be considered Vision III SL. Table 2 presents characteristics of this expanded notion of SL.

From Table 2, we see that SE expands SL by bringing in a new emphasis on social, cultural, political, and environmental issues (SCPEI). This new emphasis is to develop every citizen's skills in critical thinking, science communication, and consensus building. This new emphasis may be called Vision III in relation to Visions I and II. While SE still values Visions I and II SL, it promotes active participation in debate and seeking solutions on today's pressing issues facing the world. Examples of such pressing issues are religious beliefs and science, climate change, globalization, and security and public safety. These complex issues may not be simply perceived as scientific ones, because they involve many other aspects typically considered outside the domains of science (e.g., religion, politics, economics).

When categorizing roles of scientists in policy-making, Pielke (2007) identifies four typical or idealized roles:

- *The pure scientist*: seeks to focus only on facts and has no interactions with the decision-maker.
- *The science arbiter*: answers specific factual questions posed by the decision-maker.

Table 2 An expanded notion of SL

Emphasis	Content	Relation to Roberts (2007) vision	Orientation	Role of learner in society
Scientific content (SC)	Knowledge, skills, habit of mind, and disposition	Vision I	Within science	Pure science learner and pursuer
Science-technology societal issues (STSI)	Knowledge in action, practical problem-solving, attitude, and professionalism	Vision II	Science in relation to society	Science advocate
Scientific engagement (SE) – social, cultural, political, and environmental issues (SCPEI)	Critical thinking, communication, consensus building	Vision III	Science within society	Honest broker

- *The issue advocate*: seeks to reduce the scope of choice available to the decision-maker.
- *The honest broker of policy options*: seeks to expand, or at least clarify, the scope of choice available to the decision-maker.

Applying the above four typical roles to conceptualize the roles of learners (broadly conceived to be lifelong learners) in the society, learners in SE may take the following typical roles (last column in Table 2):

- Pure science learner and pursuer: engage in science for developing mental capacity and for preparation of a science career
- Issue advocate: engage in science for solving technological and societal problems
- Honest broker: engage in science for seeking informed/best possible solutions to complex social, cultural, political, and environmental issues

The notion of SE is not new; many researchers have argued similarly in the past. Shen (1975) proposes three types of SL that include (a) practical, possession of the kind of scientific knowledge that can be used to help solve practical problems; (b) civic, to enable the citizen to become more aware of science and science-related issues in order to participate in the democratic processes; and (c) cultural, knowledge and appreciation of science as a major human achievement and cultural heritage. In today's knowledge society, a scientifically literate person should possess all three types of literacy, and we should promote them in both school science and informal science education. Similarly, Shamos (1995) identifies three levels of SL: (a) cultural SL, a grasp of certain background information underlying basic communication; (b) functional SL, to not only know the science terms but also be able to converse, read, and write coherently using these terms in nontechnical contexts; and (c) true SL, understanding the overall scientific enterprise and the major conceptual schemes of science, in addition to specific elements of scientific investigation. While Shamos's above three types of science literacy are all desirable today, cultural SL may not be considered a lower level of SL than the other two anymore. On the contrary, it could be argued that cultural SL may be even a higher level of SL than the other two.

SE is also consistent with a distinct research program called socio-scientific issues (SSI) (Sadler 2004; Zeidler 1984; Zeidler and Sadler 2011; Zeidler et al. 2005). SSI seeks to involve students in decision-making regarding current social issues with a purpose to develop moral reasoning, ethical consideration, and character development. For both SE and SSI, the orientation is science within society, i.e., complex issues that face the society and the world today. The difference between SE and SSI is that, first of all, SE has a broader scope than SSI. For example, SCPEI would be involved in defending science when it is misused or distorted by politicians (Mooney 2005). Second, SE is intended to be a more comprehensive framework to subsume other issues (e.g., cultural, political). Third, while SSI is primarily framed within the school science context, SE intends to encompass both formal and informal science education.

Both SSI and SCPEI are examples of a more general approach called humanistic perspectives on science education, which has existed for over 150 years (Aikenhead 2006). As an alternative to Vision I SL, humanistic approaches to SL intend to make science relevant to students. Relevance to students is "usually determined by students' cultural self-identities, students' future contributions to society as citizens, and students' interest in making personal utilitarian meaning out of various kinds of sciences – Western, citizen, or indigenous" (Aikenhead 2006, p. 23). Thus, humanistic perspectives focus on Visions II and beyond.

Participation and actions are key characteristics of SE; both have been promoted in the science education literature. Roth and his collaborators call for "community and citizenship-based scientific literacy" (Roth 2002; Roth and Calabrese 2004; Roth and Lee 2004) in which students co-construct science with science experts and the general public in solving specific local problems. Hodson (2003) proposes scientific literacy to include four broad domains: (a) learning science and technology, (b) learning about science and technology, (c) doing science and technology, and (e) engaging in sociopolitical actions. It is the last domain that is what SE is calling for. Following Hodson (2003), more recently, Bencze and Carter (2011) propose an activist science and technology education program based on principles of holism, altruism, realism, egalitarianism, and dualism. Taking actions is central to this program, and examples of actions include educating others about issues, developing better products and systems, lobbying "power brokers," boycotting harmful products/services, protesting against sources of issues, disrupting socio-enviro problem situations, and changing one's own practices (Bencze and Carter 2011, p. 658).

Communication is an essential component in all visions of SL. Researchers in science education have recognized the importance of communication in achieving SL. For example, Hand et al. (2003) point out the important role of language uses in science. SL with a consideration of language values both formal literacy learned in school and informal literacy practiced outside school. It also values a variety of ways of communications in science, particularly in reading, writing, and speaking in science. In this sense, SL is a public good; it is a civic duty for all citizens. Similarly, Norris and Phillips (2003) distinguish two emphases of SL – the fundamental sense in terms of reading and writing in science and the derived sense in terms of knowledgeable and competence in science. They further claim that current notions of SL often focus on derived sense while ignoring the fundamental sense. No doubt, strong literate skills are foundational to science communication, thus science engagement.

However, communication in SE differs from science communication and public engagement in public understanding of science (PUS) (Bauer and Jensen 2011; Christensen 2007). While the major purpose of science communication and public engagement in PUS is to help the uninformed to understand a certain scientific topic or issue, communication in SE also values the process of interacting with the others. While science communication in PUS gives the privilege to the informed, communication in SE gives an equal status to all participants involved. Further, different from science communication in PUS that is limited to only the scientific community, communication in SE can take place at any age and anywhere.

The any-age-and-anywhere nature of SE implies that SE is a lifelong process, going beyond school science. This view of learning reflects the fact that school children spend far more time outside schools than inside schools. According to an estimate (Bransford et al. 2000), an individual spends only 18 % of his or her life in schools, 5 % before kindergarten, and 77 % out of school years. During a typical school year, assuming 180 school days a year, 6.5 h per school day, a typical American child spends 53 % of time in home and community, 33 % sleeping, and only 14 % in schools. It is clear that expecting children to achieve SL before they leave high school is too ambitious, because it overlooks much larger learning resources and potentials outside schools and beyond high school. Adopting the notion of SE helps tapping such large resources and potentials beyond formal science education.

It must be pointed out that the three visions of the expanded notion of SL, although distinct, overlap to some degree. The three visions of SL depend on and enhance each other. Focusing on one vision while ignoring others is undesirable. The relationship among the three visions can be represented in a Venn diagram in Fig. 1.

The above-expanded notion of SL is consistent with current theories of learning. Anderson (2007) summarizes current perspectives on science learning into three groups of theories: conceptual change, social cultural activities, and critical theories. The conceptual change learning theories emphasize the changing processes and final outcomes from naïve informal scientific ideas to formal scientific understandings, the social cultural activity theories emphasize participation in cultural and discourse communities, and critical theories emphasize empowerment and transformation of current social structures of power. Vision I SL is compatible with the conceptual change learning theories, Vision II SL is compatible with the social cultural activities learning theories, and Vision III SL is compatible with critical theories.

Fig. 1 Relationship among
three visions of SL

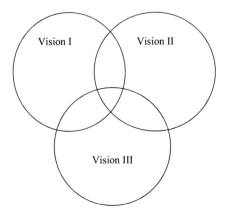

Specifically in terms of SE in the expanded SL, critical theories are directly relevant. This is because, first, the purpose of SE is to promote a more harmonic and progressive society by developing better mutual understanding of diverse views about complex issues (e.g., global warming). Second, SE requires active participation in various science-related activities inside and outside schools. Third, a prerequisite for SE is an adequate appreciation and understanding of scientific theories, principles, approaches, as well as science as an enterprise.

Science Engagement Curriculum Policy Images

Like all notions of SL, the expanded notion of SL is also a curriculum policy because it expresses a selected point of view about what counts as science education, is value-laden, and could be enacted into various forms or images of curriculum (e.g., institutional, programmatic, and classroom, Deng 2011; Roberts 2011). Given that a large body of literature on curriculum images in terms of Visions I and II SL is already available (see Roberts 2007, 2011), and a distinction between current notions of SL and the expanded notion of SL is in SE, below I provide three sample SE curriculum images as illustration.

Science and the Public: An Online Graduate Program

A notion of SE requires that people of all ages actively participate in science activities. For example, research scientists, although may be highly proficient in terms of Visions I and II SL, may not necessarily be well equipped with the skills to communicate science to the general public. Professionals with an undergraduate or graduate degree such as science reporters, government employees, museum staff, and so on, may have adequate initial education in sciences or other fields, thus meeting Visions I and II SL, but they may not possess current understanding related to such issues as science and religion, science and policy, science and humanism, and science and secularism nor may they have adequate knowledge and skills in engaging in science education. In all these examples, SL in terms of Vision III is needed.

In response to the above need, faculty from State University of New York at Buffalo (UB), in conjunction with professionals from the Center for Inquiry (CFI) – Transnational, have developed a curriculum focusing on Science and the Public. This curriculum, delivered by Graduate School of Education at UB in the form of an EdM in Science and the Public (EdM SAP), works to advance research and education concerning the public understanding of science and its intersections with public policy, culture, and values. It is offered completely online.

EdM SAP is designed to (a) prepare professionals to better engage in public activities and debates related to science, (b) to promote SL and understanding in the public at large, and (c) to promote scholarship in science and humanism, science and public policy, and science in the political, religious, and secular environments. So far, there have been over 60 students enrolled in the program, and 18 have graduated with the degree. These students have come from many US states and from other countries (e.g., Canada, Japan, Ireland, and France). These students are professionals in various fields including research scientists and engineers, public relations officers, science filmmakers, primary care pediatricians, freelance writers/editors, university professors, lawyers, veterinarians, school science teachers, and science museum educators. These professionals have a minimum of a bachelors degree; some have more advanced degrees (e.g., M.D., Ph.D., JD, DVM, MBA, and MA). After graduating from EdM SAP, it is expected that they will become leaders in science engagement in their own professions.

Science engagement takes various forms in the program. For example, in one of the courses, the following public online presentations were made on a public blog site:

1. Science and Religion: http://www.omniscopic.com/ScienceandReligion.htm
2. Energy Efficient Automobile: http://estrahle.blogspot.com
3. When the Public Become Scientists: Science in the Court: http://paradiggm. blogspot.com/
4. Suppression of Dissents: The Influence of Political Appointments on Science Policy: http://mjray.blogspot.com/
5. From Quacks and Nostrums to WHCCAMP: 50 Years of Thinking about Complimentary and Alternative Medicine Policy in America: http://www.slideshare. net/tdonnelly/lai531-presentation/
6. Anti-science: Public Ignorance, Rejection and Denial: http://www.slideshare.net/ idoubtit/antiscience-slidecast

Completing a thesis is also a requirement of the program. Through completing this requirement, students actively engage in social, cultural, political, and environmental issues related to science. The following are the sample completed theses:

1. Relationship between prevalence of science education and religious affiliation across Ireland: An analysis based on census data
2. Direct-to-consumer advertising: Its television audience and country comparative prescription drug sales
3. Energy: A public discussion
4. The attitudes of science center visitors toward a human body exhibition
5. When science meets lifestyle journalism: A look at how the news media reports on complementary and alternative
6. Being scientific: Popularity, purpose and promotion of amateur research and investigation groups in the US

The above theses address current and important social, cultural, political, and environmental issues. Students completing the above theses actively participate in debates of the issues by conducting comprehensive literature reviews, designing a study to collect and analyze data to identify patterns, and disseminating findings in various formats (e.g., print media, website, blog). Their roles in relation to the issues are honest brokers instead of pure science advocates. The purpose of their engagement in the issues is to promote understanding among various positions/viewpoints.

The Inconvenient Truth: A Documentary on Global Warming

An Inconvenient Truth is a 2006 documentary film directed by Davis Guggenheim about former United States Vice President Al Gore's campaign to engage citizens about global warming; it received two Academy Awards including the Best Documentary Feature and Best Original Sound. Gore effectively used his personal profile and multimedia presentations (PowerPoint, film, songs, etc.) to engage the general public and convince them that global warming is real and it is man-made, and immediate action by every global citizen and government is in order. The entire film was around Al Gore's keynote presentation on global warming with full facts, personal stories, and of course humors; it engages audience, young and old, in a casual way, although the underlying message is a serious one. The skillful engagement of audience through all of the above modes gradually leads audience to believe that climate change is real based on the current consensus within the scientific community, that it threatens not only future economic development but also world peace, and that without immediate actions, human catastrophe is inevitable. The film presents global warming not just a scientific issue but also a technological, social, political, cultural, economic, and moral one.

An Inconvenient Truth exemplifies Vision III SL because it deals with a contentious political and environmental issue, i.e., global warming, in the USA and around the world. It promotes evidence-based reasoning and active dialogue among competing views in order to build a consensus. Al Gore's role in the film is mainly an honest broker, although at the same time a strong science advocate. The impact of the film and his engagement of the public on global warming goes far beyond communicating science and technology to the public; it also promotes world peace and a better future planet, as evidenced by the 2007 Nobel Peace Prize co-shared by Al Gore and the IPCC (Intergovernmental Panel on Climate Change). The award citation states

> ... While the IPCC has laid the scientific foundations for our knowledge about climate change, Al Gore is in the opinion of Norwegian Nobel Committee the single individual who has done most to prepare the ground for the political action that is needed to counteract climate change. He is the great communicator. He reaches people all over the world with his message. ... No one can charge Gore with lacking concrete guidelines for what individuals can do. *An Inconvenient Truth* contains sixteen tightly-packet pages of advise on "what you personally can do to help solve the climate crisis. (Dec. 10, 2007, Oslo, Nobel Prize Presentation)

Oceanside Community Science Project (Roth and Lee 2004; Roth and Calabrese 2004)

While school science is often conceived within the boundaries of school settings and involves only teachers and students, the Oceanside community science project involved middle-school students and teachers in not only learning science and technology but also actively participating in finding solutions to a local environmental problem – upgrading a local creek to make it suitable as a trout habitat. Students participated in the project alongside with research scientists from a local university, community leaders, local residents, and so on. Students not only actively engaged in a variety of scientific inquiry practices (e.g., developing research questions, design procedures to collect and analyze data) but also presented research findings and communicated them to community leaders for actions. Thus, science learning originated from the community and also ended in the community. All learning activities were community-based participatory activities. Roth and Lee categorize this project in three principles: (a) because society is built on division of labor, scientific literacy is a collective property emerging from sharing of labors; (b) scientific literacy is not privileged but one of many resources to draw to solve problems; and (c) scientific literacy is a lifelong participation in community life.

The Oceanside community project exemplifies Vision III SL in that students actively engaged in social and environmental issues. Science learning is action-oriented and students acted as honest brokers in seeking solutions. Students learned and practiced critical thinking, communication, and consensus building as they participated in the community-based project. Scientific literacy was emergent from participating in the project because no prescribed curriculum or activities were available.

Conclusions

SL has traditionally been considered as a state to achieve or commodity to possess in other words, as being extrinsic to individuals. It has also been based on a deficit- and one-way transport model. In addition, there has been continuous divide between the scientific community and school children as well as the general public. Science education literature has accumulated a large body of conceptual and empirical studies arguing for alternative approaches to SL. This chapter proposes an expanded notion of SL by incorporating science engagement (SE) as a conceptual framework to integrate various alternative approaches to SL. SE is action-oriented, i.e., what to do with science, instead of accumulation of science knowledge. SE is both a state and a lifelong process, a personal choice and an economic necessity, and a personal enhancement and civic or collective participation. This new notion implies that SE is a task of both formal and informal science education; it creates a demand for students of all ages and all professionals to become both science participants and learners.

Adopting this expanded notion of SL has potential to improve both formal and informal science education. It is unrealistic to expect that an equal emphasis is placed on all three visions of SL for all science education settings; differential emphases on different visions at different lifelong times are reasonable. For example, it is reasonable to place more emphasis on Visions I and II in formal K-12 science education, while more emphasis on Vision III may be appropriate for adult continuing science education. It is necessary to emphasize that all three visions of SL are applicable to both formal and informal science education along the entire lifetime span, and maintaining an appropriate balance on all three visions for a specific science education program (e.g., K-12 science education) is a major challenge for developing curriculums.

The ultimate goal of the expanded notion of SL is to promote a more democratic and harmonic society through increased engagement between the scientific community and the rest of the society. This goal is particularly valuable and greatly needed in today's US political and social climates. The gridlock between the democratic and republic congressional delegations on almost all major social and economic issues exemplifies the need for more science engagement in social and political decision-making processes. The unnecessary confrontation between science and religion intensified by such prominent scientists as Richard Dawkins (e.g., *the God's Delusion*) is another example of how SE is desirable. We need more people like Carl Sagan who, through his skillful scientific engagement activities such as public talks, books, movies, and TV series, had enthused a generation of people of all ages to pursue science and engage in science-related issues and activities.

References

Aikenhead, G. S. (2006). *Science education for everyday-life: Evidence-based practice*. New York: Teachers College Press.

Aikenhead, G., Orpwood, G., & Fensham, P. J. (2011). Competing visions of scientific literacy. In C. Linder, L. Ostman, D. Roberts, P. Wickman, G. Erickson, & A. MacKinnon (Eds.), *Exploring the landscape of scientific literacy* (pp. 28–41). New York: Routledge.

American Association for Advancement of Science [AAAS]. (1989). *Science for all Americans*. New York: Oxford University Press.

Anderson, C. W. (2007). Perspectives on science learning. In S. K. Abell & N. G. Lederman (Eds.), *Handbook of research on science education* (pp. 3–30). New York: Routledge.

Apple, M. W. (1992). Do the standards go far enough? Power, policy, and practice in mathematics education. *Journal for Research in Mathematics Education, 23*(5), 412–431.

Bauer, M. W., & Jensen, P. (2011). The mobilization of scientists for public engagement. *Public Understanding of Science, 20*(1), 3–11.

Bauer, M. W., Allum, N., & Miller, S. (2007). What can we learn from 25 years of PUS survey research? Liberating and expanding the agenda. *Public Understanding of Science, 16*, 79–95.

Bencze, L., & Carter, L. (2011). Globalizing students acting for the common good. *Journal of Research in Science Teaching, 48*(6), 648–669.

Bransford, J. D., Brown, A. L., & Cocking, R. R. (Eds.). (2000). *How people learn: Brain, mind, experience, and school*. Washington, DC: National Academy Press.

Christensen, L. L. (2007). *The hands-on guide for science communicators*. Munich: Springer.

Council of Ministers of Education of Canada [CMEC]. (1997). *Common framework of science learning outcomes*. Toronto: Council of Ministers of Education.

Deng, Z. (2011). Scientific literacy: Content and curriculum making. In C. Linder, L. Ostman, D. Roberts, P. Wickman, G. Erickson, & A. MacKinnon (Eds.), *Exploring the landscape of scientific literacy* (pp. 45–56). New York: Routledge.

Fensham, P. (2011). Globalization of science education: Comment and a commentary. *Journal of Research in Science Teaching, 48*(6), 698–709.

Friedman, T. L. (2006/2009). *The world is flat: A brief history of the twenty-first century*. New York: Farr, Straus and Giroux.

Garrson, J. W., & Lawwill, K. S. (1992). Scientific literacy: For whose benefit? In S. Hills (Ed.), *Proceedings of the second international conference on the history and philosophy of science and science education* (Vol. 1, pp. 337–349). Kingston: Queens' University Press.

Gilbert, J. (2005). *Catching the knowledge wave? The knowledge society and the future of education*. Wellington: New Zealand Council for Educational Research.

Hand, B., Alvermann, D. E., Gee, J., Guzzetti, B. J., Norris, S. P., Phillips, L. M., et al. (2003). Guest editorial: message from the "Island Group": What is literacy in science literacy? *Journal of Research in Science Teaching, 40*(7), 607–615.

Hodson, D. (2003). Time for action: Science education for an alternative future. *International Journal of Science Education, 25*(6), 645–670.

Koballa, T. R., & Glynn, S. M. (2007). Attitudinal and motivational constructs in science learning. In S. K. Abell & N. G. Lederman (Eds.), *Handbook of research on science education* (pp. 75–102). New York: Routledge.

Laetsch, W. M. (1987). A basis for better public understanding of science. In GIBA Foundational Conference (Ed.), *Communicating science to the public* (pp. 1–18). New York: Wiley.

Layton, D., Jenkins, E., Macgill, S., & Davey, A. (1993). *Inarticulate science? Perspectives on the public understanding of science and some implications for science education*. Nafferton/Leeds: Studies in Education Ltd./University of Leeds.

Linder, C., Ostman, L., Roberts, D., Wickman, P., Erickson, G., & MacKinnon, A. (Eds.). (2011). *Exploring the landscape of scientific literacy*. New York: Routledge.

Liu, X. (2009). Beyond science literacy: Science and the public. *International Journal of Environmental & Science Education, 4*(3), 301–311.

Miller, J. D. (1983). Scientific literacy: A conceptual and empirical review. *Daedalus, 112*(2), 29–48.

Miller, J. (1987). Scientific literacy in the United States. In GIBA Foundational Conference (Ed.), *Communicating science to the public* (pp. 19–40). New York: Wiley.

Miller, J. D. (1998). The measurement of civic scientific literacy. *Public Understanding of Science, 7*, 203–223.

Mooney, C. (2005). *The republican war on science*. New York: Basic Books.

Mooney, C., & Kirshenbaum, S. (2009). *Unscientific America: How scientific illiteracy threatens our future*. New York: Basic Books.

National Academy of Engineering. (2008). *Grand challenges in engineering for 21st century*. Washington, DC: National Academies Press.

National Research Council [NRC]. (1996). *National science education standards*. Washington, DC: National Academy of Sciences.

National Research Council [NRC]. (2007). *Rising above the gathering storm: Energizing and employing America for a brighter economic future*. Washington, DC: National Academy of Sciences.

National Research Council [NRC]. (2011). *A framework for K-12 science education: Practices, crosscutting concepts, and core ideas*. Washington, DC: National Academy of Sciences.

Norris, S. P., & Phillips, L. M. (2003). How literacy in its fundamental sense is central to scientific literacy. *Science Education, 87*, 224–240.

Osborne, M. D., & Calabrese-Barton, A. (2000). Science for all Americans? Critiquing science education reform. In C. Cornbleth (Ed.), *Curriculum politics, policy, practice: Cases in comparative context* (pp. 49–75). Albany: State University of New York Press.

Osborne, J., Simon, S., & Collins, S. (2003). Attitude towards science: A review of the literature and its implications. *International Journal of Science Education, 25*(9), 1049–1079.

Pew Research Center for the People & the Press. (2009). *Scientific achievement less prominent than a decade ago: Public praises science; scientists fault public, media*. Washington, DC: The author.

Pielke, R. A. (2007). *The honest broker: Making sense of science in policy and politics*. New York: Cambridge University Press.

Roberts, D. A. (2007). Scientific literacy/scientific literacy. In S. K. Abell & N. G. Lederman (Eds.), *Handbook of research on science education*. Mahwah: Lawrence Erlbaum Associates, Publishers.

Roberts, D. (2011). Competing visions of scientific literacy. In C. Linder, L. Ostman, D. Roberts, P. Wickman, G. Erickson, & A. MacKinnon (Eds.), *Exploring the landscape of scientific literacy* (pp. 11–27). New York: Routledge.

Roth, W.-M. (2002). Taking science education beyond schooling. *Canadian Journal of Science, Mathematics, and Technology Education, 2*(1), 37–48.

Roth, W.-M., & Calabrese, A. (2004). *Rethinking scientific literacy*. New York: RoutledgeFalmer.

Roth, W.-M., & Lee, S. (2004). Science education as/for participation in the community. *Science Education, 88*(2), 263–291.

Sadler, T. (2004). Informal reasoning regarding socioscientific issues: A critical review of the research. *Journal of Research in Science Teaching, 41*(5), 513–536.

Shamos, M. H. (1995). *The myth of scientific literacy*. New Brunswick: Rutgers University Press.

Shen, B. S. P. (1975). Scientific literacy and the public understanding of science. In S. B. Day (Ed.), *Communication of scientific information* (pp. 44–52). Basel: S. Karger AG.

Snow, C. P. (1959). *The two cultures and the scientific revolution*. New York: The Cambridge University Press.

Snow, C. P. (1960). Science and government. In C. P. Snow (Ed.), *Public affairs* (pp. 99–149). New York: Charles Scribner's Sons.

Snow, C. P. (1963). The two cultures: A second look. In C. P. Snow (Ed.), *Public affairs* (pp. 47–79). New York: Charles Scribner's Sons.

Suleski, J., & Ibaraki, M. (2010). Scientists are talking, but mostly to teach other: A quantitative analysis of research represented in mass media. *Public Understanding of Science, 19*(1), 115–125.

Wynne, B. (1995). Public understand of science. In S. Jasanoff, G. E. Markle, J. C. Petersen, & T. Pinch (Eds.), *Handbook of science and technology studies* (pp. 361–388). Thousand Oaks: Sage Publications.

Zeidler, D. L. (1984). Moral issues and social policy in science education: Closing the literacy gap. *Science Education, 68*(4), 411–419.

Zeidler, D., & Sadler, T. (2011). Competing visions of scientific literacy. In C. Linder, L. Ostman, D. Roberts, P. Wickman, G. Erickson, & A. MacKinnon (Eds.), *Exploring the landscape of scientific literacy* (pp. 176–192). New York: Routledge.

Zeidler, D. L., Sadler, T. D., Simmons, M. L., & Howes, E. V. (2005). Beyond STS: A research-based framework for socioscientific issues education. *Science Education, 89*(3), 357–377.

Activity, Subjectification, and Personality: Science Education from a Diversity-of-Life Perspective

Wolff-Michael Roth

In this chapter, I develop a perspective on science education that contextualizes it within the totality of learners' lives consistent with the idea that I proposed to take the fullness of life as the minimal unit of analysis (Roth and van Eijck 2010). The episode that today epitomizes for me my interest in this approach is related to a student, Tom, who attended two physics courses that I taught in 1990–1992. But it was only in 2008 – while writing a book on a Bakhtinian perspective on learning in which I used transcripts from those courses (Roth 2009) – that I came to develop a better understanding of the phenomenon that I present here. It was especially while working on the project that would provide the kinds of stories that I present in the second part of this chapter that it became obvious that a theoretical approach different from the going emphasis on "science identity" was required. In my view, science *identity* focuses too much on the individual and very little on the cultural-historical aspects of who we can be in and through participating in society.

As a high school teacher, I had been interested not only in students' learning of physics but also in allowing them to develop better understandings of epistemology, the nature of science, and the nature of their own learning. The philosophical texts that I asked my students to read were produced by authors such as Gregory Bateson (an anthropologist and philosopher), David Suzuki (a Canadian geneticist, broadcaster, and environmentalist), or Bruce Gregory (the associate director of the Harvard–Smithsonian). These authors presented epistemologies that stood in contrast to the realist and objectivist ideas students brought to the classroom. Students wrote reflections about what they read and we discussed the readings in class. Moreover, in addition to reading about science as it is really done, students spent much of their time in the laboratory (about 70 %) designing and conducting their own investigations within the context of what the provincial curriculum had foreseen. Back then, I believed that such an approach would allow students to

W.-M. Roth (✉)
University of Victoria, Victoria, BC V8W 3N4, Canada
e-mail: mroth@uvic.ca

N. Mansour and R. Wegerif (eds.), *Science Education for Diversity: Theory and Practice,*
Cultural Studies of Science Education 8, DOI 10.1007/978-94-007-4563-6_3,
© Springer Science+Business Media Dordrecht 2013

develop not only a better appreciation of how science works, but also to develop a different approach to their own learning. Much to my surprise, at that time, there were students who remained strongly committed to a realist perspective on science. Even more surprising to me as a teacher, there were students who did not at all like laboratories – even though the science education literature had shown that laboratory activities were "motivating" students or at least seen as a form of time-out from the lectures that they normally attended including the other classes at my school.

Tom was one of those students who not only stated his aversion to laboratory activities but also who provided extensive explanations about his position (e.g., Lucas and Roth 1996). In short, Tom wanted to become an engineer and had set as his main goal to enter one of the country's foremost engineering schools. To get into the school, he had to take physics at the high school level and receive provincial academic credits of sufficiently high standard to make it into the school of his choice. He explained to me that this meant that he had to do well on examinations and that he had to be well prepared to do well in his introductory-level courses at the university. This is why he wanted to know and learn the scientific canon. He said that he completely bought into the constructivist argument about the individual construction of knowledge – and precisely for this reason, he rejected laboratories. He explained that the laboratories would allow him to develop ideas and understandings based on his prior knowledge. These ideas and understandings might not be consistent with the scientific canon that was the measure against which his own performance would be held in year-end, provincially controlled high school examinations and future examinations at the university. Considerably more negative than the class average in his appreciations, he disliked especially the negotiation of meaning, student autonomy, and student centeredness that characterized the classroom learning environment that I had organized. He said he wanted me to lecture, tell him the right answers, and help him do well on examinations and end-of-chapter textbook problems.

At the time, I was baffled. As a teacher, I had always been concerned with the well-being of my students. I designed my classroom environments to involve them in decision-making and provided students with the freedom to allocate their time according to their needs – as long as major milestones were met. The students did their work at the time that they felt at their best. Why would Tom have developed such a negative attitude toward a learning environment that was designed to give him greater control over his own learning? Why would Tom desire to abandon a high degree of self-determination in favor of an external locus of control over his activities and the evaluation thereof? Today I understand that in those days I was thinking about these issues from *within* science education rather than from a total-life perspective that I developed only more recently (e.g., Roth and van Eijck 2010). Tom's overarching goal was to become an engineer. He took physics not because he particularly liked the subject – in fact, he disliked the subject to a considerable extent as much as the approach I had chosen for teaching it. It turns out that he also played the piano; this he liked to a much greater extent than doing physics. But in the academic context of the province at the time, he needed the physics course to get into engineering.

Today I understand much better how to approach this and similar issues. In the course of writing my book on Bakhtinian perspectives on learning, I realized that much of what concerns us in our lives is suppressed and repressed in school (Roth 2009). Thus, in our lives, we participate in multiple *activities*[1] in the course of a single day or week. Each of these activities is characterized by a collective object/motive – farming produces resources for making foodstuff, manufacture produces clothing for keeping warm, and schooling reproduces societal structures and labor resources. To understand Tom's stance with respect to physics, I must not attempt to understand it through terms such as "science identity," "science motivation," or "interest in science." Rather, I realized that I needed to take a perspective of his total life. As part of his life, he comes to participate in different activities with different object/motives. To understand Tom, we need to look at all the activities and *collective* object/motives, which, for him, stand in a highly individual, singular, hierarchically organized relation. Thus, becoming an engineer was on the top of this hierarchy, and playing piano was also somewhere near the top. Doing or knowing physics was much less important and only subsidiary to his main personal goal: becoming a member of the engineering community. In fact, this hierarchical network of object/motives may be used in redefining the concept of *personality* (Leontjew 1982).[2] It is this position that I articulate here because it allows us to understand science education from the perspective of the diversity of an individual's life across a diversity of activities, involving a diversity of relations to other people from equally diverse backgrounds. In the course, I develop a complementary concept, *subjectification*, which is used to denote the developmental process of becoming – as the subject of a specific activity – the context and relations of which we also find ourselves subject to and subjected to. *Personality*, therefore, integrates the different forms of subjectivity that we are and experience while participating in different activity systems.

Cultural-Historical Activity Theory

Cultural-historical activity theory is the result of efforts to develop a Marxist psychology – a psychology concerned with real, living human beings in flesh and blood, their needs, interests, and emotions rather than with abstract subjects constructing their minds and knowledge about the world, who relate to others and

---

[1] *Activity* is understood throughout this chapter in the manner that the concept was developed in the German and Russian languages of the founders of activity theory. Thus, these languages make clear distinctions between *Tätigkeit/deyatel'nost'* (activity) and *Aktivität/aktivnost'* (activity). The first term refers to a specifically *societal* formation designed to meet a collective need (food, tools, shelter), whereas the second term refers to being busy without a collective object/motive (predmet).

[2] When a Russian author's name appears in the text, I consistently use the English spelling of the name. When I reference an original or a translation into another language (e.g., German), then the name appears as printed on the book cover.

the world only through representations. It is a psychology that has no interest in the Cartesian opposition of body and mind, the Galilean (constructivist) distinction between mind and world, or the opposition between individual and collective (e.g., Vygotsky 1989). There are two schools of thought that have developed. The first emphasizes the structural, synchronic aspects of an activity system, which is viewed from a god's eye perspective emblematically symbolized in triangular representations; the other emphasizes the dynamic, diachronic nature of activity from the perspective of the subject (cf. Roth and Lee 2007). Here I follow the second approach, which is more consistent with the declared intents of the founders of what today we call *cultural-historical activity theory*.

Activity and Actions

Fundamental to cultural-historical activity theory is the concept of human labor, purposeful activity (*Tätigkeit, deyatel'nost'*) that transforms nature into the means that satisfy human needs (Marx and Engels 1962). Since Marx, the simple moments[3] of activity are recognized to include labor itself, objects, means of production, and the anticipated result, which exists ideally already in the imagination of the worker. Labor not only transforms the material but also realizes the goal, literally objectifying in the process of making some object product. As humans change nature by working with and upon it, they change their own nature as well. Other important moments of activity are the rules and laws governing property and human relations, community, and the division of labor (Marx and Engels 1963). Activities are collectively motivated, serving to meet the generalized needs of members of society, which can be thought of as a network of activities. Because of the existing division of labor, individuals may participate in the activities of their choice and, in exchange for their income, meet their needs – not only those that sustain their lives, like food and shelter, but also those that meet their extended needs related to leisure and pleasure. That is, by participating in the *collective* control over conditions and in the *collective* production of provision to meet the needs of humans generally, members of society expand their *individual* control over their conditions and the production of the means to meet their personal needs.

Activities do not realize themselves: Goal-oriented *actions* do. The relation between the two – activity and action – is mutually constitutive. An action is performed in view of the activity that it realizes in a concrete manner; but the activity exists only in so far as it is realized through a series of actions. Whereas activities

---

[3]In dialectical materialism, a *moment* is a structural aspect of a phenomenon that cannot be understood on its own but only in its part/whole relation with the entire phenomenon and, thereby, in its relation to all other moments that can be identified. The moments do not add up to yield the whole, because, among others, they may in fact stand in a contradictory relation to other moment in the same way that particle and wave nature do not add up to yield the phenomenon of light.

are motivated collectively, oriented toward the transformation of specific (concrete, ideal) objects into results for meeting general and generalized needs, actions are oriented to realize specific goals on the part of the subjects of *this* activity.

In labor, there are two concurrent and interrelated dimensions that are completely separate and independent in other epistemologies: *material* praxis and its *ideal* reflection in consciousness. In human beings, material reality comes to reflect itself in ideal, generalized form (Vygotskij 2002). Each aspect of the activity system therefore has to be understood as existing and appearing on two levels: the material and the ideal (Leontjew 1982). This takes into account a fact initially articulated by Marx that human beings do not just produce something to meet an immediate personal need – in the manner chimpanzees fashion tools to fish for termites – but they produce to meet a generalized (i.e., collective) need in exchange for something that allows them to meet their personal need. Such anticipation is possible only if reality exists a second time, ideally, that is, in consciousness, so that the meeting of individual needs can be anticipated and deferred.

Activity is directed at the transformation of some object. Marx's German and Leont'ev's Russian again offer two different terms where English has only one. The term *Objekt/ob'ekt* (object) refers to something material or ideal that the person is actually working on, whereas *Gegenstand/predmet* (object) denotes something generalized at the material or ideal level that also represents a need/motive.[4] To capture the presence of both levels in any concrete instance of human praxis oriented to the transformation of an object into a result – i.e., the motive of activity – I use the notion *object/motive*. That is, the two moments of activity, its inner (ideal) and outer (material) form, constitute a single unit. This is so because when we look at and analyze any concrete activity, humans are involved in transforming something into something else. They do so *in order to* achieve something, and this *in-order-to* is as much aspect of concrete reality as the *for-the-purpose-of*, the *what-with*, the *who/what-for*, and the *for-the-sake-of-which* that characterizes everyday attention to the world in the manner that it offers itself to the subject.

Subjectification

Central to understanding activity is the (individual or collective) subject of action who, using means of production (tools), transforms some object into an outcome. In the course of so doing, the subject itself is transformed in multifarious ways. First, the subject expends energy and therefore is materially transformed. Second, as a result of repeatedly producing the same form of movements (that realize actions),

---

[4]This is why the products of human activities can create new needs, for example, the production of the cell phone created the need for cell phones so that today many people "cannot live without it." The need did not just exist; it is not a basic need that "must" be filled for humans to live. It is a need that is the result of productive human activity (Leontjew 1982; Marx and Engels 1962).

the body or bodies of the subject are transformed, becoming increasingly practically competent. Third, in praxis, the comprehension of the subject is changed, as it increasingly comes to understand praxis on the ideal level. Fourth, with increasing practical and ideal competence, the changes of the subject are recognized within the collective (community) writ large (i.e., not only within a specific group that might constitute the collective subject of activity but also within all those who are the subjects in other concretizations of the activity). We may therefore understand the transformations that an individual undergoes in the course of participation in activity in terms of a trajectory of legitimate peripheral participation (Lave and Wenger 1991). Alternatively, we may understand the process as one of *subjectification*. By this term I mean, drawing on an articulation of the concept likely meant very differently, "the production of a constitutive body and of a capacity for enunciation not previously identifiable within a given field of experience" (Rancière 1995, p. 59). This production occurs through a series of actions; and the "identification [of the body and the capacity for enunciation] comes with the reconfiguration of the field of experience" (p. 59). I read, and use, this description of subjectification in the following way. In labor I use actions to realize the activity that orients what I do. Through my actions, both I, as a constitutive body, and my capacity for enunciation are changed. The new state of my body and the new capacity for enunciation were not previously visible in my field of experience (activity). The production of my body, therefore, and the manner in which it is identified in discourse, is part of the reconfiguration of the field (activity). The advantage of such a formulation over others is that it escapes the structure–agency opposition, where the agent is the source of what happens. In Rancière's approach, the subject is not the antecedent of the action but is subject to and subjected to the field. Rather, the acting body is identified with a reconfigured field of experience. It is not the intentional action itself that brings about the change but change is a collateral of acting in a field, which changes, and these changes allow the identification of the body and the capacity for enunciation.

Thus understood, the term subjectification, therefore, allows me to denote the changes that the subject undergoes in and through its participation in activity. Of course, the subject is subject of activity – doing what it has decided to do. Learning, however, occurs not only when there is learning-oriented action but also whenever there is practical activity. Even when a person engages in the most routine, perhaps most boring actions, the subject is transformed. We could show this in the case of fish culturists, where during some parts of the year a person might be required to throw by hand, using a scoop, up to 200 kg of fish feed. Whereas newcomers often describe the task as tedious, repetitive, and hard on the body, old-timers do distinguish from afar whether a fish-feeding individual is experienced or not (Roth et al. 2008). Thus, even though it appears to be a routine and repetitive job, fish feeding changes the individual who does it. In fact, we could show that as the feeding individuals watch the fish they feed, they become better at feeding, for they begin to stop when the fish no longer take the food. The required perception is developed in and through the feeding process itself, however boring it might appear.

The term subjectification also allows me to theorize other essentially passive forms of the experience that come with being a subject. Because activity is collective, involving material aspects, tools, division of labor, and rules/laws – all of which are the results of cultural-historical developments of the activity – the subject also is *subject to* the determinations that come with any field. That is, only in an ideal world is some action or activity accomplished unproblematically. In the real world, the agent is not only the subject of but also *subject to* and *subjected to* real-world conditions and societal/material relations. As a result of this subjection, participation in societal relations also implies the dialectic of discipline. It is through the disciplining of my body that I develop an intellectual discipline (Foucault 1975; Roth and Bowen 2001). Becoming an increasingly competent member in a community of practice also means being increasingly subject to its determinations. Participation in a field therefore also is subjection to the field.

The upshot of this situation is that there are inner contradictions for diversity. Being – and becoming as – the subject of activity must not be conflated with agency. There is an inherent passivity that comes with any form of participation. This passivity is captured in the adverbial formulations "to be subject to" and "to be subjected to." A community is defined by its membership. Members also recognize nonmembers – e.g., by their nonstandard material and ideological practices. Diversity must be thought through this dialectic of renewal and change of a community as it deals with the inherent diversity that comes with the incorporation of *any* new member not just with the incorporation of members from a visible or non-visible minority. The term subjectification allows us to understand the issue of diversity in the tension of an agency | passivity dialectic.

Personality

> But the human essence is not an abstractum inherent in the singular individual. In its reality it is the ensemble of *societal* relations.[5] (Marx and Engels 1958, p. 6, my translation, emphasis added)

In the wake of Marx's dictum, cultural-historical psychologists including Lev Vygotsky and Alexei Leont'ev viewed human beings in terms of the ensemble of societal relations that they entertain in the course of their lives. This leads to the understanding that "any higher psychological function was external; this means that

---

[5]Here, as elsewhere, my translation takes into account what Marx has written rather than what translators into English produce – perhaps for political reasons. Marx writes about *societal* (Ger. *gesellschaftliche*) rather than social (Ger. *soziale*) relations. In the original text translated as "Concrete Human Psychology" (Vygotsky 1989), the authors quote Marx using the Russian equivalent for societal (*obshchestvennyj*) rather than the one for social (*sozial'nyo*) (Vygotskij 2005). Similarly, the Russian and its German translations of *Thought and Language* use the same equivalent of societal as distinct from social, whereas the English translation only uses the adjective social.

it was social; before becoming a function, it was the social relation between two people" (Vygotsky 1989, p. 56).[6] As a result, "the personality becomes a personality for itself by virtue of the fact that it is in itself, through what it previously showed is itself for others" (p. 56). He makes direct reference to Marx and Engels (1962) who stated that it is only through our relations to *others* as human beings that we relate to ourselves as human beings. Both Self and Other are concrete realizations of the genus *man*. The concept of personality was subsequently developed in terms of the activities (activity systems) that a human being participates in the course of its societal and material life (Leontjew 1982).

In cultural-historical activity theory, personality is understood in terms of the category *activity*, its inner structure, existing mediational relations, and the forms of consciousness that activity produces. This allows an articulation of the stable basis of personality and the aspects that pertain to it and those that do not. In this conception, "the real basis of human personality is the totality of the, by nature societal relations man entertains with the world, precisely those relations that are *realized*. This occurs in/through his activity, more precisely, in/through the totality of his manifold activities" (Leontjew 1982, pp. 175–176). Personality transcends the traditional oppositions of individual and collective, inter-psychological and intra-psychological, or inside and outside – because it is interested in those "transformations that derive from the self-movement of the subject's activity in the system of societal relations" (p. 173). Leont'ev understands the subject in terms of life forces that can operate only via the outside. It is there, in the outside, that the life forces *concretely realize* themselves and thereby constitute a transition from cultural possibility to concrete material reality. The real foundation of personality, therefore, does not lie in a set of preprogrammed genetically determined routines, natural capacities, knowledge, and competencies. Rather, personality is founded "on a system of activities that are realized through these knowledge and competencies" (p. 178).

In this approach, then, personality is not defined in terms of the individuality or singularity of the person but through the totality of societal relations that occur within collectively motivated activities. This immediately leads us to understand the stable basis of personality: society and societal relations. The basis of personality, therefore, is not, as in constructivism, the individual Self that produces itself and its cognitive structures based on its biology. The basis is that which is specifically human about our species: society, its historically evolved culture, and the kinds of relation that both enable and constrain the interactions with the others and the material world. What remains to be worked out next is what is different between different personalities if the basis of all personalities is the same: society and the forms of activities that guarantee its reproduction and transformation. This under-standing of personality, therefore, goes well with the position on subjectification articulated above and understood as a process resulting from the association of a

---

[6]In this quote, Vygotsky does indeed use the adjectival forms *sozial'nyi* and *sozial'nim*, social but which may also be translated as societal.

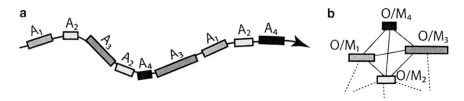

Fig. 1 Cultural–historical activity theoretic perspective on personality. (**a**) In the course of its everyday life, the individual participates in many different activities, with different *collective* object/motives and subjectivities. (**b**) Personality is the result of the hierarchical organization of collective object/motives into "knot-works"

constituted and constitutive body and forms of enunciation, inherently related to participation in collective life.

In the course of a day, week, month, and so on, an individual participates in many different activities (Fig. 1a). These activities have collective object/motives: They arise from generalized collective needs that are met with the end results of the productive activity. Within these different forms of activity, the individual constitutes different forms of subject, with different forms of subjectivity undergoing different forms of subjectification. Thus, an individual, who is a science teacher during the day, may be a (graduate) student in the late afternoon, a shopper in the early evening, a parent somewhat later, a hobby beekeeper attending a bee club meeting, and finally a husband and lover. It is still the same individual (body) but the forms of subjectivity are different and so are the forms of development that occur in each form of activity. For the individual, the different object/motives and activities come to be tied together into a hierarchical "knot-work" (Fig. 1b). It is precisely this knot-work that defines personality. The core of personality, then, is a hierarchy of *collective* activities or object/motives, which is the result of its own development. That is, "the 'knots' that combine the individual activities are tied not by the biological or mental forces of the subject, which lie in him, but by the system of relations into which the subject enters" (Leontjew 1982, p. 179, my translation). These relations are characteristically *societal* in nature rather than social. This is so because there are other social animals. But the human form of consciousness is specific to society. It is precisely this consciousness that distinguishes, for example, even the worst builder from the best bee: It allows the human "to build a cell in his head in advance of building it in wax" (Marx and Engels 1962, p. 193).

Once we take the perspective of personality as the ensemble of societal relations an individual has participated in, we come to understand that we no longer are able to investigate something like "science identity." For whatever happens in a science classroom is, from the perspective of the individual person, only part of a larger, stratified knot-work of activities and object/motives. Thus, for Tom, the physics course that I was teaching had a much lower priority than the object/motives of other activities, those that he engaged in – including playing piano – and those that he anticipated to become part of (engineering). Playing the piano is not some singular interest that characterized Tom; rather, it has constituted a form of activity since its invention with collective object/motives that are concretely realized in his playing of the piano. The specificity of his individuality arises from the specificity in which the various collective object/motives available in his society come to be knotted together and hierarchically organized. Moreover, as soon as some activities are connected, this knot-work constitutes the driving force of its own development, continuously reinforcing or re-arranging the position of the different object/motives within the overall hierarchy.

On the Way to Become a Doctor

> I haven't done any of these things, so I can't really say I don't want to. (Katie)

To better understand the relation of activities, subjectification, and personality, my research team and I followed three individuals over a 4-year period. One of these individuals was Katie, who, at the time of first contact, was enrolled in an 11th-grade biology course and a career preparatory course taught simultaneously by the same teacher. She participated in an internship in a scientific laboratory that my research center had organized for high school students. Katie participated in interviews before and after the internship and then again over the course of several semesters while enrolled in a local college. During the period, she took further science courses, worked in a lingerie store, took a course in holistic health and healing, job-shadowed a doctor and a respiratory technician, and, still during 12th grade, volunteered once a week in a hospital. Later, she took a course as home support and resident care attendant, which included 3 months of taking classes – such as "Health and Healing," "Lifestyle and Choices," or "Personal Care Skills" – and 3 months of practicum. She continued working in the lingerie store but also worked as a casual in a resident care facility. She then enrolled in a science program, taking physics and mathematics courses while working as a resident care assistant during the summer break. We interviewed her again nearly 4 years after starting the research in which she participated, and we videotaped her in two of her college courses: microbiology and organic chemistry.

Near the end of her second year in the premedical program at the college, Katie suggests that she did not decide to pursue the idea of becoming a doctor because of the sciences. In her early life, the sciences only played a minor role. There were many other formative experiences and activities in which she participated that would make becoming a doctor the primary object/motive according to which all other object/motives and activities would be organized.

In the following, I account for different activities Katie has engaged in and the relations that she has entertained in the process. In her case, the extended amounts of time that she spent in the hospital and the relations with her mother, a hospital worker, became formative early on. Katie made a decision to become a doctor; and this object/motive became the central organizing feature in the hierarchy of activities and related object/motives that defines her personality and the development thereof.

Even within the various activities, she sees her subjectification in terms of the overall goal of becoming a doctor, which encourages her to do things even though these are not her preferred activities (e.g., 11th-grade biology, organic chemistry) and to leave aside others, even though they might come easier to her (mathematics, physics) or that she might prefer (going out with friends). To give readers an appreciation of the kinds of activities she engaged in, some of which became formative, others playing a mediational role, and again others not playing a central role in her development, I present her participation in five areas: (a) early activities and relations, (b) high school science, (c) a science internship in a university biology laboratory during 11th grade, (d) college science, and (e) training as a resident care assistant.

In each of the accounts, readers are encouraged to attend to the dual aspect of development. First, as Katie comes to participate in an activity, she undergoes a process of subjectification, which involves becoming an increasingly competent agent, on the one hand, and being a person subject to and subjected to the context, on the other. The individual activities and the associated object/motives are not all valued equally; they take a different place in a developing hierarchy, the topmost feature remaining fairly constant in a relatively early part of her life: becoming a medical doctor. Katie engages in a variety of activities for the purpose of increasing the likelihood to get into medical school and of realizing the goal she has set herself for herself. That is, we can account for her personality completely in terms of the societal relations in which she engages, which provide her with the forms of experiences and discourses on which the ultimate interest and decision is founded. Nothing Katie can tell us is singular – everything we find in the accounts is constituted by forms of discourse, the ideological resource par excellence.

Early Activities and Relations

Ever since she was 15 years of age, Katie has wanted to be a doctor. Katie's mom works in a hospital. She now is a porter, bringing patients from the X-ray department to other departments, but she used to be a nurse. When Katie was a child, from the time between 5 and 10 years of age, she and her sister visited their mother in the hospital. It was the children's principal way to spend some time with their mother, who "would be working a lot." While in the hospital, Katie got to talk to the nurses and spent time at the bedside of patients. She recounts "lik[ing] the smell of it, too." In the hospital, she "talked to the workers." She went "a couple of times actually where [her] mom works and [she] talk[ed] to people around there and the environment is just so friendly and people are all there for the same reason to help people." These people included doctors and X-ray technicians in addition to the nurses and patients. As a result, Katie came to "just feel really comfortable" in the hospital and around the people whom she encountered there. During the interview, she describes her relation to the hospital as weird: "It was weird. A lot of people hate hospital, but it was very comforting for me, like it was like a welcome feeling, like I feel that peace there."

An important aspect related to her hospital experiences has to do with time. Katie noted that her mother "would be working a lot"; it is for this reason that she changed her job and became a porter, where "she would have less hours." Katie explains, "She'd have to work three shifts on and they're like eight hour shifts. Three shifts on and then she'd get three shifts off. Three shifts on but it's sometimes she would work from like eleven o'clock at night to seven in the morning. So I'd always have babysitters and stuff. Like she was always there for me but it just wasn't the same. So I don't want to, like I'm not saying I had a bad childhood. I obviously had a really good childhood, but I want to provide for my children." Katie would enjoy becoming a pediatrician. But here, too, the amount of time required mediates her choice: "there are way too many hours for that so I wouldn't be able to have a family and no family life." On the other hand, "if the hours are good, one hundred percent that is what I would do." Thus, for example, specializing in dermatology would be of interest, "because it is still in the medical field and there are good hours; and it gives you lots of money too." She would be able to work Monday through Friday. This would differ not only from what her mother has done during her early years but also from a specialization in general practice, where "you are still attached to your patients outside of the hours." As a dermatologist, in contrast, "you don't have to deal with anything outside of the hours and you can be done by four in the afternoon." There are shortcomings, however, with the job of a dermatologist: working in an office or clinic rather than in a hospital. Moreover, she feels that the dermatologist really helps: "Dermatologists help people but they don't help them if they are hurt but only [on the] surface."

Although the early experiences have allowed her to place the object/motive of being a doctor very high in her hierarchy, it is not the highest priority: Having a family is even higher. The kind of hours that her mother has worked would interfere with having a family. She says, she wants to have "good hours, so it enables me to still have a family, which is a main factor in choosing a job, too." When the interviewer insists on the job that she had selected as her most favorite career choice, Katie responds again that the favorite career is "what my heart wants, but these [have to] enable me to have a family. So there are different factors to why I pick [the careers]."

Katie not only has experiences in activities that she is necessarily part of as a young member of society – e.g., going to school or being part of a family unit – but also actively creates opportunities for new forms of relations and the experiences that come with it. For example, she signed up for a youth volunteering program as part of which she intended to spend time in a hospital, which would allow her to engage in further relations with patients and hospital workers. At another time, she arranged for shadowing a female doctor, that is, to engage in a particular kind of relation, following which she changes the hierarchy of career possibilities. She now ranks family practitioner on the top of her career choices. Katie explains: "[I changed] because of my job shadow experience. Because I was unsure if I wanted to become a family doctor because of the weird hours, cause it's like an eight to four job. But you're always on call twenty-four hours as well. Like at the hospital, if one of your patients needs you or something, you have to be there. And there's not really a question about it. So that's why I was unsure. But now that I've had the opportunity to shadow a doctor it shows me that no matter what the hours are, you make sacrifices for any job and those are the sacrifices you're going to have to make if you want to be happy as well."

Another important form of activity for Katie is sports. She has participated and continues to be active in different forms of sports. For example, she started to play soccer in seventh grade and played on a soccer team in a league. She plays a range of other sports – including basketball, baseball, and hockey – and sometimes goes to the gym. Here she has had, "at one time, a personal trainer." It is in the relation with the personal trainer that she learned about how "they show you and say what you want and they target muscle you want to do." She lists personal trainer as one of the possible careers that she could have been be interested in pursuing: perhaps "running like a local gym" where she would teach people about the body and food. She is keenly attuned to her body, in part because of relations with her mother, where she learned to detect that "if you eat bad food the next day you are tired, you don't want to do stuff, you are out of breath all the time. But if you balance all your meals and everything then you just have so much energy, and you sleep well. And just I have always found interesting how what you eat can relate to everything in your body."

High School Science

In high school, Katie did not like the sciences very much. She had a special aversion to chemistry, and she did not do well in her final mathematics course, even though she repeated it to improve on her grade. "In school, you're sitting down in a class for like an hour and twenty minutes at a time. You're listening to a teacher talk to you, or you're reading." Katie preferred "hands on." So even biology, which she attends at the time, she is not really having the kinds of learning experiences that she prefers: "Like in biology we did dissections and stuff like that like there's some hands on – but a lot of it is bookwork and a lot of it is memorization." She does understand much more and much more quickly when she has opportunities to "do hands on": "I learn a lot of things in a short amount of time in the lab, whereas in biology eleven I almost forgot, like I forget a lot of things because you're just reading it and you're doing it for the purpose of just to read it to get a good mark on your test." In terms of subjectification, the laboratory aspect of science is formative, whereas the lecture parts has tended to turn Katie off. Many years later, she notes that she did not like science during her high school years.

Perhaps unsurprisingly, then, we find Katie always involved in the laboratory tasks that the teacher has prepared for the students. Thus, even though she says, "I don't know, I don't like researching animals at all" or "I'm not really interested in any of the plants," she is the first to take the eyedroppers to squirt water onto a planarian to see how it reacts to the stream of water. Katie mounts the Petri dish on the microscope to study the reaction of the planarian to different conditions of

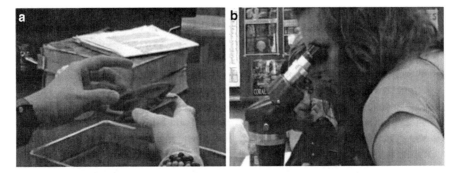

Fig. 2 Katie often is at the forefront of the task engagement in the school biology laboratory. (**a**) She is the first to handle the clam that the students are to investigate. (**b**) She is also the first to investigate the planarians, which students are invited to investigate behaviorally under a number of conditions

Fig. 3 Katie dissects a clam for her group while the others look on. (**a**) Katie uses surgical scissors to cut the tissue from the shell. (**b**) While Katie picks to get at the adductor muscle, John points to the upper part of the diagram of a clam

light (Fig. 2b). Three of the five students in her group merely look on and listen to what she has to say. Katie is also the first to take the clam to be dissected in her hands, though wearing a glove at least on one hand (Fig. 2a). Katie is even more explicit in saying, "I don't like mollusks and stuff like that." She looks forward to her 12th-grade biology class, which is concerned with the human body. This is the aspect of science that she likes, whereas there are many other things that she does not like. That is, her interest in the topics of the biology course is mediated by the same kind of priorities that also mediate her career choices.

As the students do the lab, Katie takes the clam into her left hand and begins to cut through the tissue, following the teacher's instruction. Three of the other four students in Katie's group are looking on. John has the textbook opened at the page where there is a drawing of a clam. He has oriented the book so that Katie can see the diagram with its inscriptions. Katie begins by using her surgical scissors to cut the tissue from the shell to which it is attached (Fig. 3a). She places the open clam in front of herself in the dissection tray.

K: ((*looks into the textbook at her left*)) Where is the heart? Oh so this is the posterior and adductor muscle. ((*turns to the clam in front of her, points to the top left part of the mollusk*)) This is the adductor posterior muscle. ((*She pulls some tissue aside with the tweezers*)) ((*Nobody talks; Katie picks away on the clam, Fig.* 3)) Wait, this little tube, there is a little tube ((*she looks toward the biology textbook to her left, then continues to pick away at the tissue around the posterior adductor muscle*)).

We can see Katie here, as in many other laboratory situations during her high school years, taking the lead. She is taking the lead even in those situations that she does not appreciate all too much. She comments on the smell, "It smells clam," with an intonation that shows less than appreciation for this aspect of their laboratory work. It turns out that she appreciates this aspect of science much more so than other aspects and that she develops something of a "knack," as we would observe some 4 years later in her college laboratory courses as well as during the internship in a leading laboratory in the biology laboratory of the local university.

Science Internship

One of the courses in which Katie enrolled during her 11th-grade year was "Career Preparation." The course was organized by the same teacher who also taught biology. As part of the course, the students are required to provide evidence that they have had workplace-related experiences, such as trailing a professional for a while, doing an internship, or volunteering in a relevant context. During the early parts of the course, the teacher talked about a variety of opportunities for career preparatory students to get a better understanding of science. Katie selected to do an internship in a large, world-renowned university biology laboratory concerned with all aspects of water quality. During an interview conducted prior to the start of this internship, she indicates that she is "sort of interested in it, but if it was on the body, I would be very excited." She adds by saying that she is still excited about doing the internship and that she anticipates it to be interesting.

On that day, the students participated in isolating bacteria from water collected some time ago. Each student gets a turn at placing the filter paper, moving a measured amount of water into the filtering device using a pipette, and running the pump that draws the water through the filter paper (Fig. 4). Nikki stands behind her on the left watching over her every move. She prepares a glass container into which the filter paper with the filtrate is placed. Cass makes some comment, but Nikki says, "No, you don't want to sterilize them now, because it will kill all the bacteria that we're trying to isolate." She takes over all the samples that the high school students have prepared and explains while turning about: "These samples are going into the incubator." After they had walked through several other laboratories to the one with the incubator, Nikki placed the tray. Just as she begins to explain what they would do on the next day, Katie expresses interest in the fact that they are all wearing gloves all of the time.

Fig. 4 Katie is gaining confidence as she does her part in the ongoing scientific experiment – here preparing a filtration

Katie: just a question are we wearing the gloves to protect ourselves from the bacteria or to protect the bacteria from ourselves. cause if we = re touching?

Nikki: y::yea; (0.32) well, we = ll change them before we do anything (0.66) drastic again (0.56) its to protect them from us. (0.77) tomorrow (0.35) theyll look something like this.

Cass: cool.
         (0.45)

Nikki: and you can see the blue and the red.

Katie: uh hm.
         (0.88)

Nikki: great fun
         (0.82)

Cass: the blue is e coli.

Nikki: the blue is E coli.
         (0.56)

Katie: and the red is just random bacteria?
         (0.44)

Nikki: uh:: the red is something actually completely different. ≪p≫ well not completely different> they are fecal coliforms (0.40) which is just other bacteria that live in your intestines.

This transcript shows the relations with their supervisor Nikki oriented not only to better appreciate what they are doing but also to better understand the context within which they work. This interest here pertains to the interaction between those conducting their analyses and the organisms that they are working with. In Katie's question, we see an intuitive understanding that the affect of bacteria and worker is mutual, and the gloves may protect one agent from the other. As her classmate and best friend Cass, Katie is also interested in the contents of the glass dish, including those contents that are not related to their current investigations: the fecal coliforms that show up as a red signal.

As part of the internship, Katie and her peers also had the opportunity to attend a DNA seminar. It was their first science seminar ever. When asked about this experience, the following conversation unfolded.

Katie: It is almost a scary experience.
  PL: Scary?
Cass: It is kind of like that is what I need to learn one of these days?
Katie: Yea, like I have to know and like there are so many acronyms, so many huge words that are like accepted in that lecture. They were accepted like saying "hello," like it just second nature for them to understand what that word means. For me, I'd have to look it up or okay then he'd already be moving on to next thing like you're always behind.
Cass: Yea, like it almost shows you like a window of what you're gonna have to learn, like the level of education you are going to be at in a while. And it's like, "Wow," like if I go down that path I'm gonna be like.
  PL: But [does] that scare you?
Cass: Yes very.
  PL: But somebody studies that, what I will know in the future?
Katie: It is exciting but scary.

"I feel uncertain and scared but I never thought I'd feel foreign, like stereotypically foreign. Like I'm Canadian and you're just going into the U-Vic lab you're not going to feel foreign. But you feel different and almost excluded from everybody else because you're not wearing the same clothes. You don't know what they know. So you're just drawn into a situation that is not everyday life, so it's very foreign." Katie continues to explain: "It's just so different and so out of the ordinary. I never would have thought of going into career prep biology I would go work at a U-Vic lab; like you never think of that because this is an actual project that the scientists are working on, so I think that's why it's *so* foreign. It's not just me 'cause I jobshadowed a doctor. You see doctors every day. I don't know that you see scientists every day, so it's very foreign."

Almost 4 years later, we interviewed Katie again about her experience during the internship. At that time, she had forgotten the details of the experiments she had participated in at the time – even the fact that it had occurred during 11th rather than 12th grade. But in the course of the interview, she talked about how she was personally affected and about the relation with Nikki, the laboratory technician. This account has a great family resemblance with what Katie had said right after the experience. She says: "the main thing I learned is probably just to be more confidence, because you always think even just coming to U-Vic, and being able to work in a lab was something. Like clearly we weren't doing it as aseptically as Nikki was, as clean, as precise. But we were still able to do it. Just knowing that you can do something where somebody has worked so many years toward being able to work there. You cannot do the same thing, produce the same work. But you have the ability to be able to do it. When you are accepted out of high school, it makes the work experience less scary. Because before, when you were in high school, everything is just unknown: 'After high school,' you know, 'I am going to U-Vic or Camosun [College]' and it's so big: 'What am I going to do?' But when you get here, you realize it just more people, doing what they have to do everyday kind of thing. I think it is the exact the same thing as high school, just different materials, you know. And so confidence was something that I learned from there. And then also just even techniques, like I'm doing a lot of same techniques that we've done, like dealing with pipetting and proper cleanliness, and isolating colonies, and subculturing. Like I am doing all of that now and I've already seen it. So it's not new and it has allowed me to be more comfortable."

Katie summarizes her experiences in the university research laboratory in this way: "We got a lot out of it, we did everything she did and we made new friends, so it's an experience that not a lot of people get to do, like hardly anyone." Throughout her account, she attributes much of the positive experience to the laboratory technician and the kind of relationship she has had with her. Katie experienced an increase of her control over the laboratory conditions, and we would be able many years later to observe this tremendous confidence in the laboratory compared to her college peers.

College Science

For Katie, the sciences constitute the knowledge from which "everything else branches." Because science is in everything, and it explains everything, one lives science every day. Thus, "it even explains people who are working in fast food, why the food is the way it is; or, if they are unhappy with their job, it's neurological, too. Science just explains absolutely everything. So when you have the base of absolutely everything, you are powerful." She summarizes: "knowledge is power." Knowledge allows her to "be higher up in life" because "knowledge allows her to be able to figure things out and where to go and how to save money." Studying science allows her to create career opportunities. Although Katie describes herself as being "way better in math and physics, she is still pursuing biology, just because she wants to go into medicine." Mathematics and physics are logical, "Like, 'how does it not make sense?'"

Getting ready to take the MCAT and preparing her portfolio for application to a medical school organizes everything that Katie does: her study and work habits as well as the subjects she chooses. With respect to her studying, Katie says: "You've got to do this. You've got to work hard. You don't really have an option to get bad grades so you have to study. You have to choose studying over going out with your friends or you have to choose." Here, she points out the passive aspects of participating in a form of activity, where there are particular constraints that do not tend to be selected by the subject of activity. To become a doctor, she does have to work hard; and learning to work hard, staying at home rather than going out with friends are aspects of the process of subjectification that is part of the trajectory she has chosen and that organizes her life. Katie views the subjects she takes in college through the lens of becoming a doctor or through the lens of her most important interest: the human body. For example, it is precisely when they cover sugar and its chemistry that Katie can relate organic chemistry to the body, which makes it "more applicable" and therefore more interesting to her. She studies the subject and finds it useful in as far as it is "helpful because it explains the reactions within the body."

As her time in college went on, she came to appreciate science more than she has had in high school or during the first year. She describes: "In high school, everything is laid out for you on the board, everything single piece of information you need to know is laid out on the board. And then you are tested on that. And you're generally given an outline on the test. Then in first year science, I feel like they almost give you everything you need, almost everything. But you still need to do a bit reading. You still need to go to your lab and apply that knowledge to it, to reiterate that knowledge. And then second year, I find they don't give you nearly as much information. Which is fine, because you need to go out and get that yourself. But the labs are *way* more exciting in the second year. Like the lab procedures in grade eleven, twelve, and first year, it was sort of like you go, "Okay," you know like it's nothing really interesting. Whereas I feel like second year, because you know more, you can use your knowledge; you do a lot more interesting things. And even though you are trying to teach yourself more, you always learn more because it's interesting. So it's easier to learn them and want to learn them more, 'Oh that's why they do that, oh that's why,' you know."

On this particular day, near the end of the semester, Katie comes to the microbiology laboratory together with Marsha to conduct another test for their environment isolate – to observe the motility of the bacteria using the microscope (Fig. 5). Compared to Marsha, Katie appears to be so much more familiar with the use of the equipment. Marsha is asking many questions, about how to use the microscope. They needed to use the 100X lens to see the bacteria and so needed to make sure that they put the bacteria in the middle of vision. It takes some effort and time to adjust the microscope. Yet, Katie manipulates the equipment with great ease. After a while, Katie sees her bacteria quickly moving about. Katie then notes that her unknown bacteria – she has picked the sample from the dust of her picture frame – is "very motile." Katie then brought her results of a series of tests to the instructor (Jeremy) and wanted to find out what her bacteria is. The instructor first guided Katie going online to a website, entered all the results that Katie had into the website, and then the website generated four possible bacteria candidates. Katie then wrote down names of these possible bacteria. One of them is 90 % likely. In this manner, Katie has found and identified the bacteria (i.e., achieve the project of environment isolate). In the case of Marsha, she could not "see" her bacteria properly. One possibility is that she did not use the microscope properly; the other possibility is that her bacteria is not motile at all and so it is more difficult to "see" her bacteria. Both Katie and the instructor tried to help her; but they were not 100 % sure if they "saw" the bacteria either.

Fig. 5 Katie is in the process of preparing an inspection of her bacteria for motility. Here, as throughout the course, she exhibits a great deal of self-confidence in doing what she has to do

Throughout this episode, Katie expresses great confidence in what she is doing. There is no question about the fact that she knows *what* she is doing. The video shows no signs of hesitation here as in other parts of the course. When she has a question, Katie does not hesitate to ask the instructor. That is, over the course of the 4 years that we have observed Katie doing laboratory work, she not only has demonstrated interest but also has continuously expanded her competencies. And with these competencies, she also has expanded the level in which she has control over the laboratory environment. At the same time, this competence coincides with a process of increasing discipline, which is the result of being subjected to the (required) discipline in a laboratory. Following her earlier internship in the university laboratory, Katie pointed out her admiration for Nikki, a highly competent and organized laboratory technician. Now, during her second year in the college, she herself exhibits these competencies – e.g., cleanliness, keeping the workplace aseptic – that she noted in Nikki. Perhaps unsurprisingly, Katie began to consider working as a technician at least for a few years – in case that something goes wrong with her MCAT or the medical school application.

Resident Care Assistant

Katie did not go straight through her premedical science program at the college but actually interrupts her science program and enrolls in a co-op program to be trained as a resident care assistant. The program is intended to prepare its students as frontline care providers in long-term care facilities and a variety of community settings. As the students work through the curriculum and in their clinical placements, they are anticipated to acquire the skills, knowledge, and values needed to provide professional care to the elderly and other individuals confronting health challenges.

Katie registers in the program even though she wants to be done in her second year and has written the MCAT examination. However, she became aware of the program as a health-care assistant through a fellow student in a psychology class who also signed up for the course. Even though she had not initially planned to enroll in this program and even though it is not something "that she wants to be," it would provide her with an opportunity to interrupt the science program, "work on her own without school, grow up a little bit," and then "return to appreciate school for what it is." Moreover, she described access to medical school as being very competitive. In "trying to be a more organized person," Katie "figured that this would get her into the medical aspect of things." Moreover, she hopes that this course and the experiences that came with it would "look good on her application because she will have had some hands-on experience." Even though she had to pay $2,500 for the course, the fact that she could use the certificate to "get a job anywhere after this" mediated any concerns about the expense.

After completing the program, Katie describes: She "took what she needed from that experience" and now "she wants to get back to what she wants to do." Working as a registered care assistant now provides her with extra money, which she has never had after high school or by working in the lingerie store. Now, throughout her semester breaks and over the weekends, when she has a lot of requests because "people always want their weekends off," she could work because of her status as a "casual." This gave her the flexibility to say "no" when requested to fill in for an attendant who has called in sick. As a job, it is both more flexible than the one she had in the lingerie store, and it provides her with more income ($21 for the day shift instead of $8 as a salesperson).

While working as a registered care assistant, Katie learns a lot as she relates to different members in these settings. As a registered care assistant, she has to "do pretty much the dirty work for the nurses and for the doctors." Many of her peers on the job say that "doctors don't understand what a R-C-A is doing," that it is "a hard job," in which there is "no potential to learn more [because] you are what you are." The other students in the program, who really wanted to be working as attendants suggested, when hearing about her goal to be a doctor, that she will have a better understanding of what attendants really do. Although Katie feels that she is not progressing as a person, she has come to understand that the registered care attendants develop a close relation with the patients in their care on a daily basis, whereas doctors see the patient only "once a month" and do not see what the attendants see every day. But she definitely learns that it is not the kind of job that she would want for the future. Thus, even "showing up every day is a chore, it is not where I want to be."

Much in the way she talks about doing the program to become a registered attendant care worker and her work experience, she talks about the internship as an important learning experience that contributes to her personal development. But it is not something she would "want to do for the rest of my life, just because that's not what I want to do. But I would definitely work there as a stepping stone you know, like I've just been thinking about that today." Because she needs a backup plan in case she does not immediately get into medical school, she wants to be able to find work. The training as a registered care assistant provides her with options to earn a good living while awaiting further opportunities. Because she does have the laboratory experience together with her undergraduate courses in the sciences, working in a laboratory is something she "would enjoy, like for a couple of years or something." In this situation, too, she undergoes a process of subjectification, which has two sides. On the one hand, she increases her room to maneuver, both instantly, by having a certificate that allows her to make nearly three times as much money than if she was working in as a salesperson. She also hopes to increase the quality of her portfolio that she is planning to submit to the medical schools of her choice. On the other hand, there are hardships that come with working in the particular profession, including being subject to the treatment that resident care assistants get from others working with the patients in their care (nurses, doctors). But all of these emotionally more strenuous aspects deriving from additional coursework and the job site are subordinate to the ultimate, anticipated payoff that she will derive once in medical school and even more so once she will practice as a medical doctor.

Coda

> The more you learn the more you realize the less you know … I don't know where I am going to be twenty years from now, right, maybe I will be a house painter and maybe I will be happy. (Katie)

In the course of our research, we come to know Katie as a very savvy individual, and she may have realized, in her own terms, what I attempt to make salient in this chapter. As she says in the introductory quotation to this section, she is learning continuously, including how little she knows. As a result, she cannot know where she will be 20 years from now, at which time she might be a house painter and (maybe) happy at it. That is, Katie has learned that through the kinds of relations we have with others and through our participation in various forms of activity with their different object/motives, we change, together with the hierarchy of our priorities: These two changes are mutually constitutive as the personality changes with changes in the hierarchy, and the changes in hierarchies are constitutive of changes in personality. In Katie's case, this is exemplified in a profession that she anticipated not to be particularly interesting jobs. Knowing and relating to people, participating in activities, and realizing different object/motives transforms who we are. "I think these things are just based on my knowledge at that time because I knew people who were house painters, and construction workers, and teacher and I knew people who did these things right and then I mean yeah I am not the one to judge them but that's what I saw I didn't want to." Overall, Katie considers herself lucky to have experienced a developing interest in a particular profession: "I think I was very lucky because I knew what I want to do and people are like forty-five or forty they still don't know what they wanna do I feel lucky to even have a passion about something so it is nice that I didn't change."

The narratives from her life show that in addition to the processes of subjectification and personality, emotion plays an important role in the hierarchical organization that constitutes personality and its development. The anticipated long-term goal of being a doctor is signed highly positively in terms of the emotional payoff it would yield. This anticipated payoff is so large that Katie engages in activities on shorter terms that are in themselves not or not always rewarding but that bring her closer to the chosen goal. In her case, all of this reinforces at least the topmost aspect of the hierarchy, becoming and being a doctor as the leading object/motive around which everything else in her life is organized.

In the process of living toward achieving the goal, Katie undergoes continuous development within the activities that she chooses or has to engage in. With increasing engagement, she not only becomes a more competent subject – e.g., regarding the technical aspects of laboratory work, as a health-care provider – but also she is subject to and subjected to the particularities of each activity. Each form of participation develops competencies and is formative, thereby closing off, even if momentarily, other opportunities for developing (as) a personality.

The theoretical framework that I offer here has advantages over other frameworks, especially those that emphasize the distinct nature of activities and the boundaries between them. It takes the diversity of everyday life as the fundamental starting point of theorizing knowing, learning, and development. It is because of the diversity of life that we come to observe other diversity issues that are of pertinence to science education. Diversity of life inherently means hybridity, and hybridity can be modeled only through non-self-identity. Such frameworks operate with concepts such as "boundary crossing" and "third spaces" that are used to theorize how individuals cobble together the cultural practices characteristics of their root culture and those of the culture to be learned. From these perspectives, the individual is required to cross boundaries and come to be confronted by different practices, forms of subjectivities, and processes of subjectification. As we see in the case study provided here, the category of personality is an integrative one, as it recognizes the continuity of the individual across the discontinuous forms of subjectivity and subjectification. The framework is integrative because it articulates the object/motives of the different activities in their hierarchical relations within a "knot-work" of activities and corresponding object/motives. To me, this approach is much more consistent with the continuities that we live on a daily matter, where we experience ourselves as a person whether we are subject to a subordinate role at the job, the relative superordinate role in the family, or the differential relations that we entertain as customers, for example, at the bank (e.g., while seeking a loan) or as a buyer especially of a big-ticket item. At the same time, this approach accounts for the diversity of experiences in our everyday lives that derive from the multiplicity of activities we engage in, the related diversity in the (institutional) relations we entertain, and the corresponding forms of subjectivity, knowledge, or competencies and object/motives.

Acknowledgments This research was supported by grants from the Natural Sciences and Engineering Research Council of Canada and from the Social Sciences and Humanities Research Council of Canada. I am grateful to Pei-Ling Hsu, who, supported through these grants, collected the data as part of her dissertation and postdoctoral work in my laboratory. I thank the participants, who have agreed to be part of our research program over such a long period of time.

References

Foucault, M. (1975). *Surveiller et punir: Naissance de la prison*. Paris: Gallimard.

Lave, J., & Wenger, E. (1991). *Situated learning: Legitimate peripheral participation*. Cambridge: Cambridge University Press.

Leontjew, A. N. (1982). *Tätigkeit, Bewusstsein, Persönlichkeit*. Köln: Pahl-Rugenstein.

Lucas, K. B., & Roth, W.-M. (1996). The nature of scientific knowledge and student learning: Two longitudinal case studies. *Research in Science Education, 26*, 103–129.

Marx, K., & Engels, F. (1958). *Werke Band 3*. Berlin: Dietz.

Marx, K., & Engels, F. (1962). *Werke Band 23*. Berlin: Dietz.

Marx, K., & Engels, F. (1963). *Werke Band 42*. Berlin: Dietz.

Rancière, J. (1995). *La mésentente: Politique et philosophie*. Paris: Galilée.

Roth, W.-M. (2009). *Dialogism: A Bakhtinian perspective on science and learning*. Rotterdam: Sense Publishers.

Roth, W.-M., & Bowen, G. M. (2001). Of disciplined minds and disciplined bodies. *Qualitative Sociology, 24*, 459–481.

Roth, W.-M., & Lee, Y. J. (2007). "Vygotsky's neglected legacy": Cultural-historical activity theory. *Review of Educational Research, 77*, 186–232.

Roth, W.-M., & van Eijck, M. (2010). Fullness of life as minimal unit: STEM learning across the life span. *Science Education, 94*, 1027–1048.

Roth, W.-M., Lee, Y. J., & Boyer, L. (2008). *The eternal return: Reproduction and change in complex activity systems. The case of salmon enhancement*. Berlin: Lehmanns Media.

Vygotskij, L. S. (2002). *Denken und Sprechen: Psychologische Untersuchungen* [Thinking and speaking: Philosophical investigations]. Weinheim: Beltz-Verlag.

Vygotskij, L. S. (2005). Konkretnaja psikhologija cheloveka [Concrete human psychology]. In *Psykhologija razvitija cheloveka* (pp. 1020–1038). Moscow: Eksmo.

Vygotsky, L. S. (1989). Concrete human psychology. *Soviet Psychology, 27*(2), 53–77.

Reflexivity and Diversity in Science Education Research in Europe: Towards Cultural Perspectives

Michiel van Eijck

Introduction

Recent figures on the state of science education in Europe show a dismal picture. The more a country is economically advanced and scientifically sophisticated and hence needs a workforce of scientists, the more it struggles with engaging students with the advanced study of physical sciences (Osborne and Dillon 2008). This is "a phenomenon that is deeply cultural and the problem lies beyond science education itself" (p. 14). Indeed, specifically those children who move away from science belong to cultural groups already underrepresented in advanced studies of the natural sciences. Hence, a cultural perspective is required to understand complex diversity issues in science education like these.

Cultural studies of science education emerged in response to the recognition that diversity issues like the above can be explained by the given that science education is a cultural phenomenon and as such part of the amalgam of movements and processes in society. This research field is becoming increasingly pivotal in countries such as Australia, Canada, and the USA, thereby opening up opportunities for new, significant, and viable lines of research in science education. In this chapter, I argue that the development of this research field in Europe is falling behind, particularly in regard to its present potential and need. Indeed, in European science classrooms, there are many cultural issues that interfere with the teaching and learning of science, such as those related but not limited to language, globalization, and immigration. In response to this problematic, I reflect on the state of the art of cultural studies of science education in Europe and discuss some directions for much required future research.

M. van Eijck (✉)
Eindhoven School of Education, Eindhoven University of Technology, Gebouw Traverse 3.41, Postbus 513, 5600 MB Eindhoven, The Netherlands
e-mail: m.w.v.eijck@tue.nl

N. Mansour and R. Wegerif (eds.), *Science Education for Diversity: Theory and Practice*, Cultural Studies of Science Education 8, DOI 10.1007/978-94-007-4563-6_4,
© Springer Science+Business Media Dordrecht 2013

In meeting this aim, this chapter will unfold as follows. First, I will provide a brief introduction to cultural studies of science education and their relevance for the teaching and learning of science in general and diversity issues in particular. Next, I briefly reflect on the current state of this research field in Europe. This brief reflection reveals that cultural studies of science education in Europe are falling behind. I argue that this is in part due to theoretical barriers imposed by dominant research traditions in Europe. Assumptions underlying these research traditions frustrate the integration of cultural perspectives in science education research. Based on this argument, I propose two directions for future research.

Cultural Studies and Issues of Diversity in Science Education

Cultural studies of science education examine science education as a cultural, cross-age, cross-class, and cross-disciplinary phenomenon. The research field is driven by forms of scholarly activity that explicitly connect the theoretical frameworks to social and cultural perspectives and the explicit linkage between the theories employed and the data to be explained and rallied in support. Examples of such social and cultural frameworks that have proven to be useful for science education are cultural-historical activity theory, critical frameworks (feminism, postcolonialism, etc.), discursive psychological frameworks, and actor-network theory. As such, the research aims to establish bridges between science education and social studies of science, public understanding of science, science and human values, and science and literacy. By taking a cultural approach and paying close attention to theories from cultural studies, this new research field reflects the current diversity in science education. In addition, it reflects the variety of settings in which science education takes place, including schools, museums, zoos, laboratories, parks, aquariums, and community development.

In part, the research field has its roots in studies that focus on science education in non-Western countries or in indigenous societies, or science education for groups in industrialized countries that are underrepresented in the professions of science and technology (women, ethnic minorities) (e.g., Hodson 1993; Ingle and Turner 1981; Maddock 1981; Swift 1992). Until today, cultural studies of science education are still known by and effective in particular contexts and settings that contrast with "mainstream" science education in Western countries. However, due to globalization and increased immigration, many of such settings are becoming the norm rather than the exception in "mainstream" science education today. For instance, especially in urban regions, many European science teachers increasingly find themselves in science classes with students from many different nationalities. As such, teaching science in culturally diverse settings is a challenging scientific problem with international relevance. Arguably, tackling this scientific problem requires a perspective by which processes in science education can be understood in a wider cultural frame of reference (Bryan and Atwater 2002; Moore 2007).

However, research in this field showed that even extraordinary cultural settings can be highly informative for "mainstream" science education. This is so because such extraordinary cultural settings provide an external referential frame that allows the researcher to compare it with the "normal" and hitherto dominant, seemingly universal and hence unquestioned frames of thought in a discipline (Feyerabend 1975). For instance, one foundational study conducted in the 1990s focused on the reform of science curricula in order to make these accessible to Aboriginal people in Canada (e.g., Aikenhead 1997). In this study, it was shown that Aboriginal students experience issues of social power and privilege in science classrooms by which they drop more easily out of trajectories that lead to science-related careers (Aikenhead 2001). The theoretical construct of *cultural border crossing* developed in this study described how students move between their everyday life-world and the world of school science and how students deal with cognitive conflicts between those two worlds (Aikenhead and Jegede 1999). Because students generally reject assimilation into the culture of Western science, they tend to become alienated in spite of it being a major global influence on their lives. However, these attempts of assimilation and the resulting alienation appeared not to be exclusively experienced by Aboriginal students. The construct of cultural border crossing was equally applicable to "mainstream" students to describe the trajectories that students experience from the subcultures of their peers and family into the subcultures of science and school science. Rather than being exclusively a problem experienced by Aboriginal students, it appeared that this alienation is only more acute for Aboriginal students whose worldviews, identities, and mother tongues create an even wider cultural gap between themselves and school science.

The ultimate finding that cultural border crossing is equally applicable to "mainstream" students confronts science educators with the inconvenient question what can be considered "mainstream" and what cannot in science education. Underlying this question is a theoretical perspective on diversity common among scholars of cultural studies. This perspective starts with the premise that diversity is inherent to human life. That is, the ontological difference of everything with every other thing is taken as the starting point of theorizing human life. Difference and heterogeneity are the norm, the starting point, and the prior condition that preceded any Being not something less than sameness and purity (Levinas 1998). Diversity is not only a matter of specific static characteristics of individuals that can be compared and which turn out to be different, but all human practices continuously reveal the diversity inherent to human life. This theoretical position implies that whatever categories we can think of in which two Beings are the same, we can always think of another category in which those same two Beings differ—which leads to the conclusion that two Beings are never the same *a priori* and that sameness is always a construction *a posteriori*. Accordingly, diversity issues are not thought of exclusively in terms of deficiencies located in particular groups of students that are falling out of the ordinary (i.e., are not the same as the ordinary) due to these characteristics and that need particular treatments to be solved (the so-called deficit perspective). Rather, underlying diversity issues are always processes of power and hegemony. These processes determine what is considered "mainstream" or the "same" and what is not and to what extent the posteriorly constructed sameness is positioned as a prior condition (cf. Foucault 1979; Spivak 1988). From such a perspective on diversity, alienation in science education is not exclusively located in minority groups. Rather, the phenomenon of alienation is the outcome of a process of power and hegemony that ultimately determines the acceptable subculture in science classrooms. In this chapter, I take this particular cultural perspective on diversity of human life to frame not only issues of diversity on science education. Rather, as I argue in this chapter, this perspective is equally applicable to frame current developments in *research* on issues of diversity in science education and hence is reflexive for the practice of science education research.

During the 1990s, cultural studies of science education became increasingly reflexive for the practices of "mainstream" science education. And, in turn, the research practices employed to study "mainstream" science education are enriched by highly relevant cultural perspectives. As a result of this movement, recent issues of major science education journals repeatedly feature cultural studies that have far going implications for science education worldwide (e.g., van Eijck and Claxton 2009; van Eijck and Roth 2007, 2011). But this relevance not only follows from the opportunity to better understand the processes and patterns going on in the teaching and learning of science; cultural perspectives also have methodological consequences that question the very tenets underpinning common practice in research on science education such as interviewing. For instance, by taking perspectives that are common in cultural studies such as phenomenology and cultural-historical activity theory, recent studies have shown that the practice of interviewing in science education needs rethinking (e.g., Roth and Middleton 2006; van Eijck et al. 2009). Particularly, these studies showed that data on what participants in a study know and believe obtained by interviews are not necessarily as stable as many studies on science education would suggest. Hence, they cannot be detached from either these interviews or the processes preceding them and then be treated as independent contributions.

In short, research on science education from a cultural perspective has lead repeatedly to groundbreaking results that are highly relevant for both science education and science education research. Thus, taking such perspectives provides many opportunities for new, significant, and viable lines of research in science education. As a result, this branch of science education research grew rapidly during the past decade. One of the hallmarks of this research movement in science education is the establishment of the journal Cultural Studies of Science Education by Springer in 2006. Nevertheless, as I will illustrate in the next section, these opportunities for new, important, and viable lines of research are not yet taken up substantially by the community of European science education researchers.

The State of the Art of Cultural Studies of Science Education in Europe

There are several indications that European researchers are underrepresented in the field of cultural studies of science education. In this paragraph I briefly address and discuss some of these indications. In so doing, my aim is not to provide an exhaustive and comprehensive overview of the state of the art of cultural studies of science education in Europe. Rather, the purpose is to argue for a closer look on the reasons behind this underrepresentation. As well, the nature of these reasons becomes clear by discussing some of these indications.

The first indication that European researchers are underrepresented in the field of cultural studies of science education follows from a brief investigation of the origin of the contents of the journal Cultural Studies of Science Education. Taking research articles from the past years reveals that researchers in cultural studies of science education from Europe are underrepresented in this journal as compared to researchers from regions such as North America and Australia. This indication is reflected by the actual uptake of cultural studies of science education during meetings of the European Science Education Research Association (ESERA). In 2007 in Malmo, Sweden, for instance, three of the six keynote speakers, Phil Scott, Eva Krugly-Smolska, and Cathrine Hasse, pushed for the uptake of cultural perspectives by the science education research community. This underscores the motivation for this study, that is, the need for cultural studies of science education in Europe. However, the imminence of such calls is clear as they contrast sharply with the actual uptake of cultural studies by the research community present at the ESERA. Traditionally, less researchers present in the annual meeting's strand "Gender, Class, and Culture" as compared, for instance, to comparable strand called "Cultural, Social, and Gender Issues" of the conference of the National Association of Research in Science Teaching (NARST). This observation resonates with findings from a recent study on the link between research and practice in science education in Europe (EACEA/Eurydice 2011). Often researchers in science education are employed in the same institutes that are responsible for science teacher education. Because of this link, the practice of science teacher education is to some extent indicative for research on science education. Regarding the extent to which the issue of diversity is being addressed in science teacher education institutes in Europe, however, the state of the art shows a dismal picture. This follows from a recent report on national policies, practices, and research on science education in Europe:

> Meeting the needs of a diverse range of students and the different interests of boys and girls are important for motivating students to learn. However, 'dealing with diversity' was the least addressed competence in both the generalist and specialist teacher education programmes according to the survey responses received. In particular, competences relating to dealing with diversity and gender were less frequently addressed in generalist teacher education programmes than in specialist. (EACEA/Eurydice 2011, p. 118)

Besides such quantitative differences, there is also an important qualitative difference. Presentations by European researchers in the strand "Gender, Class, and Culture" at the ESERA conference often present works without referring to any cultural theory. Interestingly, the aforementioned observations contrast with a closer look at research articles published in Europe's leading science education journal, International Journal of Science Education. During the past decade, there has been a steady increase of cultural studies of science education among European researchers in science education research community. That is, there is an increase of studies that explicitly adopt sociocultural or cultural-historical frameworks. Notably, this counts for researchers from Holland, Sweden, and the UK. These figures, however, contrast with the representation of cultural studies at the ESERA conference. Apparently, the ESERA conference does not function yet as an appropriate forum where researchers in the field of cultural studies publishing in the journal can meet, discuss their work, and organize their discipline, in short, can manifest. More so, the numbers of cultural studies on science education in the International Journal of Science Education are still low as compared to studies within the dominant constructivist framework in science education research community.

In what follows, I argue that these latter phenomena can be explained from a deeper, conceptual level. That is, the dominant frameworks prevailing among the European research community that is manifest, for instance, at the ESERA conference imposes theoretical barriers to cultural studies of science education that frustrate more practically academic progress in the field.

Reflexivity: Theoretical Barriers and Horizons

Particularly in Europe, diversity issues in science education are commonly known from large-scaled studies that focus on comparing particular groups of students with respect to the learning of science. Characteristically, studies like these theorize central concepts such as knowledge and human cognition without taking into account their cultural foundations. Of course, such "sampling" studies are required to lay bare diversity issues at stake in science education in Europe. Often, however, such studies exactly stop where cultural studies of science education come in. This is so because such rather quantitative studies lack a cultural frame of reference by which the differences detected between groups can be explained in a way that is reflexive for the culture and practice of science education. Here, *reflexivity* has a special meaning derived from cultural studies in the field of anthropology. Accordingly, theories in a particular academic discipline are said to be reflexive once they apply equally forcefully to the discipline itself. Reflexivity about the research process followed from the postcolonial critique of the methods of anthropology in the academia of hegemonic Western cultures (e.g., Clifford and Marcus 1986; Asad 1973).

Currently, the issue of reflexivity plays in several studies on diversity in science education. Particularly, the use of nonreflexive frameworks, especially those in quantitative studies, cause analytical limitations since they do not allow researchers to interpret their results at the level of atypical students. Ironically, atypical are often those students who find themselves at the margin of the practice of science education and who should be at the center of studies on diversity. For instance, in one well-known study in Europe that focuses on cultural diversity and gender equity, the Relevance of Science Education (ROSE) project (Schreiner and Sjøberg 2004), this limitation is explicitly admitted:

> It is in the nature of quantitative research to compare *groups* of students rather than individuals. In such studies, students are categorised according to, for example, sex, age, socioeconomic status of the home, religion, race, language, school type and urban/rural place of living, All research based on groups entail a loss of information at the level of the individual. This means that quantitative data facilitates characteristics of the *typical*—but inevitably at the expense of the particular. In the present study, too, *groups* of respondents will be the unit of research. The individuals are categorised into gender categories. Characteristics of boys and girls, represented by mean scores for all students of the same gender, will unavoidably do injustice to the individuals. The focus of this study is on the *typical* rather than on the particular. Thus, this injustice is a compromise this study will make. (Sjøberg and Schreiner 2006, p. 5)

In order to explain such differences detected between groups, cultural studies of science education, in contrast, aim at explaining *how* culture plays into the practice of science education. In doing so, there is a focus on *atypical* individual students and the way in which the collective diversity of their mundane, everyday (school) lives contributes to the production of culture in and out science classrooms. Yet, the theoretical frameworks dominant in the European science education research community cannot explain how the diversity of the mundane, everyday life is the very condition and resource of/for people in their ontological development and hence culture in its evolution (Husserl 1939/1973). For instance, the constructivist approach generally and the conceptual change approach specifically fail to articulate some fundamental contradictions that come with their own presupposition including the roles of commonsense knowledge and culture. Such contradictions are clearly observable in regard to the role of the tools of professional scientists that make or do not make their way to science classrooms. A closer look to these tools and their contradictions reveals that schools teach by and large for yesteryear and as such fail to make an impact on students' current lives. One example is doing longhand division given that there is an abundance of calculators. Thus, students are still required to calculate speed when in fact there are many different devices that can be used to measure speed—microwave motion detectors, laser radar, speedometers, and so on. The cultural-historical approach tells us that former goal-directed actions are first reduced to invisible, conditioned operations and subsequently crystallized into tools (Leont'ev 1978). Because of the tools, we no longer have to know what our forefather had to know but are now faced with much more complex actions unimaginable when our forefathers were alive (for a case study, see Roth 2008). As a result of such cultural-historically determined processes, human cognition is deeply embedded in the tools we use and the artifacts and objects we are surrounded within our lives and which we use to perform the actions that are required for making a living (Hutchins 1995; Lave 1988). This aspect of tools, however, is inherently ignored in constructivist frameworks.

In contrast to cultural theories, theories of science education currently dominant in Europe often take only a set of different dimensions—conceptions, affect, motivation, interest—that apparently have very little to do with how the real person is part of and contributes to his/her culture and how this relates to the culture of science (education). However, Vygotsky showed that cognition (thought) cannot be thought independently from the culture in which one participates and that manifests itself by the language of its participants (Vygotsky 1986). Yet, science educators continue to conceptualize cognition separately from culture. This is perhaps not surprising given that Western researchers, in the classical science tradition, tend to take systems apart, gaze at and theorize the parts, and then make conclusions about what these parts mean with respect to the whole. Thus, in science education the learning of science concepts during a teaching experiment or what someone says in response to interview questions about career motivations is theorized and used to make inferences without the consideration of the cultural system as a whole of which one is and always has been part (for a critique and an alternative approach, see Roth 2009). Aspects that are deeply cultural, such as affect or motivation, then come to be tacked on to the main cognitive theory, much like epicycles came to be added on to the cyclical conception of the universe (when really a scientific revolution was needed). However, the issue is to derive a theory in which motivation is a core feature rather than tacked on, a feature already possible within cultural-historical activity theory (Roth 2007).

There is thus a theoretical divide between cultural studies and mainstream studies of science education that are dominant in the European. The deepness of this divide is clearly observable in a recent special issue of the journal of Cultural Studies of Science Education (Volume 3, Number 2, July 2008). The issue focused on conceptual change theory and sociocultural theories that might be complementary or alternative in an attempt to bridge the theoretical divide. One paper of the journal laid out an historical, social-cultural framework for science education and in so doing provided a critique of conceptual change theory, identifying its inadequacies, at least as perceived by Wolff-Michael Roth (Canada), Yew Jin Lee (Singapore), and SungWon Hwang (Korea). In another paper, David Treagust (Australia) and Reinders Duit (Germany) wrote a parallel review of research on conceptual change theory. Peter Hewson (USA, South Africa), Andrée Tiberghien (France), and Stella Vosniadou (Greece) prepared review essays based on their critiques of the Roth et al.'s paper. All of these scholars are internationally renowned as leaders in science education on the topic of conceptual change theory. Finally, Neil Mercer (UK), Gordon Wells (USA), and Regina Smardon (USA), eminent scholars from outside of the conceptual change tradition, critiqued Treagust and Duit's paper. The editor expected to provoke researchers in the field to reconsider their theoretical standpoints and felt there were numerous kernels within it around which productive conversations could emerge (Tobin 2008). However, unfortunately, the dialogue between the proponents of the two frameworks did not provide much evidence of academic development or changing theoretical directions (Dillon 2008).

Theoretical barriers imposed by the dominant canon in science education research in Europe are at times not easy to cope with for researchers on cultural studies of science education. As explained by Tobin (2008), such a situation can even be damaging for academic progress:

> We do not insist that others adopt our commitments and we are determined to learn from researchers and research that incorporates diverse theoretical perspectives. Ironically, while I embrace others' perspectives and seek to learn from them, my experience in applying for funding and publication is that those in power often require me to change my theoretical frameworks to align with their perspectives. The adoption of a one-size-fits-all perspective can be damaging to progress in educational research as scholars experienced throughout the twentieth century until the present, with the dominance of behaviorism (Watson 1913) and the associated philosophy of positivism (Laudan 1996, p. 229)

Thus, these theoretical barriers between cultural studies of science education and dominant constructivist frameworks in the European research community do well explain the phenomena featured in the previous section. Therefore, for the sake of academic progress in the field of science education, it is required that the ESERA conference comes to function as an appropriate forum where researchers on cultural studies of science education can manifest in order to start fruitful discussions and collaborations. On the long run, this may contribute to increasing the numbers of cultural studies on science education in the *International Journal of Science Education* as well as the number of European researchers publishing in *Cultural Studies of Science Education*.

The need for better organizing researchers on cultural studies of science education in Europe does not necessarily mean that theoretical barriers between two research streams are problematic for academic development per se. On the contrary, controversies are the very condition for scientific progress (Latour 1987). Thus, once theoretical barriers are present in a research discipline, many opportunities lay bare for new and viable research lines. In what follows, I conclude this chapter with setting out some of these lines for future research in this field in Europe.

Towards Cultural Studies as a Unifying Research Paradigm

In this chapter, I make case for the uptake of cultural studies of science education in Europe. In regard to its present potential and need in Europe, the uptake of this discipline is currently frustrated. Researchers working on cultural studies of science education are underrepresented and poorly organized in the science education research community in Europe. In part, this is due to a deep theoretical divide between cultural studies and the theoretical frameworks in dominant, traditional science education research. This frustrates the academic development required to understand the diversity in the practice of science education in European classrooms. From this observation, two imminent themes for future cultural studies in Europe can be formulated.

One theme for further research is to understand how cultural diversity plays in the teaching and learning of science in current European classrooms. Due to globalization and migration, science classrooms are no longer the homogeneous groups they used to be but consist of different kinds of students who each responds to science education in particular ways. Many different aspects of culture play collectively in these particular ways students respond to science teaching, such as, but not limited to, language, religion, gender, and ethnicity. By taking the diversity of European science classrooms as the norm rather than the exception, cultural studies on science education are thus required for understanding the ongoing dynamics of the relationships between culture, gender, and science education in the process of the teaching and learning of science. One example of such research is a project in which I participate, called "Science Education for Diversity" and funded by the Science in Society initiative of Seventh Framework Program (FP7) of the European Commission. In order to respond more effectively to the new cultural diversity of students in Europe, universities from the Netherlands and the United Kingdom aim at learning about diversity in science education in collaboration with international partner countries from Turkey, Lebanon, India, and Malaysia. We propose that understanding the dynamics of the relationships between cultures, gender, and science education in the diverse contexts offered by the different project partners will give us a good basis for designing new flexible and diverse approaches to science education that will appeal to all students within Europe and the world. Thus, we take diversity as a reflexive methodology for the setup of our project.

Another example is a project undertaken in the context of primary science teacher education program offered at the Universitat Autònoma de Barcelona (Espinet and Ramos 2009). Some of the subjects of these programs are taught within a CLIL (content and language integrated learning) political framework of the European Union whose aim is to integrate both the subject matter and language contents in the same learning environment (European Language Council 2006). Given the importance of language in science teaching and learning, this project aims at analyzing discursive interactions within preservice science teachers' small group work while undertaking experimental work. Here, three basic cultural fields are identified through which primary science teacher education takes place: science, multilingualism, and teaching science. These cultural fields develop along the process of teacher education since they are dynamic and changes continuously occur over time and space. Approaches like these, in which central concepts are theorized as spatiotemporally dynamic fields rather than static entities, are required for addressing diversity in science education in future research on cultural studies of science education in Europe.

A second theme follows from the theoretical divide between cultural studies of science education and traditional frameworks dominant in science education research. Apparently, this divide is currently deep and especially for established scholars difficult to overcome. Yet, although the latter frameworks are incapable for explaining cultural issues in science education, debunking them *a priori* makes little sense. On the contrary, the cultural movement in cultural studies of science education owes much to constructivist frameworks. This is so because the acceptance of the constructivist framework by the European science education research community in the 1980s, including recognition of qualitative research as a viable form of inquiry, paved the way for the application of social and cultural theories as underpinnings for research in science education. More so, especially in Europe, there is a rich and long-standing tradition in subject-specific research on the study of science education. In part due to the conceptual change movement, this research community matured in Europe as an academic discipline. This yielded a body of literature relevant for the teaching and learning of science. But more importantly, without this development it would be virtually impossible to study cultural studies of science education in the European realm. Thus, the aim for future research on cultural studies of science education is certainly not to colonize the constructivist framework and to impose cultural frameworks as master narratives. Rather, in line with a cultural approach, a future research agenda should seek to open up dialogues between the two frameworks and to explore in which ways they can fruitfully complement each other in improving the teaching and learning of science in Europe.

References

Aikenhead, G. S. (1997). Toward a First Nations cross-cultural science and technology curriculum. *Science Education, 81*, 217–238.

Aikenhead, G. (2001). Integrating western and aboriginal sciences: Cross-cultural science teaching. *Research in Science Education, 31*, 337–355.

Aikenhead, G. S., & Jegede, O. J. (1999). Cross-cultural science education: A cognitive explanation for a cultural phenomenon. *Journal of Research in Science Teaching, 36*, 269–287.

Asad, T. (Ed.). (1973). *Anthropology & the colonial encounter*. Atlantic Highlands: Humanities Press.

Bryan, L. A., & Atwater, M. M. (2002). Teacher beliefs and cultural models: A challenge for science teacher preparation programs. *Science Education, 86*, 821–839.

Clifford, J., & Marcus, G. E. (Eds.). (1986). *Writing culture: The poetics and politics of ethnography*. Berkeley: University of California Press.

Dillon, J. (2008). Discussion, debate and dialog: Changing minds about conceptual change research in science education. *Cultural Studies of Science Education, 3*, 397–416.

EACEA/Eurydice. (2011). *Science education in Europe: National policies, practices and research*. Brussels: Education, Audiovisual and Culture Executive Agency.

Espinet, M., & Ramos, L. (2009, August 31–September 4). *Multilingual science education contexts: Opportunities for pre-service science teacher learning*. Paper presented at the conference of the European Science Education Research Association (ESERA), Istanbul, Turkey.

European Language Council/Conseil Européen pour les Langues (ELC/CEL). (2006). *Nancy declaration: Multilingual universities for a multilingual Europe open to the world*. Berlin: ELC/CEL.

Feyerabend, P. K. (1975). *Against method*. London: New Left Books.

Foucault, M. (1979). Truth and power. In M. Morris & P. Patton (Eds.), *Power, truth, strategy* (pp. 29–48). Sydney: Feral Publications.

Hodson, D. (1993). In search of a rationale for multicultural science education. *Science Education, 77*, 685–711.

Husserl, E. (1973). *Experience and judgment* (J. Churchill & K. Ameriks, Trans.). Evanston: Northwestern University Press. (Original work published 1939).

Hutchins, E. (1995). *Cognition in the wild*. Cambridge, MA: The MIT Press.

Ingle, R. B., & Turner, A. D. (1981). Science curricula as cultural misfits. *International Journal of Science Education, 3*, 357–371.

Latour, B. (1987). *Science in action: How to follow scientists and engineers through society*. Milton Keynes: Open University Press.

Laudan, L. (1996). *Beyond positivism and relativism: Theory, method and evidence*. Boulder: Westview Press.

Lave, J. (1988). *Cognition in practice*. Boston: Cambridge.

Leont'ev, A. (1978). *Activity, consciousness, and personality*. Englewood Cliffs: Prentice-Hall.

Levinas, E. (1998). *Otherwise than being: Or beyond essence*. Pittsburgh: Duquesne University Press.

Maddock, M. N. (1981). Science education: An anthropological viewpoint. *Studies in Science Education, 8*, 1–26.

Moore, M. F. (2007). Preparing elementary preservice teachers for urban elementary science classrooms: Challenging cultural biases toward diverse students. *Journal of Science Teacher Education, 19*, 85–109.

Osborne, J., & Dillon, J. (2008). *Science education in Europe: Critical reflections*. London: The Nuffield Foundation.

Roth, W.-M. (2007). Emotion at work: A contribution to third-generation cultural historical activity theory. *Mind, Culture and Activity, 14*, 40–63.

Roth, W.-M. (2008). Where are the cultural-historical critiques of "back to basics"? *Mind, Culture, and Activity, 15*, 269–278.

Roth, W.-M. (2009). *Dialogism: A Bakhtinian perspective on science language and learning*. Rotterdam: Sense Publishers.

Roth, W.-M., & Middleton, D. (2006). Knowing what you tell, telling what you know: Uncertainty and asymmetries of meaning in interpreting graphical data. *Cultural Studies of Science Education, 1*, 11–81.

Schreiner, C., & Sjøberg, S. (2004). Sowing the seeds of ROSE. Background, rationale, questionnaire development and data collection for ROSE (The Relevance of Science Education): A comparative study of students' views of science and science education. *Acta Didactica, 4/2004* (pp. 1–120). Oslo: Department of Teacher Education and School Development, University of Oslo.

Sjøberg, S., & Schreiner, C. (2006). How do learners in different cultures relate to science and technology? Results and perspectives from the project ROSE (the Relevance of Science Education). *APFSLT: Asia-Pacific Forum on Science Learning and Teaching, 7*, 1–17.

Spivak, G. (1988). Can the subaltern speak? In C. Nelson & L. Grossberg (Eds.), *Marxism and the interpretation of culture* (pp. 271–313). Urbana: University of Illinois Press.

Swift, D. (1992). Indigenous knowledge in the service of science and technology in developing countries. *Studies in Science Education, 20*, 1–27.

Tobin, K. (2008). In search of new lights: Getting the most from competing perspectives. *Cultural Studies of Science Education, 3*, 227–230.

van Eijck, M., & Claxton, N. X. (2009). Rethinking the notion of technology in education: Techno-epistemology as a feature inherent to human praxis. *Science Education, 93*, 218–232.

van Eijck, M. W., & Roth, W.-M. (2007). Keeping the local local: Recalibrating the status of science and Traditional Ecological Knowledge (TEK) in education. *Science Education, 91*, 926–947.

van Eijck, M., & Roth, W.-M. (2011). Cultural diversity in science education through Novelization: Against the Epicization of science and cultural centralization. *Journal of Research in Science Teaching, 48*, 824–847.

van Eijck, M. W., Hsu, P.-L., & Roth, W.-M. (2009). Translations of scientific practice to "students' images of science". *Science Education, 93*, 611–634.

Vygotsky, L. S. (1986). *Thought and language*. Cambridge: MIT Press.

Watson, J. B. (1913). Psychology as the behaviorist views it. *Psychological Review, 20*, 158–177.

Part II
From Learning to Pedagogy

Part II explores diverse experiences of learning science within interrelated historical, cultural, institutional, and communicative contexts. The approach is socio-cultural, describing learners embedded within and constituted by a matrix of social relationships and processes.

Science Education for Diversity and Informal Learning

Loran Carleton Parker and Gerald H. Krockover

Importance of Informal Environments for Learning Science

Museums, science centers, zoos, and aquariums often serve as the "face" of science in the community where they operate. They are an important place for diverse communities to learn about and be excited by science and, subsequently, are in a position to serve as facilitators of communication, cooperation, engagement, and activism among the public, K-12 school science authorities, and science research institutions (both public and private). Informal science education institutions that adopt a sociocultural stance towards science and science education can provide an entrée to science for traditionally underrepresented communities by identifying science as a way of knowing along with other socially and culturally constructed paths to knowledge. The scientific community is beginning to recognize the importance of informal experiences to the development of appreciation for and interest in science in children and adults alike.

Dierking and Falk (2010) point out that scientific research and education communities are both interested in advancing the public's understanding of science. And they point out that people assume that children do most of their learning in school. In reality, children spend less than five percent of their life in formal classroom settings. Furthermore, people's knowledge and interest in science and the environment are shaped by everyday experiences.

L.C. Parker
Discovery Learning Research Center, Purdue University, West Lafayette, IN, USA

G.H. Krockover (✉)
Department of Curriculum and Instruction, College of Education, Purdue University, West Lafayette, IN, USA

Department of Earth and Atmospheric Sciences, College of Science, Purdue University, West Lafayette, IN, USA
e-mail: hawk1@purdue.edu

N. Mansour and R. Wegerif (eds.), *Science Education for Diversity: Theory and Practice*, Cultural Studies of Science Education 8, DOI 10.1007/978-94-007-4563-6_5, © Springer Science+Business Media Dordrecht 2013

A report published by the National Academies, *Learning Science in Informal Environments: People, Places and Pursuits* (Bell 2009), indicates that everyday experiences contribute to people's knowledge and interest in science. The report notes that experiences in informal settings can significantly improve science learning outcomes for individuals from groups that are historically underrepresented in science, such as women and minorities. Evaluations of museum-based and after school programs indicate academic gains for children and youth from underrepresented groups. In addition, Bell (2009) indicates that, "Learning is broader than schooling, and informal science environments and experiences play a crucial role. These experiences can kick-start and sustain long-term interests that involve sophisticated learning" (p. 14). Falk and Needham (2011) have demonstrated that visits to a science center have long-lasting impacts on science and technology understanding, attitudes, and behaviors. They found that some of the strongest beliefs of impact were expressed by minority and low-income individuals. Simpson and Parsons (2009) point out that minority parents' decision to participate in informal science education hinges on their perception of the curriculum as culturally congruent.

These findings illustrate both the important role of informal education for engaging children in science but also the importance of considering diverse groups and cultures when designing and researching science learning in informal environments. Learning science in informal contexts is a complex process that involves the prior knowledge of the learners; guidance from others through conversation, text, and symbols; multiple perspectives about what is to be learned and the learning process; and reflection over time about what was learned. The conceptual framework for science learning used determines "what counts" as learning. According to the informal literature reviewed, learning in informal contexts such as science centers and zoos also occurs through visitors' interactions with exhibits. Learning also occurs through visitors' interactions and reflections with informal learning staff. The conceptual framework chosen by researchers allows them to examine these characteristics. Ways of viewing science learning determine how researchers and educators define and measure learning. Considering diverse ways of viewing science learning allows for the design and study of learning contexts that engage diverse cultures.

This chapter presents current sociocultural science education research about how best to design environments that support free-choice/informal learning for diverse audiences and applies this knowledge, presenting several examples of best practices. The chapter will also identify important features of informal learning in the context of diversity and equity issues. The chapter proceeds by, first, identifying key features of science learning in informal environments; second, describing the framework used by educators and researchers in designing and studying learning in informal environments; third, providing three vignettes illustrating the key features of science learning in informal environments; and, fourth, making recommendations for creating successful informal learning experiences for diverse groups.

Key Features of Science Learning in Informal Environments

In the last 20 years, research about science learning in informal contexts such as museums, science centers, zoos, and aquariums has proliferated. Researchers have examined learning in informal contexts such as science centers, museums, zoos, and aquariums; informal learning organizations have focused on family groups, individuals, and school groups and have asked diverse questions about science learning. This work has been situated in multiple research paradigms and traditions each of which approaches the concepts of learning and research from a different perspective.

The research studies below illustrate key features of successful science learning experiences in informal settings. Foundational to successful learning is the activation of visitors' prior knowledge. One of the most common strategies for this activation is through scaffolding provided by text, symbols, or interpreters. However, research also points to the importance of designing informal learning experiences that value the diverse array of knowledge and experiences of traditionally underrepresented groups. This can be achieved by designing experiences that acknowledge the multiple perspectives that visitors bring to an experience and through encouraging learners to reflect on their own experiences, knowledge, and values.

Activation of Prior Knowledge

Gilbert and Priest (1997) studied a school group's (8- and 9-year-olds) visit to the London Science Museum. They focused on locating critical incidents in learners' discourse that played an important role in the (re)construction of students' mental models. They found that recognition by the learners of a familiar object or action—something that they had previously observed or experienced in their lives—initiated discourse among the learners. They found that unexpected experiences with an invitation to explore further also initiated discourse. The initial surprise attracted learners' attention after which the learners eagerly offered comparisons and contrasts with familiar objects and actions. Guiding questions were also found to focus learners' attention and promote discourse among the group.

Gilbert and Priest (1997) also located critical incidents that allowed for the continuation of a line of discourse building more complex models about a concept. This occurred when learners were able to link a particular activity or object at the science center to broader experience—linking the particular to the general—and when experiences with different objects or actions inside the science center were linked by the learners. Alfonso and Gilbert (2007) have built on this result and note that these connections need to be made explicitly through text or other symbols in order for a meaningful link to be made to prior knowledge. Discourse was halted when the exhibit's prompts were not available to the learners—either because of their placement, their inappropriate content, or because they were missing (Afonso and Gilbert 2007).

Brody and colleagues (2002) discuss the role of prior knowledge in learning at Yellowstone National Park. They identified prior knowledge that was common to many visitors that served as "anchors" or "bridges" for their learning during and after their visit to the park (p. 1136). This prior knowledge tended to be knowledge about science concepts but was also related to the values that visitors associated with these concepts. For example, many visitors associated deep ocean thermal vents with unique and interesting life forms—after text linked Yellowstone National Park in the United States with deep ocean thermal vents, they valued the park differently.

Bamberger and Tal (2006) have also explored the influence of past personal experiences on learning in natural history museums and mechanisms through which visitors activate these experiences. They have found that moderately structured activities (where learners had choices, but textual information and prompts were provided for them) allowed for the most connection to prior knowledge and the most complex discourse when compared with completely free-choice activities or activities where learners had no choice.

Botelho and Morais (2006) have studied the characteristics of science center exhibits and learner-exhibit interaction and determined that links to prior knowledge are best made directly. That is, none of the linking process should be left to learners' imaginations. Exhibits that serve as models for physical phenomenon should be linked explicitly through symbols or text.

Hohenstein and Tran (2007) also emphasize the importance of textual, scaffolding prompts for informal science learning. They examined the effects of guiding questions on the conversations of visitors at a science center. Their research suggests that broad, guiding questions stimulate learner discourse but that the physical nature of the exhibit is also important—it determines how much attention learners pay to the prompts provided for them. It is important to provide these prompts for learning experiences, because, as Tunnicliffe (2000) has found, they create a storyline for visitors to follow at an exhibit.

Acknowledging and Valuing Multiple Perspectives

Ash (2004) identified characteristics to determine the mechanics that allow visitors to construct meaning using exhibit features. She recommends that exhibits have multiple "entry points," or multiple ways to understand what is essentially the same concept. She suggests that one way to accomplish this is to create thematic exhibits or exhibit clusters that focus on the big picture of science. She gives one interesting example: an exhibit or cluster that addresses the question, "When is something alive?" from multiple perspectives. The exhibits would provide simple prompts at multiple levels that promote discussion among learners of all backgrounds.

Zimmerman and colleagues (2010) examined the importance of diverse perspectives in the informal science learning of families. They analyzed the interconnectedness of individual cognitive resources, situated activities, and cultural resources that support learning and processes and found that families use a wide variety of knowledge to make sense of exhibit content in the area of biology by transferring cultural resources from prior experiences and two types of scientific epistemic resources to make sense of biological exhibits.

Falk et al. (2008) have developed typologies of visitor identities, which could be described as the visitor's motivations and role in a group (or individual) visit. Falk describes that through the analysis of visitor interviews and observation, he and other researchers found that learning outcomes of the visit are strongly linked to the visitor identity. He also notes that only one or two of the visitor identity types are strongly linked to the acquisition of science content knowledge. Visitors' identities determine how they will interact with the exhibit and their social group and determine what criteria will be used to determine the relevance and power of the information and experiences offered at the exhibit. This underscores the findings discussed above that visitors come to an informal science learning context with diverse motivations, knowledge, and experiences. Valuing these diverse perspectives is critical for creating and studying successful learning experiences.

Thus, the literature review shows that learning science in informal contexts is a complex process which at its foundation relies on the activation of learners' prior knowledge or experiences. Achievement of this for diverse populations involves careful scaffolding of visitor discourse and the acknowledgement and valuing of visitors' perspectives which are culturally situated. Both designers and researchers of informal science learning environments have recognized this strong connection to culture and have applied sociocultural frameworks in their work.

Sociocultural Frameworks for Informal Science Learning

Science education researchers have defined learning in several different ways. Foremost in each of the frameworks for learning developed in the research literature are the goals for learning science. These goals determine what "counts" as meaningful science learning in each framework and are essential to studying and designing learning contexts in which learners of diverse backgrounds, cultures, and interests can thrive. Because sociocultural theory emphasizes understanding the variability, as well as commonalities of the learning process, it is particularly well suited to understanding learning in informal contexts (Schauble et al. 1997). However, two types of sociocultural views of learning have dominated research and exploration of learning that occurs in informal environments. Each values a different outcome of science learning, searches for learning at different levels, and, hence, focuses on different units of analysis. The first is the social constructivist framework.

Sociocultural Approach with Individual Science Learning Goals: Social Constructivism

The social constructivist view focuses on changes to individuals' cognitive structures, but recognizes that those changes are created by both social and individual processes. This learning framework is commonly used in formal and informal science education research and represents a shift away from viewing "learning as individual cognitive growth to learning as individual cognitive growth in social settings" (Carlsen 2007, p. 58).

Learning takes place in the mind, but it is not simply an individual process; it involves dialogue with our environment—people, places, history, and culture. Glynn and Duit (1995) state that learning science meaningfully and being scientifically literate involves socially constructing and applying "valid scientific models" of the world.

The creation of a mental model requires the use of symbolic forms—a language with which to shape the representation. Therefore, language and learning cannot be separated from one another in the social cognitive framework. As Vygosky would argue, language and learning develop together (1978). Language is a social construction. It allows us to share our mental models with others. It is rooted in culture, history, and place because it originates with our mental models of the world. Therefore, both language *and* learning are social enterprises. This way of looking at learning has important implications for answering the question: How does one learn?

If learning is the construction of a symbolic mental model, then language must mediate this construction. Learning is a product of social interaction. Vygotsky (1978) described it this way: "Every function in the child's cultural development appears twice: first, on the social level and, later on, on the individual level; first, between people (interpsychological) and then inside the child (intrapsychological)" (p. 57). This process of learning has been termed social constructivism. Learning involves interpreting our perceptions of the world and organizing them into a mental model. But perceiving and organizing are mediated by language and social interaction. Social interaction directs our perception by focusing our attention, sharing pieces of mental models, and modeling the manipulation of physical objects.

Science learning with acquisition goals through social construction emphasizes opportunities for all learners to discover scientific principles through direct experimentation, discussion, and scaffolding from members of the learner's social group or through accompanying text. The process of experimentation would be the same for all learners—because through a combination of experience and guidance from their social group or another resource, they will be able to discover and acquire the same scientific principle. Questions posed are intended to create dialogue among learners and to guide them towards a particular change in their mental models. Together the learners and guide work towards building a consensus model that resembles the scientific conceptual model. It represents a shift away from viewing "learning as individual cognitive growth to learning as individual cognitive growth in social settings" (Carlsen 2007, p. 58).

The goal of science education according to this framework is the construction of scientifically valid mental models. It views scientific literacy as property of the individual. Learning is a social process that is internalized by the individual and becomes property or part of that individual. Informal science education institutions have adopted these goals for science education and used them to develop programs, exhibits, and fieldtrip experiences (Yager and Falk 2008).

This framework may not best capture the complexity of science learning at museums, science centers, zoos, and aquariums. Unlike school science classes, the visitor determines the goals and agenda for experiences at informal science institutions, and often learning scientifically valid models of the world is not a primary goal of their visit. This does not mean, however, that visitors to informal learning institutions do not or cannot learn science during their visits. The conceptualization of science learning as change, through individual or social means, is not fruitful for research about what and how visitors learn in these contexts. As Minda Borun (2002, p. 245) explains, "The learning unit . . . is not the individual, as in a classroom setting, but the small group." Learners visiting the zoo or other institutions are not visiting with the intention of demonstrating their knowledge in an exam or other individual assessment. The science learning that occurs is the result of contributions from individuals with diverse backgrounds and prior knowledge. Thus, science learning in the context of science centers, zoos, museums, and aquariums must be examined as a product of this social interaction and be examined on the group level, rather than the individual level.

Sociocultural Approach with Community Learning Goals: Collective Praxis

Learning frameworks with participation goals for science learning represent a much broader view of learning than the acquisition view. Learning is larger than the uptake of scientific concepts and processes or the participation in a model scientific community. They focus on more than individual mastery or accomplishments and view science learning as a collective enterprise that is more than the sum of its parts. Roth and Lee (2002) describe this framework as "collective praxis." They place science learning entirely in the social realm. Scientific knowledge is greater than the individuals who collectively create it; it cannot be reduced to characteristics of individuals.

This way of viewing science learning and scientific literacy does "not have boundaries coincident with formal education" (p. 33) and can accommodate the diverse forms that science takes as it is situated in everyday lives (Jenkins 2002; Roth and Lee 2004). This framework for science learning deemphasizes the science of scientists, such as the valid scientific models and concepts discussed in the previous framework, and focuses instead on how groups of people make use of and act upon their knowledge of science and science resources (Roth and McGinn 1997).

The process of communicating, locating, and acting upon scientific knowledge requires the use of language but also broader modes of communication (Jewitt et al. 2001).

Science learning as collective praxis provides opportunities for all learners to engage science in a context that is meaningful for their group or community. There are no set procedures or steps to follow. The group negotiates meaning by drawing upon the resources at hand and is encouraged to find and utilize new resources. Questions posed are reflective in nature—intended to assist the group in making decisions about what is important in their particular learning situation.

Science learning in an informal context represents a much broader view than the uptake of scientific concepts and processes or the participation in a model scientific community. It must focus on more than individual mastery or accomplishments and view science learning as a collective enterprise that is more than the sum of its parts. Science learning in informal contexts occurs socially, but the knowledge created is greater than the individuals who collectively created it. This way of viewing science learning and scientific literacy does "not have boundaries coincident with formal education" (p. 33) and can accommodate the diverse forms that science takes as it is situated in everyday lives (Jenkins 2002; Roth and Lee 2004).

In the real lives of visitors to informal science learning institutions, science cannot be separated from other forms of knowing—it is integrated with values, morals, subjectivities, tradition, and beauty. As Feyerabend (1975) contends, there are no criteria with which to demarcate science from other ways of knowing.

This way of viewing learning has also been described in other contexts as socially situated learning (Lave and Wenger 1991). Rather than being the creation of a mental model of the world through individual or social processes, science learning is the act of participation in a community. Members of the community have different levels of experience and prior knowledge. It is the distribution of prior experiences and knowledge that allow members to collaborate and create knowledge. The process of working together to create knowledge is where learning resides—it is not located in any single individual.

Science learning in informal learning contexts has been studied using participatory conceptual frameworks. Ash (2002) has also developed an explicitly participatory conceptual framework for learning science in informal contexts. Her framework for science learning is based upon Vygotsky's (1979) zone of proximal development. This is the space where collaboration and meaning making occur between individuals with distributed expertise (Ash 2002, p. 359). "Purposefully collaborative family conversations are both process and product, and are set within a larger activity system that has multiple purposes, such as having fun and learning new ideas" (Ash 2002, p. 361). Ash has used this participatory framework for science learning and used it to examine how family groups make meaning during experiences at science centers and museums.

She reports (2003) that families visiting the Exploratorium in San Francisco emphasized a wide variety of inquiry skills including observation, questioning, comparison, explanation, interpretation, reflection, and analogical modeling. Ash (2004) has also found that families at a natural history museum used questions to create organizational patterns in which to situate their new knowledge, to invite all members of the family to co-construct meaning, and to sustain ongoing content themes deemed important for learning by the family. Thus, having multiple access points or multiple ways to understand the same concept has the ability to promote dialogue and diverse perspectives.

The Practice of Informal Science Education for Diversity

Bell and colleagues (2009) provide several recommendations to education developers in informal science learning institutions. Two of those recommendations are to "provide multiple ways for learners to engage with concepts, practices and phenomena within a particular setting" and to support learners in interpreting "their learning experiences in light of relevant prior knowledge, experience and interests" (p. 6). They reiterate that science learning for diverse audiences, and ultimately all audiences, hinges on the collaboration of a learner and a guide (adult or more advanced peer). Hence, providing space for these collaborative dialogues to take place is essential for informal science learning institutions.

What follows are three research examples of such collaborative science learning events in informal institutions conducted by Parker (2009). Each highlights the roles that the design of the environment and the social interactions among the learners play in creating such collaborative environments.

Examples of the Impact of Exhibit Design on Collaborative Talk

The family in Example 1 uses exhibit text in two related ways: to frame the exhibit and direct their observations, as a source of new vocabulary with which to describe their observations.

Example 1

1 Mom: Did you guys look at this thing over here where it says domesticated or
2   wild? Do you wanna read that?
3 Mom: Let's go over and read the sign.
. . .
4 Sam: (reading) A transformation nomadic hunting and gathering lifestyles to
5   farming took place around 10,000 to 12,000 years ago in the middle east.
6 Mom: mmhmmm
7 Gabby: (reading sign) Imagine what your life would be like without dom-
8   domesticated plants or animals.
9 Mom: Is autumn domesticated?
10 Sam: Yeah
11 Mom: Yeah

12 Mom: Is Gus domesticated?
13 Sam: Yeah
14 Mom: Yeah
15 Mom: Elsie?
16 Sam: Yeah
17 Mom: How 'bout um, Grandma's things that she has in her garden?
18 Gabby: Yeah
19 Mom: Plants can be domesticated too
20 Gabby: Yeah, there it says from weeds. (points to sign)

The mother in this family approaches the zoo learning experience as a guide with the intention of directing her daughters' attention and the conversation so that they notice how a scientific concept such as domestication applies to their daily lives. Mom also uses the text at the exhibits to focus her observations and guide the girls as they create explanations.

The family in Example 2 often shared control of the science learning discourse in ways that allowed multiple family members to contribute ideas and practice using science terms, prior science knowledge, and observations to justify their ideas and explanations. The next example displays how the family uses a collaborative discourse to build explanations.

Example 2

1 Dad: Did you see what he did?
2 Scott: Yeah, he sticked his tail out.
3 Dad: He grabbed something with his tail–he used it to grab somethin'.
4 Scott: Mulch, I think.
5 Dad: Mulch? That's what it looked like.
6 Scott: A spider monkey, these are spider monkeys (looking at sign).
7 Dad: Is that what they are?
8 Scott: Yeah, it says right there. Spider Monkey.
9 Dad: Look at his tail . . .
10 Dad: Look at the end of his tail. Look at the end of it.
11 Scott: Ohh
12 Dad: On the underneath side. There's no hair, you see it?
13 Scott: Oh yeah
14 Dad: It kinda looks like a really long gorilla finger or somethin' doesn't it?
15 Scott: Yeah
. . .
16 Scott: Dad, it says um the grasping tail that works like a fifth hand. Its tail it
17   works like a fifth hand.

In Example 2 Dad and Scott use observations and the exhibit text to learn how monkeys use their long, agile tails. In lines 1–5, Dad and Scott describe their observations of the monkey's behavior (using its tail to retrieve a piece of mulch located beyond the barrier of its enclosure). Dad makes additional observations about the characteristics of the monkey's tail (line 10). Scott seeks out and retrieves relevant information from the exhibit text that explains their observations and reports it to the group (line 16). Both Dad and Scott contribute to the building of the explanation using observations and available text.

In Example 3, Dad, Scott, Mom, and Maggie were able to fully integrate the exhibit text into their collaborative explanations. For example, when discussing an eagle's nest size (lines 21–25), Dad and Mom refer to the model to elaborate on their understanding of the actual nest size question asked by Maggie.

Example 3

1 Maggie: Look at that.
2 Dad: That's the eagles' nest.
3 Scott: Wow.
4 Dad: Wanna set in it? Wanna be a baby eagle?
5 Maggie: Nnn (shakes head)
6 Scott: I will
7 Mom: No?
8 Maggie: I'll be a mother.
9 Dad: You'll be a mother eagle?
10 Dad: That's a pretty big eagles' nest huh?
11 Scott: Yeah. So this is what an eagles' nest looks like?
12 Maggie: (inaudible)
13 Dad: Yeah, what is that called, aerie? aerie?
14 Scott: I think
15 Dad: That one–
16 Scott: I don't know. I thought it was really an um teepee thing that's in there
17   (moves to wingspan painting).
18 Dad: I saw the teepee thing in there.
19 Dad: Called an aerie (looking at sign and reading aloud) one of the largest
20   birds' nests in the world.
21 Mom: Is that the actual size?
22 Dad: I think it–(goes back to sign)
23 Scott: My arms—(moves from wingspan painting to model nest)
24 Dad: (reads sign aloud) Two feet deep and five feet wide.
25 Dad: Yes, that's the actual size.
26 Mom: Ohhhh.
27 Scott: That's an–
28 Dad: (reading sign aloud) Bald eagle uses the same nest year after year
29   continually adding materials to the nest, aeries have been found that are at
30   large as 20 feet deep with a weight of more than two tons.
31 Mom: My goodness
32 Scott: Whoa
33 Scott: Alright now–
34 Dad: 20 feet deep, my land, that's a house.

Example 3 illustrates how text is consulted as a supplement to the collaboration and in the context of Mom's (line 21) and Sam's (line 11) questions about the model's relation to a real Bald Eagle nest. During collaborative explanations, the text is consulted as a resource in the context of the group inquiry, rather than as a frame through which to view the exhibit. A science learning event at the Bald Eagle Exhibit illustrates how the design of this exhibit allowed them to access the information using multiple approaches including role-play.

Leinhardt and Knutson (2006) have shown that learners' roles change throughout their experience at museums, science centers, and zoos. They studied grandparent-grandchild groups at a natural history museum to explore the roles and identities that members of the group displayed during the visit. They can become learner, teacher, modeler, storyteller, historian, scientist, and mediator within the same visit. The exhibit itself can provide structure for these multiple and changing roles through prompts and scaffolding that support multiple roles and perspectives. Members of these families played different roles during science learning events. Parents acted as guides in parent-directed explanation—a role similar to "teacher" as described by Leinhardt and Knutson (2006). But they enacted other roles during collaborative explanation: Dad in Example 1 described one of his roles as "devil's advocate."

Text at exhibits can create a frame for interpreting observations. However, easy access to scientific vocabulary and explanations can encourage parents to use this text to "teach" the group in a directed manner and can limit the ability for the group to contribute their own interpretations and culturally relevant knowledge.

Gilbert and Priest (1997) also found that successful consultation of the exhibit text helped the groups to continue their discourse at the exhibit. They describe this consultation occurring after the group has initiated interaction and conversation at the exhibit. They then consulted the text to assist them in thinking about or explaining their initial observations.

How social groups understand or approach learning in informal environments is also a factor in how they construct explanations during their science learning events. Tunnicliffe (2000) has found that exhibit text creates a storyline for an exhibit. She studied learner conversations at a robotic dinosaur exhibit cluster at London's Natural History Museum. All facets of the exhibit related back to a primary, broad storyline (in this case, that dinosaurs had very diverse diets), and this scaffolding helped keep learners' discourse focused towards the exploration of a big picture or theme. The storyline is not meant to convey a certain set of facts (although some facts are present), but to encourage and sustain the sharing of ideas on a broad subject.

Gilbert and Priest (1997) reported that when experiences with different objects or actions inside the science center were linked by the learners, meaning-making discourse was initiated and sustained by the group. Their study focused on an exhibition in a science center that had a clear theme that was known to the visitors because it was the title of the exhibition.

In studying a video-based exhibit, Stevens and Hall (1997) found evidence that creating records of experiences at a science center allowed learners to reflect on their experience at that exhibit but also provide a catalog of learner experiences for the next learner to view and model or utilize in some way—effectively broadening the social interaction from which learning is constructed. This reflection process also allowed visitors to treat learning as a continual process that continues after the actual experience at the exhibit has ended. Creating a space for reflection over learning at each exhibit and during the visit as a whole could improve visitors' ability to connect exhibits to each other and encourage visitors to view science learning as a process rather than a collection of facts.

Getting family visitors to focus upon and discuss the "big ideas" of science during and after their visits assists informal learning centers, including zoos, in achieving their stated educational objectives. Emphasizing broad themes rather than disconnected concepts also assists visitors in recognizing science as more than isolated explanations of observations and support a view of science as one of many ways to understand and interact with their environment.

Anne Lorimer's (2007) exploration of a "hands-on" exhibit on commercial aviation at the Chicago Museum of Science and Industry raises important questions about how we approach teaching and learning science at informal education institutions. How should we portray science and technology? How are science and technology perceived by visitors? Lorimer's study showed that visitors do learn, reflect, and make connections with their own lives about science and technology during their time at museums and science centers. She explored one case where the connection made was emotional and deeply personal—a reminder of unrealized adolescent aspirations. For others the visit reinforced their broader feelings of alienation from science and technology.

The museum's goal was an admirable one: to create a sense of excitement, wonder, and infinite possibility in young visitors and their families. The visitors, however, took those feelings of wonder and connected them with their personal experiences with science and technology—associating the wonder and awe felt during the visit with something that was unattainable. They were constructing or reconstructing knowledge about science and technology that placed a boundary between science and technology and themselves. The exhibit developers did not intend for visitors to interpret their exhibit in this way. As informal educators, we cannot control the meaning that visitors construct at museums and science centers, but we can help to shape that meaning by providing the necessary context and resources to visitors and our community.

Lorimer's study is a wake-up call for informal educators. It shows that even with the best of intentions, our exhibits can have results that are opposite of those that we expect. Visitors do make deep, personal connections at the museum and science center, but simply providing them with objects to manipulate is not enough to help guide their meaning making. Science and technology presented without the context or resources that visitors need to integrate new knowledge in a positive way may end up reinforcing previous understandings of science and technology as something unattainable. Placing science on a pedestal reinforces visitors' views of science as "other," as separate from themselves. Museums and centers that focus on science and technology must change their approach—away from contextless exhibits focused on very narrow concepts, towards broad themes in science and technology that touch everyone's lives.

Involving Diverse Groups in Development of Programs and Exhibits

One way to change this is to better integrate science museums and centers into the community. Involve community groups in development and discussion about programs and exhibits. Help visitors personally connect with exhibits by basing the exhibits on local concerns, interests, and resources. Most importantly, reach out to community groups who are not typically associated with science centers. Fight the alienation from science and technology that Lorimer describes by purposefully including a diverse array of groups and interests from the community.

AAAS (1993) has suggested that the best way to promote "science for all" is to emphasize the nature of science rather than individual science concepts. Part of learning about the nature of science is thinking about the human, personal, social aspects of science and technology–connecting science and technology to our lives, our society, our history, and our culture and giving context back to science and technology. The research indicates that surrounding science and technology with context that assists visitors with connecting it to their own lives and promotes learning dialogue among visitors, their families, and friends is the way to create a meaningful informal learning experience.

What follows is a series of practical steps that informal science institutions can take to involve community groups in the development of exhibit content and programs:

- Step 1: Make connections with representatives of community groups.
- Step 2: Compare suggestions and interests from the community with a theme.
- Step 3: Research other informal learning centers for the physical context.
- Step 4: Develop learning objectives for the exhibits.
- Step 5: Draft the exhibit design and text.
- Step 6: Construct the exhibits and conduct pilot visits with the community.

To elaborate on each step, we have included specific recommendations for involving diverse groups in the development of programs and exhibits.

Step 1

Make connections with representatives of community groups—especially those not typically associated with the science and technology center (civil rights groups, church groups, cultural groups, neighborhood associations, etc.). Hold a town meeting to present the theme for the future program and solicit input from the community. Connecting with community groups that have not previously been associated with the center may take extra time and effort.

Step 2

Compare suggestions and interests from the community with the theme, the nature of science, and determine how they could be integrated into the theme. For example, suggestions might include questions about how scientists do their work such as, "Do all scientists follow the same procedures or method?" Another community group may be interested in the history of science as an accepted way of understanding the world. Both of these interests could be integrated with the issues that comprise the nature of science. Part of the program could focus on scientific inquiry and explore issues of how something is "known" in science. Another part of the program could focus on the history of science and its position in society. Ideas for portions of the programs should be created from the community suggestions then the final ideas that will become exhibits in the program will be selected by science and technology center staff. The final ideas for exhibits should be easily connected so that exhibits have continuity and so that exhibit ideas echo the representation of the nature of science in Science for All Americans, Chapter 1 published by AAAS (1990).

Step 3

Science and technology center staff and volunteers should research the development of other exhibits focusing on the nature of science at other informal learning centers to gather possibilities for the physical contexts of the exhibits. Pedretti (2002) reviews several recent examples of exhibits focusing on the nature of science: A Question of Truth at the Ontario Science Center which focused on science as a human endeavor, the history of prejudice in science, and science's interaction with the local community; Science in American Life at the Smithsonian explored the history of science and technology's positions and roles in society; Birth and Breeding at the Wellcome Institute for the History of Medicine in London explored the role of biology, medicine, and social sciences shaped the perception of motherhood in the current culture; and Mine Games at Science World in Vancouver focused on engaging visitors in open discussion about the costs and benefits of mining science and technology to society. The physical contexts of each of these programs should be examined and a list compiled of possible ways to approach the exhibits (hands-on, interactive, visitor forums, dioramas, etc.). The choice of physical context for the exhibits should focus on what selection of approaches would allow for access by the visitors with the widest range of physical and academic abilities, allow for flexibility/choice by the visitor so that each experience can relate to their personal background and interest, and allow for the simplest access to resource information (supplemental text, discussion questions, prompts, pictures, etc.) by visitors. For example, an exhibit focused on the methods used by scientists might include a hands-on section where geology/paleontology is compared with experimental biology, an interactive area where visitors construct and receive feedback on their own "scientific method" using computer technology and a video area showing biographies of scientists who approached science in diverse ways—Darwin and Rachel Carson, for example. An exhibit focusing on the history of science as a way of knowing might include an interactive timeline (which responds to touch) which discusses ways of knowing in various cultures from Ancient Egyptians, through the present day, a hands-on mystery box that lets visitors experiment with how they "know" what's inside, and a daily visitor forum that discusses the role of science and scientists in current society.

Step 4

After the exhibit ideas and approaches have been finalized, science and technology center staff should develop learning objectives for the exhibits. These objectives should echo the basic concepts of the nature of science developed by AAAS (1990). For example, a learning objective for an exhibit about scientific methods that corresponds with AAAS' description of the nature of science: Learners should be able to discuss the diverse ways that scientists interact with their world including passive observation, description and collection, active probing and experimenting. An exhibit that focuses on the history of science's role in society could have the following learning objective: Learners should value science as an important way of understanding the world, but recognize that it does not have special authority as the only way to understand the world.

Step 5

Draft the exhibit design and text. Share the exhibit drafts with the same community groups that provided suggestions at the beginning of the process and incorporate their feedback wherever possible. Be sure to consult with disability advocates to check the accessibility of the exhibit designs.

Step 6

Construct the exhibits and conduct pilot visits with the community. Use the information gathered from these pilot visits to fine-tune the design and text at each exhibit. Specifically, collect observations and data to determine if the exhibits meet objectives 1a and 1b. Use this opportunity to prepare the science and technology center staff, management, donors, volunteers, and educators for the full implementation. By involving the community in the development of the program, hopefully they will be familiar with the program's theme and objectives. The center's staff, volunteers, management, and donors will also have had the opportunity to take part in the development of the program so they should also be familiar with the program's goals and objectives. Extra preparation should be given to docents and educators who will interact with visitors at the exhibits. Familiarize them with the concepts involved in the nature of science by giving them copies of

Science for All Americans. Provide them with training on good question asking (wait time, open questions)—inform them that they will be facilitators of dialogue and not sources of information. Allow them to practice these skills during the pilot program.

In Conclusion

It is clear from the literature and from the studies conducted that informal science education via museums, science centers, zoos, aquariums, etc. can play a positive role in not only learning science, but also in providing opportunities for diverse populations. The examples provided from the literature and our own research point out the importance of multiple ways of meeting the diverse needs of informal education visitors in order to help them achieve their goals. As our examples show, the visitor has multiple opportunities to utilize exhibit information in order to answer their own questions. Visitors treasure their direct experiences and value opportunities to inquire into informal education. Visitors also need to recognize that they bring diverse and cultural experiences with them to the informal setting. Text, observations, interactions, direct experiences, and the use of artifacts and models all contribute to the opportunity to provide diverse experiences for successful informal education experiences.

References

Afonso, A. S., & Gilbert, J. K. (2007). Educational value of different types of exhibits in an interactive science and technology center. *Science Education, 91*(6), 967–987. doi:10.1002/sce.20220.
American Association for the Advancement of Science. (1990). *Science for all Americans.*http://www.project2061.org/publications/bsl?online/bolintro.htm. Accessed 17 Dec 2007.
American Association for the Advancement of Science. (1993). *Benchmarks for science education.*http://www.project2061.org/publications/bsl/online/bolintro.htm. Accessed 3 June 2012.
Ash, D. (2002). Negotiations of thematic conversations about biology. In G. Leinhardt, K. Crowley, & K. Knutson (Eds.), *Learning conversations in museums* (pp. 357–400). Mahwah: Lawrence Erlbaum.
Ash, D. (2003). Reflective scientific sense-making dialogue in two languages: The science in the dialogue and the dialogue in the science. *Science Education, 88*, 855–884.
Ash, D. (2004). How families use questions at Dioramas: Ideas for exhibit design. *Curator, 47*, 84–100.
Bamberger, Y., & Tal, T. (2006). Learning in a personal context: Levels of choice in a free choice learning environment in science and natural history museums. *Science Education, 91*, 75–95.
Bell, P. (2009). Informal science learning. *The Science Teacher, 76*(3), 14–15.
Bell, P., Lewenstein, B., Shouse, A. W., & Feder, M. A. (Eds.). (2009). *Learning science in informal environments: People, places and pursuits.* Washington, DC: The National Academies Press.
Borun, M. (2002). Object-based learning and family groups. In S. G. Paris (Ed.), *Perspectives on object-centered learning in museums* (pp. 245–260). Mahwah: Lawrence Erlbaum Associates.
Botelho, A., & Morais, A. M. (2006). Students-exhibit interaction at a science center. *Journal of Research in Science Teaching, 43*(10), 987–1018.

Brody, M., Tomkiewicz, W., & Graves, J. (2002). Park visitors' understandings, values and beliefs related to their experience at Midway Geyser Basin, Yellowstone National Park, USA. *International Journal of Science Education, 24*(11), 1119–1141.

Carlsen, W. S. (2007). Language and science learning. In S. Abell & N. Lederman (Eds.), *Handbook of research on science education* (pp. 57–74). New York: Routledge.

Dierking, L. D., & Falk, J. H. (2010). The 95 percent solution: School is not where most Americans learn most of their science. [Article]. *American Scientist, 98*(6), 486+.

Falk, J. H., & Needham, M. D. (2011). Measuring the impact of a science center on its community. *Journal of Research in Science Teaching, 48*(1), 1–12. doi:10.1002/tea.20394.

Falk, J. H., Heimlich, J., & Bronnenkant, K. (2008). Using identity-related visit motivations as a tool for understanding adult zoo and aquarium visitors' meaning-making. *Curator, 51*(1), 55–79.

Feyerabend, P. (1975). *Against method*. London: NLB.

Gilbert, J., & Priest, M. (1997). Models and discourse: A primary school visit to a museum. *Science Education, 81*, 749–762.

Glynn, S. M., & Duit, R. (1995). Learning science meaningfully: Constructing conceptual models. In S. M. Glynn & R. Duit (Eds.), *Learning science in the schools: Research reforming practice* (pp. 3–31). Mahwah: Lawrence Erlbaum Associates.

Hohenstein, J., & Tran, L. U. (2007). Use of questions in exhibit labels to generate explanatory conversation among science museum visitors. *International Journal of Science Education, 29*(12), 1557–1580.

Jenkins, E. W. (2002). Linking school science education with action. In W. M. Roth & J. Desautels (Eds.), *Science education as/for sociopolitical action*. New York: Lang.

Jewitt, C., Kress, G., Ogborn, J., & Tsatsarelis, C. (2001). Exploring learning through visual, actional and linguistic communication: The multimodal environment of a science classroom. *Educational Review, 53*(1), 5–18.

Lave, J., & Wenger, E. (1991). *Situated learning: Legitimate peripheral participation*. New York: Cambridge University.

Leinhardt, G., & Knutson, K. (2006). Grandparents speak: Museum conversations across the generations. *Curator, 49*(2), 235–252.

Lorimer, A. (2007). The Cockpit's empty chair: Education through appropriating alienation at a Chicago Technology Museum. *Teachers College Record, 109*(7), 1707–1724.

Parker, L. C. (2009). The use of zoo exhibits by family groups to learn science. *Dissertation Abstracts International, 70-11*, 4231A.

Pedretti, E. (2002). T. Kuhn meets T. Rex: Critical conversations and new directions in science centres and science museums. *Studies in Science Education, 37*, 1–40.

Roth, W.-M., & Lee, S. (2004). Science education as/for participation in the community. *Science Education, 88*, 263–291.

Roth, W.-M., & Lee, S. (2002). Scientific literacy as collective praxis. *Public Understanding of Science, 11*, 33–56.

Roth, W.-M., & McGinn, M. K. (1997). Deinstitutionalizing school science: Implications of a strong view of situated cognition. *Research in Science Education, 27*, 497–513.

Schauble, L., Leinhardt, G., & Martin, L. (1997). A framework for organizing a cumulative research agenda in informal learning contexts. *Journal of Museum Education, 22*(2&3), 3–8.

Simpson, J. S., & Parsons, E. C. (2009). African American perspectives and informal science educational experiences. *Science Education, 93*(2), 293–321. doi:10.1002/sce.20300.

Stevens, R., & Hall, R. (1997). Seeing Tornado: How video traces mediate visitor understandings of (natural?) phenomena in a science museum. *Science Education, 81*, 735–747.

Tunnicliffe, S. D. (2000). Conversations of family and primary school groups at robotic dinosaur exhibits in a museum: What do they talk about? *International Journal of Science Education, 22*(7), 739–754.

Vygotsky, L. (1978). *Mind in society: The development of higher psychological processes*. Cambridge, MA: Harvard University Press.

Zimmerman, H. T., Reeve, S., & Bell, P. (2010). Family sense-making practices in science center conversations. *Science Education, 94*(3), 478–505. doi:10.1002/sce.20374.

Diverse, Disengaged and Reactive: A Teacher's Adaptation of Ethical Dilemma Story Pedagogy as a Strategy to Re-engage Learners in Education for Sustainability

Elisabeth (Lily) Taylor, Peter Charles Taylor, and MeiLing Chow

The Value of Sustainability

The United Nation's *Brundtland Report* defines sustainability as patterns of living that '...meet the needs of the present without compromising the ability of future generations to meet their needs' (United Nations World Commission on Environment and Development (WCED) 1987). Australia has recognised education for sustainability as an issue of great importance, a national priority. The *Australian Sustainable Schools Initiative* (AuSSI) was implemented in over 2,000 schools Australia-wide (Australian Government – Department of Sustainability, Water, Environment, Population and Communities 2011). One would therefore expect to see education for sustainable development reflected in the structure and content of the new Australian Curriculum which is currently being developed and implemented. In the new curriculum documents, sustainability appears as one of the cross-curricular priorities, together with Aboriginal and Torres Strait Islander histories and cultures and Asia and Australia's engagement with Asia (Australian Curriculum, Assessment and Reporting Agency (ACARA) n.d.a). The curriculum outlines the relationship between sustainability and English, mathematics, history and science. Science '...provides content that, over the years of schooling, enables students to build an understanding of the biosphere as a dynamic system providing conditions that sustain life on Earth'. Students are expected to gain an understanding that life is interconnected through ecosystems and that humans depend on ecosystems for their survival and well-being. Scientific understanding

E. Taylor
School of Education, Curtin University, GPO Box U1987, Bentley, WA 6845, Australia
e-mail: Elisabeth.taylor@curtin.edu.au

P.C. Taylor (✉) • M. Chow
Science and Mathematics Education Centre, Curtin University, GPO Box U1987, Bentley, WA 6845, Australia
e-mail: p.taylor@curtin.edu.au

N. Mansour and R. Wegerif (eds.), *Science Education for Diversity: Theory and Practice*, Cultural Studies of Science Education 8, DOI 10.1007/978-94-007-4563-6_6, © Springer Science+Business Media Dordrecht 2013

and scientific inquiry provide the knowledge and skills to 'forecast change and plan actions necessary to shape more sustainable futures'. A focus on sustainability allows science education to address systems change processes, their causes and consequences, thereby '... assisting students to relate learning across the strands of science' (Australian Curriculum, Assessment and Reporting Agency (ACARA) n.d.b). These statements emphasise science education's leading role in preparing learners for sustainability thinking, planning and action.

With such a strong curriculum emphasis, it would seem that the teaching of sustainability is set to become a widespread and integral component of Australian classrooms. But the presence of sustainability in science classrooms might not be easily achieved due to its 'meta-position' which embraces science, society, the environment, culture and the economy. In other words, sustainability is situated at the nexus of scientific methods and sociocultural perspectives which examine differing human interests, motivations and cultural values. Are science teachers well prepared to engage their students in debates whose resolution is not amenable to solely scientific content and processes (Robottom and Simonneaux 2012)? Robottom (2012) concludes that whilst a scientific element is undoubtedly necessary within an education for sustainability discourse, it is insufficient for a rigorous educational exploration of 'socio-scientific' (or science in everyday life) issues. A lack of preparedness to address socio-scientific issues might explain many science teachers' hesitancy to embrace education for sustainability – after all they are the product of science teacher education programmes that focus almost entirely on scientific content and processes with a high level of epistemic certainty and predictability. However, according to the Sustainability Curriculum Framework (Australian Government – Department of Water, Environment, Heritage and the Arts 2010), students of education for sustainability

> ... will be able to assess competing viewpoints, values and interests; manage uncertainty and risk; make connections between seemingly unrelated concepts, ideas and outcomes; and test evidence and propose creative solutions that lead to improved sustainability. (p.5)

We may conclude from this statement that education for sustainability has strong links to sociocultural perspectives due to its connection to human activities, interests and cultural values.

According to the Framework for Values Education in Australian Schools, our future depends on young Australians developing a solid foundation of intellectual, physical, social, moral, spiritual and aesthetic abilities (Australian Government – Ministerial Council on Education, Employment, Training and Youth Affairs 1999; Australian Government – Department of Education, Science and Training 2005). In this respect, values education has been identified as a vital ingredient of effective education for sustainability. At its heart lies the recognition that, as a national value, sustainable development starts with individual value systems which shape people's attitudes to the (natural and social) environment, which, in turn, affect their abilities to make decisions that impact the future of not only the nation but the world beyond. These decisions are clearly ethical in nature since their outcomes will affect current and future generations.

Making sound ethical decisions requires informed decision-making skills based on sound scientific knowledge of the environment, high-level awareness of the environmental impact of science and technology and an ability to engage in critical thinking and critical reflection, thus being able to distinguish between beneficial and potentially detrimental policy decisions. Gower (1992) discusses our moral obligations towards future generations and raises an interesting point: despite the rich stock of ideas as to how moral issues 'should' be examined, tested and resolved, often there is a dialectic, or opposition, between one set of ideas and another – this can lead to important insights even if the dialectic remains unresolved (p. 11). From this perspective, the task of science teachers is not only to prepare students to participate in an informed and competent manner in the public discourse on science but also to enable them to gain insights from uncertainty associated with the highly complex issues of sustainable development. The question is whether or not science teachers can cope with uncertainty and multiple potential solutions.

Science Education for Sustainable Development

With regard to solving global problems in a world of rapid change and development, much hope is invested in education, especially science education. This uncritical yet popular perspective, which assumes that education is responsible for solving the world's problems, has been critiqued and labelled 'educationalisation' by authors such as Depaepe and Smeyers (2008). Despite this critique, it seems that imbuing science education with a special role in facilitating sustainable development is still an appropriate view due to the complex nature of the environmental challenges facing us. Drawing on Polanyi's critique of markets in capitalist societies, Sharma (2012) regards science education as a central element of a societal response because major environmental issues such as climate change can be viewed as products of the commodification of nature in market-dominated societies. This view is supported by UNESCO's (United Nations Educational, Scientific and Cultural Organisation) 'Science and Technology Education' website which emphasises that science and technology education is an essential tool in the search for sustainable development. However, UNESCO also laments that current science and technology education has lost relevance and is unable to adapt to the challenges facing education systems. UNESCO therefore encourages development of curricula and policies for science education that employ multidisciplinary approaches that promote (1) gender-sensitive, sociocultural and environmental knowledge, (2) life skills and (3) scientific literacy (United Nations Educational, Scientific and Cultural Organisation 2011).

According to Van Eijck and Roth (2007), high-ranking scientists have expressed an urgent need for improvement in science education. The authors view this demand as an expression of the growing consciousness that high-quality science education is vital not only for sustaining a lively scientific community capable of addressing global problems, such as global warming and pandemics, but also for bringing about and maintaining a high level of scientific literacy in the general population. This sentiment is echoed in a report on a recent conference in South Africa where physicists from around the world discussed the role of science and science education in relation to achieving sustainable development. It was concluded that '...the problem-solving style of scientists and engineers is a mindset sorely needed for the sustainable development challenges facing developing countries and an ever-increasingly globalised world' (Moore 2006, p. 42).

Several fundamental questions arise from the expectations invested in science education described in the previous statements. Is the high hope invested in science education for helping us solve our problems appropriate? Is it appropriate to try to address socio-scientific issues such as sustainable development with the mindset of a scientist? It seems that different mindsets are at work in science labs and in science classrooms. Robottom (2012) contrasted his own experiences as a researcher conducting research for its own intrinsic value with addressing socio-scientific issues in education and concluded that 'socio-scientific issues are necessarily and irrevocably located within a community context' (p. 97), which implies shared community values.

If we accept that creative problem-solving practised by scientists and values learning grounded in shared values of communal concerns are required for preparing both future scientists and non-scientist decision-makers for managing systemic global change policies, are science educators rising to the occasion? Is creative problem-solving as a way of thinking applied by successful scientists actually taught in science classrooms?

Traditionally science has been taught as if it could and should be value-free (Allchin 1998). Science teachers adopting a technical and scientistic view of science tend to shy away from addressing values because of a trenchant belief that values lie outside the domain of science education. In our own research, we have found it increasingly difficult to find science teachers willing to collaborate on a project that seems to be 'slightly outside the norm of science education', preferring to delegate the teaching of creative thinking to colleagues in the arts and humanities. This 'silo' view of one's discipline fails to heed the growing worldwide realisation that everything taught in schools influences how students understand and shape the human culture/natural environment relationship (Bowers 1993; Orr 1992). Such an integral perspective, which promotes a broader approach to how we think about education and its relationship to society, is fuelled by a realisation that the resolution of global environmental crises requires knowledge, skills and values drawn from various disciplines: complex curriculum solutions for complex problems.

In this respect, there is a pressing need to reconsider the pivotal role of science education in fostering education for sustainability. Some might say that a 'macroshift' in thinking (Laszlo 2008) is required of science educators if their discipline is to benefit from integration with other disciplines, such as the arts and humanities. This demand for change in thinking is grounded in the insight that engagement in values learning as part of education for sustainability will contribute significantly to a scientifically literate citizenry for the twenty-first century (McInerney 1986; Zeidler 1984).

Lack of Engagement and Proactivity

In recent years, concerns have been voiced about falling student numbers in mathematics and science-related disciplines. Studies by Australian psychologists have identified disconcerting levels of hopelessness and feelings of helplessness amongst many adolescents (Fien 2001). These issues are especially problematic since young people represent the leaders and decision-makers of the future, and it is their decisions that will need to be both ethically and sustainably sound.

So, how does one educate a somewhat reluctant generation challenged by unprecedented global problems? How does one educate young people who seem to have 'abandoned ship'? Are our current methods of instruction and thinking about curriculum sufficient? Traditionally, the response to similar educational conundrums, at least in science education, has been that if students do not perform well in science, we must inject more science content into the curriculum and timetable more science classes in the school week. Environmental education curricula have been treated in a similar way. However, this approach is based on a deficit view of learners' access to knowledge: preparation of future leaders from this perspective focuses primarily on content learning. The question arises as to whether this is enough.

Uzzell (2008) points to a widespread mistaken assumption in educational circles that global crises can be resolved solely by developing children's knowledge levels. This belief is grounded in the conviction that children will assume the role of 'little experts' when they return home after school whereupon they will positively influence their parents to conserve water, save electricity and recycle – thus transforming societies and cultures. Unfortunately, for proponents of the 'content plus' approach, research funded by the European Union (cited in Uzzell 2008) has concluded that the widespread assumption that the provision of more environmental facts to students will lead automatically to enhanced concern and action in the community through passive osmosis has been shown to be false.

At this point, we would like to draw on Steven Covey's (1989) scholarly work on personal empowerment and leadership education. Covey distinguishes between our 'circle of concern', issues we are concerned about, and our 'circle of influence', issues we can actually do something about. For 'reactive' people, the circle of concern is much larger than their perceived circle of influence, thereby leading to a sense of hopelessness and lack of proactivity. In comparison, 'proactive' people's circle of influence is much larger, leading to a sense of empowerment. Applying Covey's model to young, disengaged learners, it seems that for many, there is a mismatch between the two circles. Most students in the so-called developed world lack neither concern about the environment nor do they lack content learning or opportunities to engage in learning about the environment. There is already plenty of attention paid to that in current curricula. There seems to be a correlation, however, between many students' experience of 'voicelessness' and disengagement and their experience of a diminished personal circle of influence: it rarely overlaps with their circle of concern.

Responding to student disengagement by adding more content – and thereby adding more problems without adding a sense of empowerment – is unlikely to lead to enhanced engagement. It seems that this approach can only exacerbate the problem by increasing students' circle of concern whilst failing to increase their perceived circle of influence. Based on our long-term experiences with ethical dilemma pedagogy as a way of engaging students in deep learning through moral dilemmas, we argue that ethical dilemma story pedagogy, in a supporting role for traditional science curricula, may counteract the trend of disengagement by giving students opportunities to practise decision-making and problem-solving with their voices being heard.

Our research has shown that ethical dilemma stories have the potential to engage a diverse range of learners in issues related to the use of science and technology in daily life, especially to issues of sustainability. From a pedagogical point of view, our approach addresses the requirements of successful programmes for education for sustainability, as suggested by Fien (2003) by promoting care and compassion for the environment and for stakeholders. Because students are not just passively taught but are actively involved in trying to make ethical decisions and to finding solutions to problems that relate to the curriculum of their lifeworlds, they are more likely to re-engage with science and sustainability issues. This strategy seems to sit well with what Covey argues the leaders of the future need: the new leaders born and bred in the Knowledge Age, which started with the fall of the Berlin Wall, cannot rely on the same strategies that worked for leaders of the Industrial Age. These new leaders will have to find their voice and thus enhance their moral integrity (Covey 2004).

This perspective has direct implications for how we view curriculum development for future science education – curricula that are successful at enhancing students' environmental awareness and agency and that encourage future citizens to get involved and be active rather than closing down and becoming helplessly reactive. We believe that ethical dilemma story pedagogy can help counter the chronic disengagement amongst science learners.

Ethical Dilemma Story Pedagogy: A Sociocultural Perspective

Ethical dilemma stories are stories with characters and a storyline that contain one or more ethical dilemma scenarios. The story is best told freely by the teacher who breaks the storyline at appropriate junctures to pose ethical dilemma questions. Students are instructed to engage with each dilemma question, thereby making a series of ethical decisions on behalf of the story's character. Ideally, the story has direct curricular links to specific concepts or skills as well as perceived relevance to students' lifeworlds. Examples of ethical dilemma stories, including suggestions for teaching, are available at www.dilemmas.net.au.

In our research, we use the approach to dilemma story pedagogy suggested by Gschweitl et al. (1998). According to this approach, one of the key pedagogical aspects of ethical dilemma learning is the requirement for students to reflect individually on a dilemma question and to record in writing their decision and the reason for their decision, thereby engaging with their personal values. The next step is to interact in small groups with peers to compare and contrast their decisions, thereby promoting critical reflection on their decision-making values. We have found it best to build up group sizes gradually during the course of the story as students encounter each successive dilemma question. The reason for a staged approach to group work is that discussing personal values and decisions is not often done in public, especially not amongst adolescent peers. Building up group size allows students to 'warm-up' to having this type of unfamiliar discourse in class, thereby building rapport and trust. The teacher ensures that 'rules of engagement' are clear from the beginning and that, because there is no single correct answer, it is okay to voice an opinion that is different to that of other students. Ethical dilemma story pedagogy is thus an approach to values learning that employs ethical dilemma stories as a means to engage learners in:

- Critical thinking about a dilemma problem that has no clear cut, black-and-white answer
- Critical self-reflection on taken-for-granted assumptions grounded in personal values, which involves individual reflection and explanation of dilemma decisions
- Social learning through subsequent discussions with peers
- Emotional learning through promoting active and empathic listening skills when different views are shared in class
- Problem-solving by codeveloping suggestions for possible solutions

Ethical dilemma stories date back to the psychologist, Lawrence Kohlberg, who based them on an 'ethic of justice' as a means of engaging learners in moral reasoning. Kohlberg (1984) was interested in how moral development progresses. He developed a stage theory of moral development that he applied when analysing student responses to dilemma problems. Subsequent feminist researchers, notably Carol Gilligan, added an 'ethic of care' to Kohlberg's theory, thereby opening the door to the multidimensional and multivocal nature of the moral domain (Gilligan 1982).

Gilligan's work is of particular importance for ethical dilemma story pedagogy since she recognised the importance of words, language and storytelling as a form of human discourse essential to a moral life. Parker Palmer defines discourse as what humans do every day that involves the use of language in the form of speaking (Palmer 1993). With the moral self being a shared or distributed product of social relations and communicative practices, the role of social-cultural-historical-institutional contexts has gained importance in the field of moral learning, especially with a view to human action and interaction or, as we would like to add, inaction (Tappan 2010). According to Tappan, a sociocultural, dialogical view has become increasingly influential in moral education, especially through the inclusion of the theoretical and empirical work of Vygotsky and Bakhtin. Tappan (1997) outlines areas of overlap between sociocultural theory and theoretical aspects of moral development, such as the assumption that higher moral functioning – for example, ethical decision-making – is mediated by words, language and forms of discourse such as storytelling. Furthermore, this mediation is made possible through inner speech resulting in an inner moral dialogue. There is interplay between the inner moral dialogue of the individual and processes of social communication whereby the person becomes engaged in social relations with moral implications. Consequently, moral development is always shaped by the particular social, cultural and historical context in which it occurs. Applying Tappan's outline to our work, we can say that in the context of values learning through ethical dilemma story pedagogy:

- Moral functioning in the form of ethical decision-making is mediated through the telling of dilemma stories and the discussion and discourses involved in solving the dilemmas.
- At every dilemma situation in the story there is a break where students are requested to reflect individually on how they would solve the ethical dilemma. They are then asked to write down their decisions plus their reasons why they would decide in a certain way.
- This phase of individual reflection is followed by an exchange of ideas between pairs, at first, with growing group sizes as the story progresses, culminating in a whole-class discussion at the end.
- Our research into students' responses and decisions indicates that they are influenced by their social, cultural and historical context.

Active involvement in problem-solving, where students' opinions count and where the teacher facilitates rather than determines the decision-making process, promises to provide a natural antidote to student disengagement, hopelessness and help-lessness. It is important to emphasise that we are not suggesting the replacement of traditional content learning in science curricula. Rather, we are excited by our research which is demonstrating how ethical dilemma pedagogy is being incorporated into mainstream science lessons in a variety of ways by creative teachers committed to promoting education for sustainability.

Preparing the Teachers for Ethical Dilemma Pedagogy

The key to success of ethical dilemma pedagogy is a teacher who is convinced of the importance of socio-scientific issues in science education, cognisant of ways to establish and maintain a social constructivist learning environment and prepared to take on the role of a facilitator rather than that of a traditional knowledge dispenser. As part of the project teachers received intensive professional development in ethical dilemma pedagogy which included professional learning about theoretical aspects such as the social constructivist nature of ethical dilemma pedagogy as well as practical aspects of the role of the teacher as facilitator. Teachers were not only introduced to the structure and nature of ethical dilemma stories but also encouraged and trained to write their own locally relevant stories that would sit well within their curriculum. Furthermore, teachers were encouraged to adapt the existing model of dilemma story pedagogy (Settelmaier 2009) to their needs. Teachers in the project accepted the challenge and creatively adapted and remodelled our suggested 'recipe'. It was entirely up to the individual teacher's professional judgement as to how much science content they would teach before engaging their students in an ethical dilemma lesson. For this reason, some teachers chose to use an ethical dilemma story as an introduction to a new topic, others taught content first to prepare their students for the story and yet others used a story as the culminating finale of a curriculum topic.

Student Diversity in Ethical Dilemma Story Research

In this chapter, we draw on some of the results of our 3-year research into ethical dilemma story pedagogy funded by the Australian Research Council (ARC) to illustrate its effectiveness in diverse classrooms. This project is investigating science and mathematics teachers' experiences in designing and implementing ethical dilemma story pedagogy as part of education for sustainability as well as their students' dilemma learning experiences. Of special interest is how ethical dilemma story pedagogy successfully engages students across a diverse range of academic abilities and sociocultural backgrounds by tapping into the personal values they bring from home.

For this purpose, we are less concerned with demonstrating students' attainment of dilemma-learning outcomes (such as critical thinking, critical reflection, collaborative decision-making) and more with identifying the qualities of dilemma story teaching that spark students' interest and deep involvement in science-related learning. In our research across a variety of contexts, we have found that some students 'get' an ethical dilemma whilst others do not. Sometimes students find an ethical dilemma deeply engaging compared to other students who find it mildly concerning. Local relevance seems to be one of the key factors as described in our earlier work (Settelmaier 2009). The key question that we address here is: *what makes dilemma story teaching more or less compelling for students?*

In the next section, we present a synopsis of a case study conducted at Hardbridge College (a pseudonym), a Western Australian middle school renowned for its success in educating students from a low socio-economic suburban area with a multicultural student population, including a large number of Australian Aboriginal students. Yet it's not only the students who have culturally diverse backgrounds, so do some of their teachers, such as MeiLing Chow, the young Asian science teacher with a strong Buddhist background, who was the participating teacher in this case study. At the time of the fieldwork, she was in her 6th year of teaching. The case study focused on her three Year 10 science classes and in particular on the effect of the dilemma stories on her students' thinking about and engagement in science. Other publications (Chow et al. 2011; Settelmaier 2009; Settelmaier et al. 2010) describe in detail the research and teaching methodologies and data analyses. Suffice to say that the case study results presented here were generated in accordance with an interpretivist epistemology that employed teacher and student interviews, classroom observations, a student questionnaire and analysis of student work samples, all of which were subjected to grounded theorising within the sociocultural perspective on dilemma thinking outlined above.

Sociocultural Context of the Case Study

There are several aspects we would like to discuss briefly in relation to the sociocultural context of the case study, including the impact of mining on families in Western Australia, Australian Aboriginal culture, multicultural issues and Buddhist environmental ethics.

Hardbridge College is a Year 8–10 government secondary school situated in a low socio-economic area in the Perth foothills, with wealthier suburbs occupying the hilltops of the Darling Range. The population is mostly working class and multicultural consisting of Anglo-Celtic Australians and migrants from Asia and Europe, in particular from Balkan countries. Some parents of students work in the mines, which usually means 'fly-in fly-out' work rosters which can put severe strain on family relationships. Despite the relatively good income from mining, wealth is hard to come by since family break-ups are common. At Hardbridge College, some students' families have migrated from overseas, and it can be assumed that some students have experienced life as refugees before settling in Australia, especially students from Afghanistan and Balkan countries.

There is a large Australian Aboriginal community in the local area. According to Simon Forrest (2002), Head of Curtin University's Centre for Aboriginal Studies, an Aboriginal is a person who has practices, language, behaviours, values and beliefs common with other Aboriginal people. Aboriginal identity is strongly related to the often complicated kinship system distinguishing language and culture groups. In more traditional groups, this system is referred to as 'skin group' system in which an individual identifies with family members who may not be biologically related to them as father, mother, brother, sister, uncle and auntie. What many Aboriginal Australians share is a strong connection to 'the Land'. A strong sense of connectedness and family bonds affect social structure, economic behaviour, language and spiritual values. The existence of preferred Aboriginal learning styles has been discussed in the literature, e.g. Ryan (1992), referring to Aboriginal students' preference for group-based learning, storytelling as a vital aspect of teaching and strong community involvement in education.

According to the Census Document 2006, Western Australia is the most culturally diverse of all Australian states and territories. Whilst the majority of the population is of Anglo-Celtic origin, a high number of people were born overseas and speak languages other than English at home. In Western Australia, religious diversity as well as linguistic diversity is higher than in other parts of Australia

(Australian Bureau of Statistics 2007). These trends were reflected in Mei Ling's Year 10 class which represents a snapshot of the multicultural nature of Western Australia.

As Mei Ling is a practising, devout Buddhist with knowledge of Buddhist scripture, a factor influencing her work as a science teacher is Buddhist environmental philosophy. Srivastava (2005) reminds Buddhists that there is no need for them to go to the modern prophets of environmental 'doom and gloom' but instead to revisit and reread their own ancient texts since it was the Buddha himself who, in *the Buddha Vacana*, apprehended today's impending eco-crisis. He advocated proper management of natural resources and protection of natural environments from human encroachment. He stipulated the pillars of Buddhist environmental thought as stewardship of nature and protection of nature. It is important to note that Buddhist theory of nature conservation is 'cosmo-centric': neither is a human being regarded as the master nor is nature his slave or something to be exploited for human consumption or pleasure. It is rather the abandonment of selfish thought in favour of the moral precepts of loving kindness (karuna), joy (mudita) and friendliness (metta) – which translates into eco-friendliness – and a concern for the well-being of nature (Srivastava 2005).

At Hardbridge College, teachers are required to make adaptations to the curriculum so that it is more culturally relevant and community based. Low literacy and numeracy levels plus many problems common in low socio-economic areas tend to affect student learning. The students are organised in year groups taught by teacher teams with a strong focus on pastoral care. Mei Ling is an energetic and dedicated science and mathematics teacher in her 6th year of teaching. She is always interested in finding new ways to engage her students in education for sustainability.

> As a science educator, I am passionate about issues of pollution and sustainability. I feel that students are not given enough opportunities in school to explore environmental issues. Yes, there might be the odd lesson that teaches about pollution and recycling, but how often do the students treat it seriously?

Mei Ling is concerned that sustainability does not rate highly on her students' priority list.

> For many it's just another Science topic, especially in my school where the majority of my students come from low socio-economic backgrounds, with 47 % of the school population being Aboriginal students. Science education seems to be largely irrelevant to my students. It is either too boring or the boys complain that they are not doing enough experiments that 'blows things up'. Some students somehow have the idea that they already know everything, already have it all worked out, and that Science is not going to be any use for them in their future!

Mei Ling's comments touch on one of the major concerns raised in this chapter: chronic lack of engagement in science due to a perceived lack of curriculum relevance to students' lives. Mei Ling's students are very culturally diverse, with a large cohort of Aboriginal students and a minority of White students in some classes. Many have problematic and unstable family backgrounds. We observed that

Mei Ling maintained a high degree of rapport and mutual respect with her students. We never witnessed situations where students were confrontational towards her or pushed the limits of acceptable behaviour. Mei Ling saw ethical dilemma teaching as a potentially valuable way to engage her otherwise disengaged students in meaningful learning about issues of sustainability.

> When I was first introduced to dilemma story teaching I immediately saw the potential of engaging my students in a topic that might interest them. I believed that this innovative teaching approach might help me connect with their interest in topical issues in the world beyond school, while at the same time connecting with the Science curriculum. I really liked the idea of tapping into the students' personal values in an attempt to engage them in learning science.

Like all the other participating teachers, Mei Ling received extensive professional development in ethical dilemma story pedagogy. Subsequently, she wrote two dilemma stories designed for her Year 9 and 10 science classes. *The Prime Minister Dilemma* story raises the question of how future government funds should be allocated: for fixing catastrophic environmental problems or for providing half of Australia's population with an escape to a new life on a newly discovered habitable planet. Here, we focus on the second of Mei Ling's dilemma stories: *The Mining Dilemma* which she taught in three Year 10 classes during 2011. Both dilemma stories can be viewed on our website www.dilemmas.net.au.

The Mining Dilemma

The Mining Dilemma addresses the environmental impact of the Western Australian 'mining boom' which provides jobs for many families but also threatens to destroy pristine wilderness areas of 'outback' Australia. The story is told through the eyes of a boy named Akiki whose father works in the mining industry in a remote region of Western Australia, requiring him to leave his family for weeks at a time. The father is offered a high-level job in a northern town (in the Kimberley region) that will enable him to live with his family. However, there is a sting in the tail in that the mining company requires him to open a new mine close to the town in an environmentally sensitive area. The community is confronted with the decision as to whether a mining project that is likely to damage the delicate environment is to go ahead. The main dilemma question is:

What is more important to the community – the environment or improved job prospects?

Mei Ling prepared the students for the Mining Dilemma teaching story by providing preliminary science lessons on the geology, chemistry and technology of mining for minerals, which are topics associated with the 'Natural and Processed Materials' and 'Earth and Beyond' strands of the K-10 Science Syllabus of Western Australia (Government of Western Australia – Curriculum Council 2010). She taught the Mining Dilemma story over two 60-min lessons as a culmination to this science topic. During the dilemma teaching, she directed the class to engage in a series of individual reflective thinking and small-group discussions. In order to support students who struggle with basic literacy skills, she provided a PowerPoint presentation whilst telling the story to the whole class. Her collaborating Society and Environment colleague taught the students about the major environmental effect mining can have on a community and steps mining companies can take to protect both the community and the environment from harm. The Mining Dilemma story provided a cross-curricular link to the Society and Environment curriculum strands of 'Resources' and 'Place and Space' (Government of Western Australia – Curriculum Council 2010)

Aaron

Aaron is an Australian Aboriginal boy who engaged meaningfully in dilemma-learning activity, in the process drawing upon his family values to personalise and resolve his dilemma thinking. Aaron has a relatively high literacy level and was keen to support fellow students struggling with poor reading skills by reading aloud to the whole class passages on Mei Ling's PowerPoint presentation of the Mining Dilemma story. When asked how he would resolve the dilemma if he had been in Akiki's situation, he responded as follows.

Aaron: I made the decision that his dad should stop working in the mines and stay with his family
Intvr: And why did you make that decision?
Aaron: Because if he loves his family, then he'll stay with his family and try and get another job in the Kimberleys. There might be less pay, but at least you'll be with your family.
Intvr: Family is an important thing for you?
Aaron: Yep.
Intvr: Are you close with your family?
Aaron: Yeah.

Aaron's responses are very much in line with traditional Aboriginal culture where family ties are very tight and important in everyone's lives. The link between Aaron's responses and his cultural and social background becomes clearer in the following excerpt:

Intvr: What did you find interesting about the story?
Aaron: That his father had a hard choice to pick. And we didn't really know which choice we had to pick so it was just really interesting for me to know which one he was going to pick.
Intvr: Does any of your family work in the mines?
Aaron: My dad used to, but we asked him to stop working there, and he stopped and he started working somewhere else now.

Intvr:	He stayed with the family?
Aaron:	He used to work out in the mines, but now works for people who get into the mines. Like, helps people get jobs.
Intvr:	He doesn't fly-in, fly-out anymore?
Aaron:	Nope.
Intvr:	Do you like it that way? Or do you prefer seeing less of your dad?
Aaron:	Nah, it's better with my dad here.

Many families in Western Australia are subject to a 'fly-in fly-out' regime resulting in family members being separated often for weeks at a time. Many students in Mei Ling's class have parents working in the mines. Whilst the mining boom has been responsible for making Western Australia the richest of the Australian states, it seems that this wealth does not extend to Mei Ling's school community where poverty is widespread. Nevertheless, the story has a strong resonance with the lives of many of Mei Ling's students. When asked how he had liked the dilemma story approach, Aaron said that he liked it, '...cause it kept me on task'.

Intvr:	How did it keep you on task?
Aaron:	Like, other times I would rip out my iPod and start listening to it, but this had me thinking sometimes.
Intvr:	What did it make you think about?
Aaron:	Most of the times I couldn't even sleep, just thinking about what would he do? Stay with his family or go up to the mines?
Intvr:	It affected you pretty deeply?
Aaron:	No. Not pretty deeply, just that it made me think for some reason.

Aaron's comments indicate that the Mining Dilemma story had resonated with his personal beliefs and values; it had engaged him in dilemma thinking; it had '*made me think*'.

Marisa

The Mining Dilemma story also elicited the personal values of Marisa, a European Australian girl, engaging her deeply in thinking about the family-environment dilemma confronting Akiki.

Marisa:	He [Akiki] had to decide whether he wanted his dad to be home constantly and have the mining through the town he's just moved to or keep the town as it is and have his dad still working away.
Intvr:	What was the dilemma of the story?
Marisa:	That he had to choose between seeing his dad. He was put into a bad position, he had to choose and it was a tough decision.
Intvr:	Did you guys have to place yourselves as Akiki?

Marisa:	Yeah. We were asked what we would do if we were in that position, and we had a bit of a debate about it.
Intvr:	And what did you talk about?
Marisa:	How the Kimberley was such a nice place, and putting mines there is a stupid idea because there's families living there and it would pollute the area, and noise pollution and all that, eventually people would have to leave anyway...Lots of parents have to work away, and it's just something you have to learn to deal with. And it's better off that way, rather than a year or so and they have to move again.

Unlike Aaron, Marisa was prepared to 'sacrifice' Akiki's father living with his family in order to protect the environment. She explained that she was opposed to mining in the Kimberley region because its long-term effects could cause problems for everybody.

> I think it's better that his dad can choose working away [from the family in his current job], and they don't mine in the Kimberleys. Because it's not necessary. He's got a job where he is. And that's another thing; he would get a higher position, better pay. But umm it's still ruining their home. Their family, he had a younger sibling I think, it would affect him and everything.

The personalisation of the Mining Dilemma story for Marisa becomes clear in her next comments:

Intvr:	How do you feel normally about mining?
Marisa:	My step-dad works away...like mining is something that needs to be done. It brings good income, and everything. And it sucks that he goes away but it's part of life. And I've still got my mum as Akiki still had his mum.

Her family values expressed a conflict between having a dad who is with the family and the necessary 'evil' of mining which provides an essential family income. The story seemed to have deeply touched Marisa. When asked whether she could identify with Akiki's feelings, she said:

> Yeah!...Yeah, like it's a hard decision. Like if my stepdad had the decision to work here, like we're considering moving to Collie [a rural coal mining town] 'cause we'd be right on site and he'd come home every day. But then we'd have to move and stuff, like Akiki, and we've come to the decision that we're going to stay [in Perth]. And that's what I think Akiki should have done.

Marisa's subsequent comments highlight aspects of the widespread feeling of helplessness amongst adolescents. When asked what she thought about environmental protection, she responded:

> I think the environment is important, but they're going to mine regardless. So, there's no laws against it and companies pay lots of money for the mining they do. So, it's not like they're going to stop it. So, I haven't really thought that much about it but they're going to mine anyway, and global warming's going to kill the earth anyway. So...there's not really much you can do to stop it.

Marisa's further comments confirm that she had engaged in dilemma thinking about environmental protection versus mining:

Marisa: There are not very many positive aspects of mining compared to the negative. Like with the whole environment and stuff. But again, it supplies many jobs, it employs thousands of people, so it's kind of a win-win-situation.

Intvr: Do you think it's possible to mine while looking after the environment or do you think that's not possible?

Marisa: I don't know. At the moment I don't think it is, but in the future they could come up with a way to. At the moment I just think it's destroying the environment.

We conclude that The Mining Dilemma Story case study illustrates how deeply engaged Aaron and Marisa were in dilemma thinking, with Marisa evidencing a more nuanced perspective. Aaron can be viewed as an example of a student who normally doesn't engage in science or learning but who was drawn into dilemma thinking by the values-based questions in the story. In struggling to resolve the family-vs-environment dilemma facing Akiki, the main character of the story, they arrived at different conclusions based in large part on their different family values. Although many of the Year 10 students indicated that they were in favour of protecting the natural environment, ultimately Akiki's father's job prospects were regarded by many as being more important. This resolution is not surprising given that many of these students' families face the daily spectre of unemployment and poverty. Regardless of the ways in which the students resolved the dilemma, the results confirm that this ethical dilemma teaching story successfully engaged these socioculturally and academically diverse Year 10 students in science-related learning.

Summary and Conclusion

Our research has found that ethical dilemma story has the potential to re-engage students in science learning, especially when it is paired with sociocultural issues relating to sustainable development. The goal of ethical dilemma story pedagogy is to support students' growth as responsible, autonomous, democratic citizens, capable of practising an ethic of consistency, capable of evaluating the consequences of their actions and able to practise both empathy and care in their adult lives. These character strengths are sought after in decision-makers of the future – people on whose decisions the sustainability of the world's resources will depend. We believe that science education for sustainability should have an important role in developing these character strengths in students. It is for this reason that we are working with science teachers to investigate implementation of ethical dilemma pedagogy in their classrooms.

Teachers in our research have been developing ethical dilemma teaching stories as curriculum resources aimed at the lifeworld interests of their students. In this chapter, we have focused on one science teacher, Mei Ling, whose Year 10 science class was an exemplar of academic and sociocultural diversity. Our research has shown that ethical dilemma story pedagogy has been successful in getting Mei Ling's diverse adolescent students involved in resolving ethical dilemmas linked to their normal science curriculum.

Relevance of the storyline to students is important if they are to identify with the character in the story and to engage in problem-solving on his/her behalf. In Mei Ling's case, her Mining Dilemma story seems to have struck a chord in students' minds and hearts due to the similarities between the main character and their own lives. We have learned through our work with other teachers that a teacher's enthusiasm for the content of a dilemma story, especially one she has authored, is not a sufficient guarantee of student engagement. If students cannot attach personal significance to the dilemma in the story, if they find the story too far removed from their lifeworlds, then they are unlikely to feel compelled to engage in dilemma thinking.

Mei Ling's case study supports earlier research which found that in a class of diverse students a good dilemma story can (and should) elicit differing responses, some of which might be due to cultural differences. In the Mining Dilemma Story, Marisa voiced a more nuanced understanding and personal resolution of the family-vs-environment dilemma than did Aaron, who was opting in favour of a strong family connection over financial gains, even though there was evidence that both had engaged in dilemma thinking. It is only natural, and indeed educationally desirable, that a range of perspectives is elicited and voiced in response to dilemma thinking. This is supported by Killen et al. (2010) who argue that varying application of moral principles reflects the diversity of life experiences students bring to the classroom: group functioning, group identity, cultural expectations and traditions all influence moral considerations.

Applied to our case study, it would be easy to jump to the conclusion that Marisa's comments were representative of her more individualistic European Australian cultural background since they were more clearly focused around individualism, independence and separation, whilst Aaron's ethical decision-making seemed to reflect the collectivist nature of Aboriginal culture with its strong family bonds. Yet, Smetana cautions against drawing such simplistic conclusions. The cross-cultural application of theories of moral reasoning has generated much interest and controversy, especially in relation to the distinction between individualistic and collectivist cultures (Smetana 2010, p. 143). In contrast to proponents of theories of structural moral development, such as Kohlberg who deemed that moral development followed universal principles, researchers such as Nucci (2010) have become increasingly cautious and tend to look beyond the cultural surface when exploring moral reasoning. They contend that within cultures with a recognised collectivist nature, individual decision-making can focus on individual needs and interests, whilst in individualistic cultures, individuals may opt for collectivist values. What does this mean for our scenario of culturally diverse Year 10 science students engaged in ethical dilemma thinking? We conclude that whilst students may be strongly influenced by their home cultures, individual responses to ethical dilemmas might differ from the cultural norm, and so we should be wary of stereotyping particular ethnic or racial groups, especially individuals within these groups, as having special interests or needs.

Also of importance to the quality of student engagement in dilemma thinking is the teacher's style of delivery. We have witnessed how the tone of storytelling, where the teacher abandons reading a prepared script and speaks directly from her heart, can be far more captivating as in the case of Mei Ling who told the story more or less freely. She had designed a PowerPoint presentation to support the delivery of the Mining Dilemma in order to help engage the visual thinkers amongst her semi-literate students. It was fascinating to observe an Aboriginal boy (Aaron) seize the initiative to read the story aloud to the whole class, thereby (wittingly or unwittingly) supporting his cultural peers struggling with poor functional literacy skills. Whatever the mode of delivery teachers choose, it is important that they are sensitive to what works best to capture students' interest and captivate their imagination, to ensure that the storyline and the central character resonate with students.

This is a good moment to recall that ethical dilemma pedagogy is not a search for a single best answer, even though as teachers we might be tempted to steer students towards our personal view or belief. In the case of education for sustainability, it could be very tempting to abandon the 'teacher as facilitator' role and adopt the 'teacher as instructor' role of the content expert, pointing out to the class the 'politically correct' answer. However, the pedagogical goal of ethical dilemma teaching is to enable students to develop key ethical decision-making capabilities:

• Awareness of and critical reflection on their own values and the values of their peers
• Ability to engage deeply in dilemma thinking – to struggle with the horns of a dilemma
• Ability to negotiate and justify a personal or shared resolution of a dilemma

The results of this and other case studies give hope to science educators interested in making their curricula more socially responsible and inclusive. Whilst we are not suggesting that ethical dilemma story pedagogy is a 'magic bullet' that will solve all problems afflicting science education, there are strong indicators that student engagement with important issues of sustainability can be enhanced through this approach. If socially responsible science education is to successfully meet the pressing challenge of sustainable development, then classroom discourse on values and the ethical implications of science, on the way scientific knowledge is obtained and used in daily life, and on how science and environmental issues are interlinked is inevitable.

We believe that a particular strength of ethical dilemma story pedagogy is that it supports students in finding their *voice* as critically aware decision-makers and prospective leaders in the Knowledge Age. We close with Mei Ling's perception of the value of ethical dilemma pedagogy for engaging otherwise difficult-to-engage students in science-related learning.

> The fact that my students are not very bright academically and for them to be able to think critically, to have discussions about it at home with their parents, and lastly to be able to produce posters is truly amazing. I strongly recommend any teachers to give dilemma teaching a go, to try it with an open mind, and to simply enjoy the discussions that will follow on naturally from the students.

References

Allchin, D. (1998). Values in science and in science education. In B. J. Fraser & K. G. Tobin (Eds.), *International handbook of science education* (pp. 1083–1092). Dordrecht: Kluwer Academic Publishers.

Australian Bureau of Statistics. (2007, October). *2006 census QuickStats Western Australia.* Retrieved July 14, 2011, from Australian Bureau of Statistics: http://www.censusdata.abs.gov. au/ABSNavigation/prenav/ProductSelect?newproducttype=QuickStats&btnSelectProduct= View+QuickStats+%3E&collection=Census&period=2006&areacode=5&geography=& method=&productlabel=&producttype=&topic=&navmapdisplayed=true&javascrip

Australian Curriculum, Assessment and Reporting Agency (ACARA). (n.d.a). *Australian curriculum – Cross curriculum priorities.* Retrieved February 4, 2012, from http://www. australiancurriculum.edu.au/CrossCurriculumPriorities

Australian Curriculum, Assessment and Reporting Agency (ACARA). (n.d.b). *Australian curriculum – Cross curriculum priorities in science.* Retrieved February 4, 2012, from http://www. australiancurriculum.edu.au/Science/CrossCurriculumPriorities

Australian Government - Department of Education, Science and Technology. (2005). *National framework for values education in Australian schools.* Canberra: Commonwealth of Australia.

Australian Government – Department of Sustainability, Water, Environment, Population and Communities. (2011, October 18). *Australian sustainable schools initiative (AuSSI).* Retrieved November 12, 2011, from http://www.environment.gov.au/education/aussi/

Australian Government – Department of Water, Environment, Heritage and the Arts. (2010). *Sustainability curriculum framework.* Retrieved November 14, 2011, from http://environment. gov.au/education/publications/pubs/curriculum-framework.pdf

Australian Goverment – Ministerial Council on Education, Employment, Training and Youth Affairs (MCEETYA). (1999). *The Adelaide declaration on national goals for schooling in the twenty-first century.* Retrieved 11 March 2009, from http://www.curriculum.edu.au/mceetya/ nationalgoals/index.htm

Bowers, C. A. (1993). *Critical essays on education, modernity, and the recovery of the ecological imperative.* New York: Teachers College Press.

Chow, M-L., Taylor, E. (nee Settelmaier), Taylor, P. & Hashim, J. (2011, June 29–July 2). *Enhancing engagement of students in science through use of ethical dilemma stories.* Paper presented at the 42nd Annual Conference of the Australasian Science Education Research Association Conference, Adelaide, SA.

Covey, S. R. (1989). *The seven habits of highly effective people.* Melbourne: The Business Library.

Covey, S. R. (2004). *The 8th habit: From effectiveness to greatness.* New York: Free Press.

Depaepe, M., & Smeyers, P. (2008). Educationalization as an ongoing modernization process. *Educational Theory, 58*(4), 379–389.

Fien, J. (2001). *Education for sustainability: Reorienting Australian schools for a sustainable future.* Fitzroy: The Australian Conservation Foundation.

Fien, J. (2003). Learning to care: Education and compassion. *Australian Journal of Environmental Education, 19*, 1–13.

Forrest, S. (2002). That's my mob: Aboriginal identity. In G. Partington (Ed.), *Perspectives on Aborginal and Torres Strait Islander education* (pp. 96–105). Riverwood: Social Science Press Australia.

Gilligan, C. (1982). *In a different voice: Psychological theory and women's development.* Cambridge, MA: Harvard University Press.

Government of Western Australia – Curriculum Council. (2010). *K-10 syllabus: Early adolescence science.* Retrieved February 15, 2011, from http://www.curriculum.wa.edu.au/internet/Years_K10/Curriculum_Resources

Gower, B. S. (1992). What do we owe future generations? In D. E. Cooper & J. A. Palmer (Eds.), *The environment in question: Ethics and global issues* (pp. 1–12). London: Routledge.

Gschweitl, R., Mattner-Begusch, B., Neumayr nee Settelmaier, E., & Schwetz, H. (1998). Neue Werte der Werterziehung: Anregende Lernumgebung zur Anbahnung uberdauernder Werthaltungen bei Jugendlichen [New values in values-education: Engaging learning environments for initiating values and attitudes in adolescents]. In O. Jugendrotkreuz (Ed.), *Gibt es nur einen Weg: Informations- und Unterrichtsmaterialien zur Friedenserziehung und Konfliktarbeit im Sinne der Genfer Abkommen und des Humanitaren Volkerrechts* [Is there only one way: Information and curriculum materials for peace education and conflict work in the sense of the Geneva Convention and the Charta of Human Rights] (Vol. 2, pp. 13–21). Vienna: OBV Pädagogischer Verlag.

Killen, M., Margie, N. G., & Sinno, S. (2010). Morality in the context of intergroup relationships. In M. Killen & J. G. Smetana (Eds.), *Handbook of moral development* (pp. 155–184). New York: Psychology Press.

Kohlberg, L. (1984). *Essays on moral development: The psychology of moral development. The nature and validity of moral stages* (Vol. 2). San Francisco: Harper & Row.

Laszlo, E. (2008). *Quantum shift in the global brain: How the new scientific reality can change us and our world.* Rochester: Inner Traditions.

McInerney, J. D. (1986). Ethical values in biology education. In M. J. Frazer & A. Kornhauser (Eds.), *Ethics and social responsibility in science education* (Vol. 2, pp. 175–181). Oxford: Pergamon.

Moore, K. (2006). Science and sustainable development. Retrieved April 2011, from *Berkeley Science Review*, (Spring), 40–42. http://www.sciencereview.berkeley.edu/

Nucci, L. (2010). Education for moral development. In M. Killen & J. G. Smetana (Eds.), *Handbook of moral development* (pp. 657–681). New York: Psychology Press.

Orr, D. W. (1992). *Ecological literacy: Education and the transition to a postmodern world.* Albany: State University of New York Press.

Palmer, P. J. (1993). *To know as we are known: Education as a spiritual journey.* New York: HarperCollins.

Robottom, I. (2012). Socio-scientific issues in education: Innovative practices and contending epistemologies. *Research in Science Education, 42*, 95–107. doi:10.1007/s11165-011-9258-x.

Robottom, I., & Simonneaux, L. (2012). Editorial: Socio-Scientific issues and education for sustainability in contemporary education. *Research in Science Education, 42*, 1–4. doi:10.1007/s11165-011-9253-2.

Ryan, J. (1992). Aboriginal learning styles: A critical review. *Language, Culture and Curriculum, 5*(3), 161–183.

Settelmaier, E. (2009). *'Adding zest' to science education: Transforming the culture of science classrooms through ethical dilemma story pedagogy.* Saarbrucken: VDM.

Settelmaier, E., Taylor, P., & Hill, J. (2010). *Supporting teacher, challenging students: Socially responsible science for critical scientific literacy.* Refereed paper presented at the XIV IOSTE Symposium: Socio-cultural and human values in science and technology education, 13–18 June, Bled, Slovenia. ISBN 978-961-92882-1-4.

Sharma, A. (2012). Global climate change: What has science education got to do with it? *Science Education, 21*, 33–53. doi:10.1007/s11191-011-9372-1.

Smetana, J. (2010). Social-cognitive domain theory: Consistencies and variations in children's moral and social judgments. In M. Killen & J. G. Smetana (Eds.), *The handbook of moral development* (pp. 119–154). New York: Psychology Press.

Srivastava, A. K. (2005). Buddhist environmentalism. In D. C. Srivastava (Ed.), *Readings in environmental ethics: Multidisciplinary perspectives* (pp. 138–149). Jaipur/New Delhi: Rawat.

Tappan, M. (1997). Language, culture and moral development: A Vygotskian perspective. *Developmental Review, 17*, 78–100.

Tappan, M. (2010). Mediated moralities: Socio-cultural approaches to moral development. In M. Killen & J. G. Smetana (Eds.), *Handbook of moral development* (pp. 351–374). New York: Psychology Press.

United Nations Educational, Scientific and Cultural Organisation (UNESCO). (2011). *Science and technology education*. Retrieved April 21, 2011, from UNESCO education: http://www.unesco. org/new/en/education/themes/strengthening-education-systems/science-and-technology/

United Nations World Commission on Environment and Development (WCED). (1987). *Our common future*. Oxford: Oxford University Press.

Uzzell, D. (2008). *Challenging assumptions in the psychology of climate change* (T. A. Society, Ed.). Retrieved 2011, from InPsych: http://www.psychology.org.au/publications/ inpsych/highlights2008/#s3

van Eijck, M., & Roth, W. M. (2007). *Improving science education for sustainable development*. Retrieved April 2011, from OpenAccess Plos Biology: www.plosbiology.org

Zeidler, D. (1984). Moral issues and social policy in science education: Closing the literacy gap. *Science Education, 68*(4), 411–419.

Tracing Science in the Early Childhood Classroom: The Historicity of Multi-resourced Discourse Practices in Multilingual Interaction

Charles Max, Gudrun Ziegler, and Martin Kracheel

Introduction

This chapter presents research conducted in early childhood classrooms in Luxembourg, a European country with a complex multilingual situation. A multi-layered corpus of classroom interactions, consisting of photos, videos and audio recordings, was collected over a period of 6 months and then classified, annotated and partially transcribed. Drawing from this corpus, this study sheds light on the discourse practices of 6–12-year-old children and examines the co-construction of the children's growing understanding of science in collaborative inquiries. Arguing from a context-sensitive perspective, our research approaches the learning of science as an interactional achievement in situ, one that encompasses the enactment of science as shared discourse and therefore as a cultural accomplishment. Luxembourg has a highly multilingual and multicultural population with over 40 % of students being non-nationals. Portuguese form the largest migrant community living in the country. The Luxembourg schooling system is trilingual. In kindergarten, children start to learn Luxembourgish, which is used as vehicular language across elementary school. German is the first foreign language taught in first grade and the language of literacy. French starts in elementary school in grade three. Beside the country's three official languages, Portuguese is strongly present across the schooling system. School offers (catholic) religion education classes that are broadly accepted in the society.

In examining children's interactions around the exploration of water, we frame science as practices mediated by semiotic resources, i.e. discourse formats and repertoires. These repertoires emerge from students' opportunities to engage in

C. Max (✉) • G. Ziegler • M. Kracheel
Faculty of Language and Literature, Humanities, Arts and Education,
University of Luxembourg, Walferdange, Luxembourg
e-mail: charles.max@uni.lu

N. Mansour and R. Wegerif (eds.), *Science Education for Diversity: Theory and Practice*,
Cultural Studies of Science Education 8, DOI 10.1007/978-94-007-4563-6_7,
© Springer Science+Business Media Dordrecht 2013

joint experiences, which involve multimodal as well as multilingual resources. Repertoires with related resources within a specific activity are then available to students beyond the actual situation. We contend that what counts as "knowledge" related to science education for young children needs to consider the situated ways in which science emerges as a cultural achievement. Children construct science *in*, *from* and *as* socio-cultural events pertaining to and expanding from their everyday activities. Specifically, discourse formats, which involve everyday practices bound with a particular feature of the element at stake (here, water) are captured. Excerpt 1, for instance, shows how the general characteristics of floating are explored by everyday practices, e.g. the "putting stones in a cup", in terms of available resources for the learning of science in the multilingual classroom setting. Moreover, the analysis demonstrates how processes of transformation of discourse practices by the children in a classroom can be traced in line with and sometimes in contrast to the situated emergence of new ways of "speaking about" water-related issues. Therefore, "doing science" is marked not only by the diversity of resources children draw from (e.g. different languages, different religious educational contexts) but also by the ways peers and the teachers acknowledge, ratify, reject or modify the resources brought into the "doing of science". Hence, studying learning in diversity contexts and from a perspective which carefully integrates elements of diversity as they show in discourse practices provides insight for understanding the early science classroom context in general. In fact, the tracing of discourse practices points to general co-constructed practices of science in the classroom, going beyond the actual context discussed in this paper. This study raises awareness as regards the complex nature of multiple resources, their emergence and integration in the science classroom. We show how children manage different linguistic resources and facilitate each other's contribution to the science learning activity at hand.

In sum, we explore the empirically tangible ways in which young children construct science as a gradually emergent accomplishment. We demonstrate that children's discourse practices with their relevant repertoires can be approached as enactments of science. These co-constructed enactments display representations of scientific concepts. Moreover, they point to the diversity of resources and conditions of such science accomplishments by young children, therefore allowing for tracing science learning. Hence, we emphasize the actual doing science, which – at first sight – might not be considered (relevant) canonical science. We argue that the tracing of discourse practices values science learning as accomplished by the young children and helps to elaborate on science education.

Tracing the Historicity of Discourse Practices in Multilingual Interaction

Studying learning in diversity contexts and from a perspective, which integrates features of diversity into the analysis provides insights for understanding highly diverse linguistic and cultural contexts, which are still under-researched. This chapter presents research conducted in elementary science classrooms in Luxembourg, a European country with a complex multilingual situation. Over 40 % of elementary school students are non-nationals, the largest group coming from the Portuguese speaking community. The country has a trilingual schooling system, which reads as follows. In the kindergarten, children start to learn Luxembourgish, which is used as vehicular language across the elementary school years. German is the language of literacy. French enters elementary school in grade three. Beside the country's three official languages, Portuguese is strongly present across the schooling system. School offers (catholic) religion education, broadly accepted in the society and therefore serving as a resource in discourse.

The cultural and linguistic diversity of Luxembourg classrooms and beyond calls for a research approach that is sensitive to the social, cultural and linguistic backgrounds of the children and to the ways in which they are brought into interaction through multiple mediations (social, semiotic, material). The rationale behind this research is to explore the nature of science learning as a social phenomenon that is discursively bound. In discourse construction, learners link elements of knowledge and build science knowledge in and through their interactions. Two research questions guide our inquiry:

(a) How do 6–12-year-old children combine resources in terms of interactional coordination and knowledge construction and thus generate dialogically rooted, multifaceted and shared discourse practices about scientific phenomena?
(b) To what extent do multimodal discourses (talk, pointing, gestures) facilitate changes in understanding, knowing and reasoning of physical phenomena of the world and vice versa?

Through our analysis, we highlight the apparent ("at hand") and emergent ("developed") discourse-in-interaction practices, features and formats with regard to the resources from inside and outside of school, which are mobilized by the children. Moreover, we look at the multimodal quality of the discourse practices and focus on their multilingual dimensions in terms of learning science. Specifically, we show the children's ways of imagining, representing and knowing as deployed in the activity of inquiring the specifics of the physical element "water".

Following this view, our research conceives science learning as an interactional achievement, one that encompasses the enactment of science as a particular cultural practice. More specifically, we situate "doing science" as an accomplishment of discourse-in-interaction. In order to take into account the complex interpersonal processes that are fuelling the dynamics of this accomplishment, our research has developed a multidimensional framework that blends socio-cultural, cultural-historical and interactional perspectives on learning and knowledge production in terms of co-constructing science (Fig. 1).

The following sections present this multidimensional framework, which high-lights the dialectics between processes taking place at individual and collective levels. We detail the core perspectives of our work further throughout the discussions of empirical data in later sections of this paper. Moreover, we describe our methodological perspectives which allow for turning phenomena emergent in

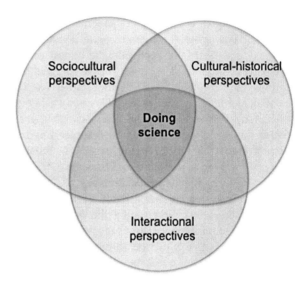

Fig. 1 Three theoretical perspectives for researching "doing science"

open science stations into analytically relevant data. In sum, this chapter provides a discussion of two key concepts, "discourse-in-interaction" and "doing science". Both relate to the emergence and co-construction of science in early educational contexts and help to understand the dynamics of science learning with a focus on the diversity context of the early multilingual classroom.

Socio-cultural Perspectives on the Co-construction of Science

Socio-cultural approaches (Cole 1996; Lave and Wenger 1991; Rogoff 2003; Roth 2006; Daniels 2008) emphasize that processes of learning and knowing are embedded in social, cultural and historically framed practices, which are distributed among people, semiotic and material mediators, time, space and physical environment. These frameworks regard learning being essential to all social practices as "learning, thinking, and knowing are relations among people in activity in the world, rather than something static that is to be internalised" (Lave and Wenger 1991, p. 51). In the water workshops we ran with the children, the social, cultural and historically framed context of being in a school, the presence of a teacher as well as the practise of learning science trough experimentation and abstraction play a crucial role. They constrain learning but nevertheless open possibilities for learning. With regard to human learning and knowledge building, the appropriate "unit of analysis" for socio-cultural research (Rogoff 1995; Valsiner 2002; Matusov 2007) has to account for the social situatedness of these processes within the discursive and local context and their instantiation through mutual interactions *hic et nunc*. Since individuals "do" science in interaction with others, they share and co-construct meaning within the situation at hand while relying on a range of familiar resources, from the classroom and beyond, in the form of conceptual framings, discursive repertoires and formats. These resources are permeated with specific, idiosyncratic or shared, meaning and mediate the actors' "doing in situ". As these resources mutually impact on each other and specific meanings interpenetrate during their situated enactment, the tools in use get re-shaped through collective use and imbued with new meanings, which emerge during the interactions at hand. In the data we analyse, a cup, for example, is labelled first as yoghurt cup, relabelled later as a cup, then as a ship, then as a boat and finally again as a cup (cf. Excerpt 1–2).

Cultural-Historical Perspectives on the Co-construction of Science

The cultural-historical school of thought in the legacy of Vygotsky (cf. Wertsch 1981; Engeström et al. 1999; Hedegard 2001; Van Oers et al. 2009; Daniels et al. 2009) stresses that the development of individual persons and social collectives can only be understood dialectically, i.e. historically and in interaction with each other. Hence, children "doing science" are continually shaped by and are shaping the collective context, e.g. the inquiry group or the class community through multiple systemic interactions.

> The uncertainty of knowledge, which can be seen as knowledge being problematical, is a powerful catalyst for children's intellectual activities. Within that trend, new and as yet vague knowledge arises, while clear and final knowledge is transformed into uncertain and unclear knowledge. Through the interaction of these processes the child develops, exploratory, trial-and-error forms of integration when several global structures that "grope" for a pattern of regularity are synthesised. That process has been poorly studied, yet it has immense significance in the thought process. (Poddiakov 2011, p. 61)

The efforts to bridge previously loosely connected or unconnected communities in the science classroom give rise to interesting boundary-crossing phenomena, where subjects and cultural tools come into contact. Specifically, they cross linguistic, pragmatic or organisational borders and allow for cultural practices to come into contact with each other. Currently, much research is directed towards diversity in hybrid pedagogic spaces and emerging opportunities to create rich zones of innovation and learning. Tuomi-Gröhn (2003, 2005) discusses processes within boundary zones, which "are not closed spaces but networked and mediated practices, which give rise to alternative framings and metaphors" (Edwards 2005, p. 5). Gutiérrez et al. (1999) qualify learning contexts as "third spaces", which are "immanently hybrid, that is, polycontextual, multivoiced and multiscripted. Thus, conflict, tension and diversity are intrinsic to learning spaces" (1999, p. 287). These zones or spaces might be conceived as in-between arenas of polycontextual practices, where cultural elements from various communities collide and intermingle according to Cole's dynamic conception of context as "that which weaves together" (Cole 1996, p. 135).

Cultural-historical activity theory does not include a specific theory of learning, but highlights that learning arises within communities as individual and collective endeavours. At the individual level, learning takes place through active involvement of a subject (or a group of subjects) in the particular interaction and within an activity. Moreover, learning is understood as the dynamic interplay of internalization and externalization processes when culturally relevant knowledge and situated action feed into each other. At the collective level, learning arises through critically questioning the object of the activity system within which the subjects are engaged. In the present case, this is the elementary science investigation. Learning cannot be reduced to a mere sum of a single subject's knowing, nor conceived as a unidirectional achievement from a less competent to a more competent user status. Learning at the collective level is the transformation and creation of the learning culture itself. This approach conceives practices beyond the situated elementary science learning context and looks at these practices in terms of a transformative dialogue. In and through the transformative dialogue, everyday and school literacy practices interact and culturally new patterns of activity emerge. The movements across and between contexts transcend the institutional boundaries of the school (Sannino et al. 2009) and promote qualitative transformations in the science learning activity itself. These movements stimulate processes of dialogical and intertextual interactions, which create enhanced opportunities to draw on available repertoires of discursive formats or practices. Specifically the repertoires of discursive formats come into play when "learners learn something that is not yet there. In other words, the learners construct a new object and concept for their collective activity, and implement this new object and concept in practice" (Engeström and Sannino 2010, p. 2).

From an epistemological account, the dialectical and materialist perspective on "doing science" continuously works to overcome dichotomies raised by idealist and rationalist accounts, which advocate for the separation of thought and material world. Specifically, a materialist point of view strives to stress the interconnections between mind and body with regard to cultural artefacts. It foregrounds the fact that "mind is not in opposition to the material world but is embedded in social activities of people and is mediated by material tools and cultural-semiotic artefacts that people use and historically produce in those activities" (Kostogriz 2000, p. 4). Furthermore, a cultural-historical perspective emphasizes that "doing science" is not to be understood as an apolitical activity. "Doing science" goes beyond the concept of an individualized cognitive process related to acquiring and processing information composed of a collection of abstract concepts, ideas and reified thoughts.

The materialist perspective stresses "doing science" in and as social practice. Knowledge is generated through an ongoing process of tension and struggle between controversial positioning and alternative discourses. "The essence of these contradictions are relations of power in the processes of its cultural-historic production. Hence, knowledge is not a thing in itself, a product but a process-object of human activity" (Kostogriz 2000, p. 10).

In diversity contexts, the dynamics between the individual and the collective are critical to framing the "doing of science". Diversity contexts fuel our interest to analyse learning processes (Ziegler 2011) within activity systems. Science is a human activity, which comes into existence as it is done within interaction, whereby interactants are continually shaping, and are being shaped by, the activity and the context under construction. The cultural-historical lens allows to move beyond traditional static and linear models to work towards a depiction of the complexity of human interaction and experience that depicts social phenomena as "multiple systematically interacting elements" (Engeström and Miettinen 1999, p. 9).

Interactional Perspectives on the Co-construction of Science

The above frameworks ground our approach towards science learning practices, which emerge from the interactional formats participants use to engage in joint activity. In the following, we look at the various ways by which learners accomplish "doing science" interactionally. Participants converge to multilingual talk, embodied action in time and space (gestures and body orientation) in combination with the use of diverse material and semiotic artefacts. Therefore, we focus on the multilingual and multicultural semiotic resources that students draw upon in their exploratory transactions in the science classroom. They carry these resources from the cultural communities they inhabit or go across. The interactional lens allows for tracing the ways in which the students deploy the interactional architecture of talk. The conversational dynamics provide clues about how the children are coming to "know" by achieving joint topical orientation (Melander and Sahlström 2009). Joint topical orientation in performing science entails processes such as comparing, predicting, verifying and so on. Moreover, these dynamics comprise the multimodal ways the children participants talk, sing or even dance their context-sensitive understandings into being (Gallas 1995; Lemke 1990; Roth 2005).

The conversations, which unfold during and around the collaborative investigations, allow for a close analysis. Specifically, our analysis focuses on the interactional dynamics through which participants are doing actions together which go far beyond talking. Some of the action, which we observe as children interact are

1. Categorizing each other as novices or experts,
2. Managing agreement or disagreement,
3. Moving into or out of specific activities,
4. Regulating role-relationships in a reciprocal manner.

For instance, they propose a topic that the other treats as the current focus of attention or attempting to understand each other's perspectives and achieving a state of intersubjectivity (Pekarek-Doehler and Ziegler 2007, p. 74). The analysis of these interactive deployments allows to look more thoroughly at how participants create and negotiate common interests. In sharing, they achieve inter-understanding and mutual learning. Space configurations and material equipment are of major relevance in these processes as learner-centred approaches foster hands-on tasks such as manipulating objects, gathering items or (de)constructing objects (Inan et al. 2010).

Discourses are seen as community-bound forms of communication. They follow certain rules, which "are not anything the participants would follow in a conscious way, nor are they in any sense 'natural' or necessary." These rules or practices are established over time, within related formats and in line with certain topics. They can be compared to "historically established customs" (Sfard 2009, p. 57). Adapting Sfard's components to "Science-as-Discourse", we integrate the following dimensions:

(a) The keywords or a standardized terminology and their use (e.g. melting, solution, floating, sinking),
(b) Visual mediators as semiotic means for identifying the object of talk (e.g. the use of an iconic gesture in reference to a ship),
(c) Routines "which determine or just constrain the patterned course of discursive action and the circumstances in which this action may be undertaken" (ibid., p. 57), e.g., the teacher asking for clarification in the context of an investigation task,
(d) Endorsed narratives that the reference community labels as correct (e.g. ice melting due to climate change).

Situating "Doing Science" as Discourse-in-Interaction

Talk that is jointly shaped and managed by all the participants has the potential to change educational reality, i.e. learning situations, role relationships, purposes and procedures. In transformative and transactional modes of interaction, individuals engage in joint interaction and negotiate a specific goal of the interaction at stake (e.g. organising the exploration task at hand). Moreover, they co-define the direction of discourse, jointly determine the relevance of contributions in dialogic response to other participants' requests and co-construct meaning (van Lier 1996, p. 179).

Interactants create and transform particular discourse practices, which are a part of the larger canonical perspectives of science and which we previously referred to as "Science-as-Discourse" (Pekarek-Doehler and Ziegler 2007). These discursive formats are both specific and general. They are specific as being the actual Science-as-Discourse, which is identifiable for instance when the children talk about "water can become ice, ice is made out of water" (cf. Excerpt 2). They are general in that they rely on nonscience specific talk as conducted in everyday interactions, e.g., when one child advises the other "don't put a big stone in the cup, otherwise it will drown" (Excerpt 1 and 2).

When science becomes relevant within a group's interaction, e.g., in the water-related workshops at school, children organize their general talk with regard to the more specific science topic at hand. In such a case, an utterance like (taken from our data) "because ice when it gets too warm, becomes water as you know"(cf. Excerpt 2) illustrates how children select, adapt and expand available discourse practices, which then pertain to the science-specific perspective emerging from the interaction.

The context-specific organization of science-related discourse-in-interaction draws from, feeds into and constitutes the larger "Discourse-as-Science", which develops over time and through science-related interactions. Young children's science-related talk can therefore not be considered as a low-quality version of standardized or established ways of science. We argue that it is important to acknowledge that these discourse formats often precede (and are at the origin of) the standardized canonical discourse formats, which function as norm-orientation in the general landscape of science, both as regards practical science as well as the theoretical elaborations of science.

Multiple Resources and Integration of Diversity in Science Learning

The outlined perspective on discourse formats, which highlights their emergent nature and development towards accepted science discourse, sheds light on the issue of learning. Specifically, looking at learning as a twofold process as has been suggested in the literature, the process of "differentiation" on the one hand (e.g. the cup vs. the ship) and "integration" on the other hand (e.g. the cup floats like a ship) (Poddiakov 2011, p. 56) is of particular interest when it comes to multiple resources available to the children in doing science in interaction. Various linguistic resources from one or more languages as well as references to non-school topics, and other learning experiences are identified. "Stable knowledge" (Poddiakov 2011, p. 60) such as established formulations or accepted patterns of explanation get challenged, when children engage in science talk. Specifically, multilingual resources such as the naming of elements relevant for the science learning activity at school with its ascribed schooling language, which is Luxembourgish in our case, are operated in another language, French, German or Portuguese for instance. Children refer to these resources when an issue of the science learning at hand is challenged or indicated as being unclear by the children (see, for instance, the code switch to Portuguese in Excerpt 1b). The functioning of such resources in a diversity context such as the Luxembourgish pre-primary context with its high percentage of children coming from various migrant and other linguistically and culturally different homes has been highlighted for other contexts as well (e.g. Switzerland, cf. Gabo and Mondada 2000). Linguistic resources act as transformative elements (Poddiakov 2011) at moments when knowledge transformation and expansion are tangible in and through the discourse of the children. Discourse patterns which are stable as regards the specific knowledge they convey as well as the language within which the knowledge is accepted (see Excerpt 2) are challenged by patterns and linguistic elements from various contexts, when children struggle with getting acquainted with these patterns. Specifically, contradictions emerge as the accepted patterns or discourse elements compete with non-accepted patterns from other, non-science contexts. Competition of elements (e.g. competing verbs from two or more languages for describing a feature of the water contexts) and negotiation as well as ratification of the expected science discourse act as drivers for transformation of the emergent science knowledge with regard to the knowledge at stake.

In order to conceive of knowledge development in interaction in the science classroom, experience as acquired through activity and communication (Poddiakov 2011, p. 60) is tangible not only in the verbal elements that children use, e.g., some specific vocabulary such as boat instead of ship (see Excerpt 1b) or topics addressed. Discourse formats as discussed above and their patterned nature as emergent are not the only discourse entities children take up, elaborate on or question during the course of the interaction. "Stable knowledge" from other school relevant knowledge contexts (e.g. religious education) comes into play as well (Poddiakov 2011, p. 60) as children draw from specific formulations or discourses of reference (e.g. creation as a religious topic) (cf. Excerpt 2). The ongoing negotiation of linguistic elements and discourse patterns from non-school context gradually moves towards the accepted norm of "doing science". This process allows for tracing the historicity of the various discourse elements and patterns which get ratified and therefore stabilized over time. This norm then, as language of norm (Luxembourgish in our case) on the one hand and discourse of relevance in a specific situation (religious discourse for instance) on the other hand, is part of the learning and knowledge construction process in the science classroom (Tomasello 2008; Poddiakov 2011).

In the following sections we explain the design and the context of our study. Then we move towards exploring the analysis of young children's interactions around science investigations.

Observing Children's Practices of Co-constructing Science in a Discursive Process: The Study

This work is part of a larger 3-year research project conducted in kindergarten and elementary school grades one and two in the multilingual country of Luxembourg. The selected vignettes examine one aspect of this study, which is investigating children's collaborative processes and constructions of science within curriculum activities addressing the diversity of their resources in interaction. The overall project incorporates five different schools with multiple early childhood classes, which organized a variety of water inquiry topics and activities during a 4–6-week period. The following table summarizes the grade levels in which data has been gathered (cycle 1 = Kindergarten, and cycle 2 = grades one and two) across the 2 years time-span of the study. The vignettes for this chapter have been selected from two data sets (colours) emphasising diversity issues (Fig. 2).

The research group worked in close collaboration with the school teachers as they were asked to organize open child-centred workshops around the physical, chemical or natural features of water. The guiding idea was that such open workshops might generate practices through which children co-construct science as a discursive

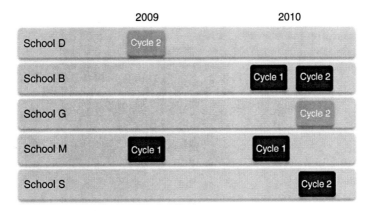

Fig. 2 Overview of the project schools in 2009 and 2010 and the cycles, which the excerpts presented in this chapter have been selected from

process. Yet, the research team was supporting the teachers in implementing learner-initiated science activities and expanding inquiry-based approaches through data-based feedback. Two to three researchers participated in all the science workshops and were documenting the children's interactions within water explorations using audio recorders, two stationary and eight handheld cameras (for the children and the researchers). Ethnographic methods of data collection included also the use of participant observation and field notes. The data corpus represents about 200 hours of video and audio recordings from 45 days of science activities and, furthermore, photographs, writings and paintings by the children. The data are organized as a searchable corpus for extended analysis with Transana (Woods 2007), an open-source software for qualitative analysis of video and audio data. Selected excerpts regarding specific research topics are transcribed for fine-grained micro-analytical work.

The investigation of the children's practices in science interactions refers to four different data sets, with data set 1 and data set 2 being the most relevant sets for this study. The following four domains of observation guide our analysis: (1) children's interactions during group inquiry activities; (2) classroom talk, teacher initiated; (3) children's self-recorded moments of hands-on group inquiry activities (their perspective); and (4) children's (retrospective) comments about the/their recordings.

Selection of Examples and Analysis

In the following sections, we discuss a selection of micro-phenomena with regard to discourse patterns and their co-construction across sequences and over time. The analysis of these examples shows how resources from various repertoires, linguistic and other, are used by the young learners and shaped in and as "doing science".

The chosen vignettes illustrate the ways in which children "do science" through their discourse-in-interaction. As discussed, the schooling context is marked by very heterogeneous population. The vehicular language is Luxembourgish, whereas the official (normative) language of instruction is German.

Excerpt 1: e schëff get awer mol ënner – *a ship can sink*

In Excerpt 1 (composed of Excerpt 1a and 1b) the students are about to fill cups with stones in order to see if (and when) the cups keep on floating or sink. The task brings together children from cycle 2 (7–8 years old) with older pupils from cycle 4 (11–12 years old). The younger children repeat an experiment that they had already performed as a group a week ago, but do it now in collaboration with the older students to improve their initial understanding. The three older children are Kim, Bob and Luc. Kim has an Asian background and Luc and Bob have a lusophone background. All three speak Luxembourgish. The younger children are Han, Ced and Ric. Han and Ric have a lusophone background and do speak Luxembourgish. Ric is new at the school. Being Portuguese, he does not speak Luxembourgish.

For reasons of clarity, we present Excerpt 1 in two stretches, Excerpt 1a and Excerpt 1b.[1]

The stretch (Excerpt 1a) starts with an explanation sequence, as the teacher (Tea) enters the room, in which all six children are gathered around the table with the equipment. The children have written instructions on how to do this experiment with plastic cups and small stones in a big container filled with water. Upon her entering, Tea asks if they know what they are supposed to do (line 02). When Luc says that they do not understand (**[mir verstin et net** line 03), the teacher explains the task. The conversation is conducted in Luxembourgish.

The excerpt allows us to observe relevant aspects of how the teacher is framing the inquiry activity using various resources, specifically multimodal ones. Furthermore the teacher points out that the activity is about swimming and sinking. These words are from the children's life-world language (Gee 2004) and supported by pointing and other supporting gestures (Fig. 3). Children's life-world language is often quite different from the accepted academic terminology and used by the children themselves and by more advanced peers when introducing the scientific mode to discuss and represent observable features. Moreover, these terms function here as the standardized terminology for the description of the physical aspect of the activity.

---

[1]Transcription conventions follow the GAT system. Translations are provided in English line by line. All excerpts are available as original recordings via http://dica-lab.org/research/data/

Fig. 3 The teacher (*Tea*) explains the task and puts a cup in the water (vignette from beginning of Excerpt 1)

Excerpt 1a: Negotiation of Discourse Practices and Multimodal Resources

002	Tea	**verstit dir?** *do you understand?*
003	Luc	**[mir verstin et net** *[we don't understand*
004	Tea	**also (.) die heiten dat sin jo sou youghurt becheren gell** *So these here are a kind of yoghurt cups ok* **((takes a cup and put in on the water surface))**
005		**äh dat [sin** *äh that [are*
006	Kim	**[(fësch)** *[(fish)*
007	Tea	**quatsch dat sin einfach sou becheren** *nonsense that are simply some kind of cups* **((pours water out of the cup and puts it again on the surface))**
008	Bob	**jo** *yes*
009	Tea	**an déi schwammen normalerweis** *and they usually swim*
010		**musst der probéieren dass se schwammen** *you must try that they swim* **((puts a stone in the cup))**
011		**dass se su riit bleiwen, gell** *that they stay upright, ok* **((holds cup with right hand, puts another stone in the cup with her left hand))**
012	Luc	**jo ((reaches over the basin to get the cup))** *yes*
013	Tea	**dir musst einfach probéieren** *you simply have to try*
014	Luc	**jo ((reaches over the basin to get the cup))** *yes*

```
015    Tea    an da musst der och kucken wéini se ennergin, gell
              and then you have to look when they sink, ok
              ((pushes cup deeper into the water))
016           well irgendwann kippen se jo=
              because at a certain point they turn over you know
              ((takes cup out of the water))
017           firwaat↑ se ënner gin
              why they sink
018    Luc    (          )
              ((wants to take the cup, but Tea takes it away – mutual gaze))
019    Tea    an dann as hei na eng fro↑
              and than there is here one more question
020           well e schëff↑ dat belueden as mat esou steng, gell
              well a ship that is loaded with stones alike, ok
021           wësst der déi sou riesegrouss schëffer
              you know these immense big ships
              ((lifts her hands up and indicates size, cf. figure 4a))
022    Luc    jo ((nodds))
              yes
023    Tea    déi si jo SOU schwéier
              they are SO heavy you know
024           déi sinn sou voll belueden mat schëff
              they are so fully loaded with ship
025           firwaat gin déi NET ënner,
              why do they NOT sink,
026           an e klenge becher dee geet ënner,
              and a small cup that goes down
027           elo musst der [dat rausfannen
              now you have  [to find out
028    Kim                  [e schëff
                            [a ship
029           [e schëff get awer mol ënner
              [a ship happens to sink
030    Tea    [firwat dat esou ass
              [why this is so
031    Ced    well de becher méi [LIIcht ass
              because the cup is [LIghter
032    Kim                        [dat hänkt dervun of wivill gewicht
                                  [this depends of how much weight is
033    Tea    dat musst der rausfannen
              that's what you have to find out
034    Kim    d=schëff aushält
              the ship is able to carry
035    Tea    ((teacher leaves the room))
```

By establishing that the cups usually swim *an déi schwammen normalerweis* (line 09), the teacher refers to both dimensions, the everyday experiences of the children and the physics of objects and shapes. In fact, the children are acquainted with the manipulation of cup-like things and their reaction on water. The canonical understanding of physics is also addressed as the teacher refers to a general truth, holding for the physics of objects when brought on water, when an object of a certain design is expected to swim. Moreover, the teacher introduces the cups as yoghurt cups, known to the children from their everyday out of school contexts: *die heiten dat sin jo sou youghurt becheren gell (These here are a kind of yoghurt cups ok, line 04)*. Simultaneously, the teacher puts the cups she refers to in the container with the water, using pointing gestures and demonstrating the actual reaction of the cups when put in water (see Fig. 3). The cups in the water are "directly available", i.e. they are part of the co-constructed context shared by the teacher and the children. Shortly after this introduction sequence, the teacher produces a self-repair concerning the origin of the cups, going beyond the very specific context of use of the cups *(youghurt becheren, line 04)*. The teacher indicates that the cups are cups in general terms *(einfach)*, sharing specific – and for the experiment relevant – features: *quatsch dat sinn einfach sou becheren (nonsense that are simply some kind of cups, line 07)*. The teacher's self-repair therefore modifies and advances the co-constructed shared context, indicating the general nature of cups with regard to water beyond the actual, concrete cups in front of the children's and the teacher's eyes. This co-construction of the shared context, which goes beyond the actual visible elements and scene, continues in the following sequence. The teacher in her talk introduces the reference to ships for the first time (line 21*): wësst der déi sou riesegrouss schëffer (you know these immense big ships, line 21)*. Tea emphasizes the introduction of the reference to ships, which are not present in the co-constructed scene, by multimodal means, supporting her explanation by an iconic gesture, indicating a ship. As we can see in Fig. 4a, she creates the space for and the shape of the symbolic size of the ship by positioning her arms. The dashed arrow indicates the teacher's gaze towards one particular student. This time, and in contrast to the former everyday-life-related presentation of the cups (yoghurt cups) and their expansion (general features of cups), the teacher does not use any concrete element at her disposal she would refer to in a direct and later more abstract way. As regards the ships, she does not refer to a directly available artefact, but she multimodally (linguistically and gesturally) constructs the ship within the visual shared space between her and the students, the co-constructed micro-context around the water container as a virtual element which shares the features of the generally presented cups. Tea indicates that the ships and their size are available for reference and gesturing by the children from their everyday life context, without however indicating a specific context (e.g. a harbour). Tea puts the first mentioned and then modified reference to the tangible cups together with the verbally and iconically created ship not only in the same shared space of activity: *firwaat gin déi NET ënner, an e klenge becher dee geet ënner (why do they NOT sink, and a small cup that goes down, line 26–27)*. Moreover, the children follow the discursive as well as scientific operation of comparing the features of the cups (at hand) to the ship (virtually referred to). Yet, the children's discursive uptake shows ambiguity as regards the shifting between the enacted shared space in front of their eyes with the concrete cups on the one hand, and the abstract but possible comparison of the behavioural features of the cups versus a ship on the other. Ced's causal explanatory reaction may represent both, the intended science-related experiment (cups as ships) and the everyday-life statement that cups are lighter than ships: *well de becher méi Llicht ass (because the cup is lighter, line 31)*. Kim, in return stays with the rather scientific discourse context when mentioning weight in general terms: *dat hänkt dervun of wivill gewicht d=schëff aushält (this depends of how much weight is the ship is able to carry, line 32, 34)*. The excerpt, taken from the co-constructed

Fig. 4 (**a**) Tea producing the iconic ship gesture (*solid lines*) while gazing towards Luc (*dotted line*). (**b**) Bob moves into the activity by taking a cup and repositioning himself at the table

situation around the water activity, shows the blending of the talk about everyday-life referable objects (yoghourt cups), general physical rules (sinks normally) and the reference to abstract, virtually enacted elements (huge ships).

Besides the verbal and gestural elements, which serve as supporting resources for setting up and enacting the science learning activity, participants co-construct the situation by means of interactional resources. These resources help to manage the interactional development in terms of next-turn management (e.g. overlap, pause). Moreover, they serve as indicators for transitions between the contributions made by the more advanced participant, the teacher and the student. For instance, the student Luc positions himself through back-channeling (*jo lines 13, 15 and 23*) to the teacher's contributions as the lesser advanced person in the interaction. He offers himself through the interactional resources at hand (e.g. backchannel, responding to gaze) as being an (or even the) addressee for the interaction. In fact, the teacher acknowledges this activity of reacting upon her actions and has Luc in focus throughout the sequence as manifest in her continued eye-gaze towards Luc (see Fig. 4a). The teacher and Luc co-construct the teacher's position as "the one who can explain and explains" by mutual gazing between her and Luc.

In sum, Excerpt 1a shows how everyday-life contexts are brought up and modified in a science learning activity in the discourse of the participants. Practices of referring to at-hand elements (here: the cups) are used in order to demonstrate general physical properties (here: conditions of sinking). Diversity is tangible in terms of varied context of reference and ways of referring to objects and talking about an observable action. The close analysis of the deployment of the interaction demonstrates the gradual blending of the contexts of reference across the participants within the co-constructed shared space of joint learning. Discourse practices (e.g. reference to the cups at hand) are modified across the course of interaction, allowing for tracing the historicity of – in the end – accepted discourse practices which emerge from the shared context (here, the elicited explanations by the teacher). Participants use a wide range of resources for establishing and managing the science learning activity. Moreover, they position themselves as more/lesser advanced participants (e.g. gazing by teacher), orient to each other in their discourse practices (e.g. self repair, repetition) and establish a mutually negotiated way of doing science over time (e.g. task to be carried out, targeted abstraction).

Excerpt 1b: Linguistic Diversity and Peer Positioning in Interaction

The subsequent stretch (Excerpt 1b) allows for tracing the diversity that the children participants bring in when engaging in the science learning activity as a peer activity (e.g. various multilingual elements). Once the teacher leaves the previously constructed shared context, the children start focusing on specific activities and manipulate the cups, which have been introduced and discursively made available by the teacher in her explanatory discourse. The excerpt allows for analysing the various ways in which the children engage in science learning with peers by relying on their resources. They position themselves as more knowledgeable participants by use of interactional means, they phrase and modify their discourse practices as they become more acquainted with the science issue at hand, they bring in and adapt their multilingual linguistic resources.

```
037   Luc   komm mir kucken
            let us have a look
            ((Luc takes the blue cup and Ced puts a big stone in it))
038   Ced   [(          )
039   Luc   [net sou vill steng [dran
            [not so much stones [into it
            ((Luc takes the stone out again))
040   Kim                   [<<French> trois>(1.0)
                            [<<French> three>
041   Luc   komm mir kucken
            lets have a look
042         ((Luc puts the empty cup on the water surface))
043   Bob   ((Bob takes the orange cup and moves to the container))
044   Luc   komm mir machen (.) nëmmen ee becher
            let's make              only one cup
045   Bob   ((Bob puts the cup back on the table))
046   Luc   also de becher as sou wann der léist da fällt en
            so the cup is like that if you relase it it turns
            [ëm
            [over
047   Kim   [(          ) mir mussen [(…)
            [(          ) we have to [(…)
048   Luc                         [also (.) musst der
                                   [therefore you have to
                                   ((puts a stone in the cup))
049   Kim   [(          )
            [(       .)
050   Luc   [jo an also musst der eppes voll an d'mëtt maachen
            [yes and so you have to put something right in the centre
            ((puts a stone in the cup, shakes it to move the stone into the
            center))
051         eppes voll an d mëtt ewéi dat heiten
            something right in the centre as this one
052   Bob   <<Portuguese> consegue fazer>
            <<Portuguese>you can do>
            ((to Ric who nods and laughs, leans forward to the basin))
053   Kim   (            ) laughter
054   Luc   jo en ass net richteg an der mëtt
            yes it is not exactly in the middle
055         dofir maache mer nach eens (.) een eng
            that's why we put another one (.) one one
056   Ced   ((puts another stone in the plastic cup))
057   Luc   <<Port.> Por isso temos mais uma>
            <<Port.> so we have one more>
            ((looks at bob))
058         <<Port.> bob, tu tudo traduz em português>
            <<Port.> bob, you translate everything into Portuguese>
059   Bob   [<<Port.> (                )> ((talking to Ric))
```

Luc establishes his leading position within the group, suggesting that the group investigate the cups (lines 37, 41 and 44) after the teacher left. Kim produces a code switch when proposing in French how many stones, three, they should put in a cup (line 40). French is one of the three official languages in the Luxembourgish school system. The counting, which is operated in French by Kim, is accepted by the peers, the switch in the linguistic resources is not referred to by the others, none of the children addresses this code switch. Indeed, French represents the language of academic work and success as the children move on in their schooling life. The children display mutual engagement around the same science learning activity, with changing linguistic resources (Luxembourgish, French, Portuguese) and by using relevant discourse practices for the science learning activity, such as counting, formulating causality ("so") and consequence or eliciting others to move on with the experiment. All children engage jointly in the activity, yet, they rely on different resources for doing so: Luc and Kim, for instance, engage by verbal means in Luxembourgish, addressing the others three times verbally (lines 37, 41, 44); Bob moves into the activity by taking the physical object of the cup and repositions himself at the table (line 43); Ric, Ced and Han change their body positions, lean forward on the table, towards the joint activity. Specifically, the verbal means demonstrate how scientific discourse is emergent from the situation. Luc in particular refers to discourse practices which mark the advanced science learning (*so, sou wann*).

But when the interaction unfolds in Luxembourgish, Ric disengages spatially and looks at the recording camera and around in the room. As mentioned, Ric is new in the school and he only speaks Portuguese. When Bob addresses him in Portuguese (line 52), inviting him to take action in the activity, Ric reacts immediately, turns his gaze and his body to the container with the water and leans forward on the table. The code switch addressed to him allows Ric to move into the ongoing activity. Interestingly, Luc follows up on this code switch in using the Portuguese language when addressing Ric (line 57). Furthermore, he makes the choice of language resources and the way these should be used an issue to be mentioned, telling Bob that he should translate for Ric (line 58). The reference to the choice of the linguistic resources available points to the awareness of the children as regards participation in the science learning activity. Two participants, Bob and Luc, display their understanding of the importance of the choice of the linguistic resources, in order to make sure that Ric gets the possibility to participate in the activity (e.g. his actions). Moreover, the issue of the linguistic resources to be used becomes an element in the positioning of the participants in the science learning activity. Luc takes the role of an expert by eliciting actions of others, allocating a specific role to another participant (e.g. Bob to translate), ratifying actions, making sure that all the participants have access to the shared activity space (e.g. gazing) and addressing the resources (e.g. linguistic) to be used in order to keep the activity going and the participants involved (e.g. lines 37, 41, 44, 48 and 50). Finally, both switches of the language show how the participants bring up and develop their discourse practices with regard to the enactment of the science learning activity: counting, for instance, is done in French (line 40), commenting the scene of the water experiment with the sinking cups is done in Luxembourgish. Interestingly, the issue of language choice for one particular participant (Ric) is embedded in the commenting sequence of the observable science learning, mainly operated by Luc. In fact, Luc engages in providing causal explanations with regard to the observable elements by relying on complex discursive practices: Luc specifies and actually over-specifies the action to be taken (line 50), by indicating and repeating the fact that the stones have to be placed right into the "middle" of the "middle" (*voll an d'mëtt, lines 50, 51*). Deictic reference is then used to demonstrate this over-specification (line 51) by using a comparison device. Furthermore, Luc relies on an advanced discourse practice related to the science learning activity. He not only describes what is observable. He actually provides the reason why a specific phenomenon of the science experiment is not functioning (line 54), followed by the verbal indication of what needs to be done to make the experiment work (*dofir, line 55*) while still demonstrating the actions to be carried out (here: acting on the cups).

In sum, the analysis of the above excerpt provides insights into how diversity (e.g. linguistic) emerges both as an issue to be dealt with and a resource for accomplishing the actual activity. Participants address the choice of language when organising participation and positioning themselves within the co-constructed context. Moreover, the navigating between the linguistic resources at hand is embedded in highly relevant sequences of science learning with regard to the discourse practices which need to be carried out in terms of accepted ways of doing science, for instance, providing evidence for claims through demonstration, being specific (and over specific) in discussing an element, providing causal relations. Tracing these phenomena of "doing science" amongst peers across the micro-historicity by looking at the discourse practices allows for grasping both, the situated embeddedness of emergent discourse practices (e.g. explanatory sequence with increased awareness of language choice for carrying out the activity) and the potential moments of development, when, for instance, the need for being specific (and over-specific) is made relevant by the participants in doing science in an evermore general way.

The following excerpt provides additional evidence for the ways in which available resources for managing the science learning activity emerge and are subsequently modified by the children with regard to gradually accepted discourse practices for doing science.

Excerpt 2: a vu wou ass dann den REEN komm – *and where does the rain come from*

This excerpt shows an interaction in a science learning activity in a second grade class with fifteen children. The teacher organizes the class in four groups and invites the children to gather questions, familiar facts, ideas and interesting propositions for an upcoming classroom project about water. The groups are asked to write their ideas on a sheet of paper for the classroom discussion following this initial exchange phase in small groups. The research team is recording the discussions of all the groups, which last for about 40 minutes and the following teacher-led

class talk. The excerpt below represents the interaction of one of the groups, the
group of three girls – Altina, Lynn, Lea (Alt, Lyn, Lea). The beginning of the
exchange is marked by pauses and short utterances for about seven minutes. Then
they identify and co-construct a potential topic to discuss in line with the given
task. Altina (line 001) raises a question which is taken up and further developed
by the two co-participants. The beginning (Excerpt 2) is of particular interest for
tracing both, the micro-development of the discourse practices and the elements
the children draw from. In fact, the children refer not only to everyday context
related with the question about the origin of water. Interestingly, they refer to, use,
recycle and modify discourse elements stemming for other discourse contexts, for
instance, religious education. Over the short time of the excerpt presented below, the
three participants produce evermore specific and sequentially related utterances and
explanatory devices. Their exchange on the key questions shows how interacting and
speaking gradually constructs the activity as a science learning activity, calling for
appropriate discourse elements of science with a certain degree of generalization.
Therefore, we see how the children refer to, use and then co-develop discourse
practices as needed by the actual topic at hand, moving from a diversity of discourse
contexts (e.g. religious) towards the normatively expected words and elements for
"doing science".

Excerpt 2: Diversity and Development of Discourse Practices

```
001   Alt   firwaat' um: ·hh
            why um: hh
002         nee ↑wi ass d´waasser iwwerhaapt hei op d erd ʌkomm:
            no ↑how did water come here to earth at all
003         (1.2)
004   Lin   h (1.0) m:a (1.0) ma well: ·h(0.5)
            h (1.0) we:ll (1.0) well because: ·h(0.5)
005         hhh (0.4) wéngst ·hh
            hhh (0.4) because ·hh
006         (8.0)
007   Lin   hh (0.8) ·hh ma well (* déinië *) mmh (0.7)
            hh (0.8) hh well because (* déinië *) mmh (0.7)
008         wou mier nach net op der WELT' waren do: ·hh
            when we were not on the EARTH' ye:t ·hh
009         (1.0)
010         do huet den:: (0.4)
            at that time did the:: (0.4)
```

```
011              de JESUS eis de WAAsser matbruet
                 JESUS brought with him the water to us
012              (3.0)
013    Alt       :hh ech me:ngen well et déck déck oft gereent' huet ·h
                 :hh i thi:nk because it was pouring' very very often
014              dofir ëmmer d´waasser mi héich a mi héich
                 therefore continuously the water got higher and higher
015              a mi héich gaangen ass (0.8)
                 and got higher (0.8)
016    Lin       jo mee ·hh (0.3) awer: e' den eh (0.8)
                 yes but hh (0.3) but: e the eh: (0.8)
017              den=ehm (1.0) de reen ass och waasser (0.6)
                 the=ehm (1.0) the rain is also water (0.6)
018    Lea       [jo
                 [yes
019    Lin       [a vu wou ass dann den REEN komm
                 [and where did then the RAIN come from
020    Lea       a jo, do kret en och d waass[er
                 o yes, there you get the wat[er as well
021    ??:                                   [hehe (laughter)
022              (1.3)
023    Alt       villäi:cht ·h
                 maybe ·h
024              (1.4)
025              well ↑äi:s wann et ze WAArm gëtt
                 because ↑i:ce when it gets too wAArm
026              gëtt et jo WAAsser (0.8)
                 becomes WAter as you know (0.8)
027    Alt       an emmer ass den ÄIs méi kléng jo gin
                 and all the time the ICe became smaller
028              (0.4) do
                 (0.4) there
029              (1.5)
030              op der welt ass ëmmer mi KLÉNG gin (0.5)
                 on the earth ice became SMALLER and smaller (0.5)
031              den [äis
                 the [ice
032    Lea           [oder huet et eng kéier gereent
                     [or did it rain one-time
033              (1.0)
034    Lin       .hh awer wann tschk (0.9)
                 .hh but if tschk  (0.9)
035              a wou kréie mer dann den ÄIs 'hir
                 and where do we get the Ice 'from
036              (1.8)
037              dann brauche mer ↑och waasser
                 then we need    ↑also water
038              well ·hh      [(*bringst*)
                 because hh [(*bring*)
039    Alt                    [ma nee daat ass jo schon(.)
                               [but no that has already(.)
040              alles laang do vun               [(inc.)
                 been there for a long time from [(inc.)
```

Altina introduces her inquiry by the following formulation, marked by a self-repair as regards the question tag (why): *firwaat' um: ·hh nee ↑wi ass d´waasser iwwerhaapt hei op d erd ∧komm: why um: hh no ↑how did water come to earth at all* (lines 01–02). The first question tag (why) marks an inquiry, whereas the second question tag (how) in the chosen combination with the formulation "came on earth" refers to the religious education discourse, which discusses how the earth came into existence and which people or creations came to earth by which movement or action by some religiously identified agent. Both elements, the finally selected question tag (how) within the same utterance, refer to a discourse practice which introduces certain explanatory sequences and formats of story narration in religious education. In formulating her inquiry, Altina draws from this discourse practice (i.e. how, came on earth) by relating it to the actual earth (here) and the observable phenomenon. The transformation of the format of the questioning, helping and actually doing the inquiry is manifest in the changing questioning tag, which the children transform as they work on the initial question, marked by the self-repair on the questioning tag. In fact, the narratively bound "how" is transformed into questions referring to the place of origin by introducing the question tag "where" (*a vu wou ass dann den REEN komm, and where did then the RAIN come from, line 19*) or "from where" (*a wou kréie mer dann den Äls 'hir, and where do we get the Ice, line 35*). These questions are more specific. The participants manage to develop an exchange around these more specific questions by bringing in possible candidate answers to these "where" and "where from" questions (line 019 and line 035). Interestingly, the transforming move from one question tag (how) to another one (where, where from) affords two actions completed by the children. Firstly, they manage to come up with candidate answers to the specific "where" question. The discourse practices which help to provide answers to the "where" question are available to the children and allow for discussing, accepting or rejecting one option after the other. The interactional development shows that the participant who drives the interaction by the "where" questions keeps this type of questioning up until or a response is found or it becomes evident that no suitable answer can be found at the moment. Secondly, this inquiry-bound advancement by the children allows them to respond to the initial task. They start to write up their guiding questions and potential answers on the sheet, using the "where" and "where from" questions as guiding and structuring discourse devices in their writing. In sum, the transformation of the initially presented questions and question tags shows how the actual way of managing the inquiry activity by means of talk-in-interaction (e.g. mutual uptake) relies on available discourse practices (e.g. narrative religious discourse) but gets discussed, rejected, modified and ratified in line with the objective of the interaction, namely, to conduct a science-driven inquiry into issues of water.

Another feature to observe with regard to the discourse practices mobilized and modified by the children across this excerpt is the marking of closeness to the topic. In fact, when a topic is developed, participants mark their closeness towards the topic at talk by indicating specific reference (Ziegler 2006). In this excerpt, the children start by marking their closeness to the actual earth here and now as they talk about *here on earth (hei ob d erd, here to earth, line 02)*. The "world" *(der welt, line 08)* is addressed as an overreaching context, which precedes and goes beyond the participant's, Lin's, personal presence *(mir)* in the "here and now" (line 08). Moreover, the children rely on the narrative format of religious discourse to refer to a potential personified agent by addressing *de Jesus* (line 11) in relation to themselves as "us" *(eis, line 11)* to whom he specifically brought the water. The way in which Lin develops her response to Altina's question *(lines 07–11)* displays hesitation (pauses, lines 07, 10) as she mobilizes a discourse practice stemming from another discourse context, the religious context involving Jesus and his doings on the earth the children indicate as being the one they inhabit. The rather long pause (3.0 s., line 12) shows the ongoing change of the participants' means of discourse as it unfolds. In fact, the marking of closeness, by reference to themselves and the "here and now", as well as the taking up of the narrative perspective of religious discourse, is not available in the subsequent section of the excerpt (lines 13–33). To the contrary, several phenomena demonstrate how the discourse practices change throughout this subsequent section as the children mark stance towards the arguments they discuss.

Firstly, Altina uses epistemic stance markers with lengthening of sounds, indicating her point of view by discursive means *(ech me:ngen, line 11; villäi:cht, line 23)*. The discursive indication of one's own point *(ech me:ngen, line 11)* of view refers to the discourse context of informed discussions or argument-driven discussion rounds. Both instances of discourse markers *(ech me:ngen, line 11; villäi:cht, line 23)* are placed at the beginning of Altina's turn, followed by the indication of a causal element *(well, because, lines 13; 25)*, which is put forward by Altina to account for the water issue under discussion. More causal discourse elements are used in this stretch *(dofir, that's why, line 16)*.

Secondly, the children do not use the "we-voice" in the same way they do up to this point of the interaction. They do not refer to themselves in this stretch (lines 13–33) of interaction. Rather, they refer to the elements in third-person voice or in impersonal forms *(et déck déck oft gereent' huet ·h, it was pouring' very very often, line 13; dofir ëmmer d´waasser, that's why always the water, line 016; de reen ass och waasser, the rain is also water, line 017; huet et eng kéier gerent, it had rained once, Line 32)*. Speaking of and about the elements, e.g. water, rain, in third person voice is in line with a more science bound discourse practise, the elements are given, their actions happen and are described in a non-personal way, using the non-personal pronoun: *(a jo, do kret en och d waass[er, o yes, there one gets/you get the wat[er as well, line 20)*.

Thirdly, the children rely on discourse elements, which indicate the type of conditions related to a specific water element. Children use elements such as "always", which describe the normal conditions of a water phenomenon (line 27, 30). The children even come up with conditional correlations in an "if x–then y" format, indicating how the state of water changes under which conditions: *well ↑äi:s wann et ze WAArm get gëtt et jo WAAsser, because ↑i:ce when it gets too wAArm becomes WAter as you know (lines 25–26)*. Also, they refer to candidate answer options *(oder, or, line 34)* and formulate more specific questions, e.g. where does the water come from *([a vu wou ass dann den REEN, [and where did then the RAIN come from, line 19)*.

In sum, the participants clearly display discourse practices from other contexts, e.g. religious, when they discuss the characteristics of water in an evermore science-oriented way. The different discourse practises emerge as the interaction develops; the practices change over the course of the interaction. In the present stretch, the ways that discourse practices are mobilized change, the difference in the way that "doing science" discursively is done at the beginning of this interaction (01–12) as compared to the later sequence is tangible (lines 13–33). The transformation of discourse practice is subtle as the children do not move from extreme discourse practices (e.g. real religious quotes) to other equally highly normative discourse practice (e.g. standardized science discourse). Rather, we see a gradual movement from a narrative I-/we-voice, which relies on religious elements towards a argumentative, fact-oriented third-person voice. Discourse elements such as the double or triple use of comparative forms *(déck déck oft, pouring' very very often, line 13; mi héich a mi héich a mi héich, the water got higher and higher and higher, lines 14–15)* are observable at the transition areas within the interaction development. In fact, when starting the reasoning (line 12), the participants leave the discourse practices used so far behind (e.g. religious) and co-construct a more science-bound argumentative way of looking at and discussing the water phenomenon. They manage this transition by means of resources, which they draw from a diversity of elements. Therefore, in doing science as is observable from this stretch the children draw from their diversity of resources and transform the available discourse practices. The discussed micro-phenomena observable in the discourse practices (e.g. first-person vs. third-person voice) allow for tracing the historicity of this "learning". This transformation is operated in interaction and across the participants. They work together in moving towards the more science-related discourse, thus not one single party (e.g. Altina, Lin or Lea) maintain discourse practices pertaining to one reference discourse only. Rather, they propose options, reach common ground on the practices that are mobilized, ratify or reject them and carry on working on the topic by bringing up, expanding and reworking available discourse practices of high diversity. In Excerpt 2, the emergence of both a rather religious bound discourse context and a more science-related discourse context, which are confronted with each other (cf. Poddiakov 2011) and actively worked into each other, is an active accomplishment done by the children in dialogical interchange (cf. Wegerif et al. 2010). These micro-actions can be traced across the sequence of interaction.

In this respect, it is interesting to mention the way in which the children finalize their task (see above) and in which way they write down their reached outcome of the interaction on "water", even though the writing sequence is not the core of this paper. In fact, the transformation of the questioning practises as discussed in the analysis shows in the actual writing of the outcome done by the children. They write the following questions in the following order (see Fig. 5): Why does water exist? Where does the water come from? Why do some people live very close to the water? When fish are on land they cannot breathe, when we are under water we cannot breathe?

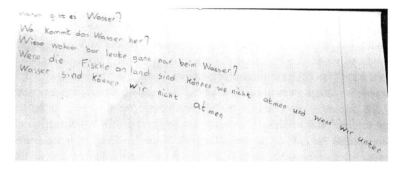

Fig. 5 The written questions (Adapted from Max and Ziegler 2009, see text for translation)

The "how" question gets transformed during the interaction, carried out in Luxembourgish (Excerpt 2, line 2) and is finally written down in German in the following wording, taking up the transformed questioning (expert 2, line 19): *Wo kommt das Wasser her* (where does the water come from)? During the writing-up sequence, the participants push for most accurate, answerable questions they agree to present their questions to more knowledgeable peers and the teacher. It is interesting to notice that they replace the initial "how" question altogether by two questions which they write down in the end: the "where from" question and a "why" question, which focuses on the reason of the existence of water (*Warum gibt es Wasser?* Why does water exist?). The "why" and "how" questions which have been developed by the children throughout their interaction are the major topic of the follow-up classroom, transforming the discourse practices even more, closer to the normative science discourse, gradually operated by the teacher (Max and Ziegler 2009).

Discussion: Observations in the Light of the Framework

The following graph illustrates schematically how children mobilize different discourse practices which they then gradually modify and adapt towards "Science-as-discourse" as the inquiry of the children is developed in interaction. More fact-driven, science-bound discourse is emerging in the early science classroom from the different resources that children bring into play across a variety of contexts, transgressing boundaries in their talk-in-interaction. Discourse practices, which are reasonably connected with specific discourse contexts (e.g. catholic religious context), engage in a productive process. Other discourse might also serve as resources, for instance, comic-discourse (Max and Ziegler 2009).

The participants shape the various practices and elements as well as their positioning throughout the interaction. Children's emergent and continuously developed discourse practices are not seen as something individual they own, rather

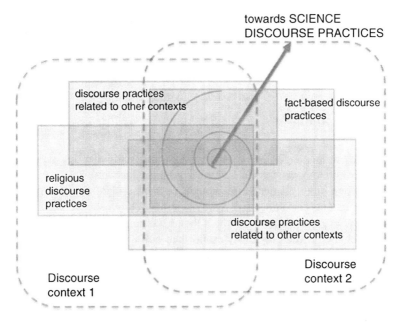

Fig. 6 Emergent, gradually established discourse practices during the children's talk

they collaboratively bring up resources which are then exchanged, ratified and modified in a mutual process which can be traced. The analysis carefully traces their micro-actions, what they do and what they use as they interact with multiple semiotic and discourse means in science-related activities. In short, "doing science" within interaction is carefully organized, complex, and inherently multidimensional (Fig. 6).

Looking at the discourse practices in interaction from a sequential point, different discourse practices are sequentially brought into play by the children, which are then accepted, modified, rejected, taken up as the inquiry of the children carries on. Given the specific indicators – first/third-person voice, argument structure, reference to specific topics – the historicity of the development of these resources and practices can be traced. Rather than showing which relevant discourse of reference is selected by the children in absolute terms, the aim is demonstrate the diversity of resources and their ongoing modification within the interaction. As discussed, the analysis shows how science is done in and as practices mediated by semiotic resources, i.e. discourse formats and repertoires. These repertoires emerge from students' opportunities to engage in joint experiences, which involve multimodal as well as multilingual resources. The fact that the discourses of reference vary (e.g. catholic religious discourse), exist alongside of each other, contradict each other and disappear and reappear in the children's discursive doings confirms the actual work of moving towards a science-bound discourse. Also, the individual participant does not establish one discourse or another. Rather, the co-construction across and through repertoires is developed in interaction as science learning takes place. Historicity in the line with the socio-cultural perspective on science learning then is a moment-by-moment accomplishment, which marks the goal-driven, gradually more science-bound activity of the children.

Concluding Remarks

The diversity of languages, home and group cultures and other discourse elements are the driving elements for the science learning which can be traced along the instances when other elements are brought up by the children in their discourse practices. Elements pertaining to diversity in terms of linguistic elements or ways of talking about everyday elements, which then become part of the emergent science discourse, challenge the previously established norm with regard to language and contexts of reference. They allow for testing against the norm in a highly efficient way within the classroom. This testing challenges stabilized knowledge which is presented to the children at school and expands or transforms the previously stabilized elements – of the science classroom discourse but also of other contexts such as the home context.

In sum, the issue of diversity in terms of resources and languages brought into the science classroom (Ziegler 2011) are highly relevant in order to understand the construction of science as a situated process in and as discourse practices. Focusing on the diversity in terms of students backgrounds and, more importantly, multiple linguistic and discourse resources as they are brought up in science learning sheds light on the otherwise poorly studied processes of transformation of science learning elements and integration of newly elaborated knowledge (Poddiakov 2011). When children address and negotiate problematic issues, the interactional deployment as well as the discourse elements which are brought up allow for tracing science as emergent and stabilized in such moments. Practices in discourse which might seem separated from "doing science" per se, such as translation from one language to another, are brought in by the participants as devices which actively transform knowledge elements available to the children into the science-related inquiry as some of these resources then are maintained (for instance when writing up results) and others rejected by the children.

These transformations then are traceable in the discourse practices as they become historic for the children in the particular learning context (Siry et al. 2012). Putting the issue of multiple resources from diversity at the centre for understanding processes of knowledge construction and its stabilization from a micro-analytical perspective, two elements are tangible from the analysis above (Excerpt 1, Excerpt 2): Firstly, children display awareness of relevant discourse elements and related topics. Given the position of religious education in the pre- and primary school context of Luxembourg and its verbal presence ("Jesus"), children rely on these resources, evoke meaning and knowledge elements by confronting different discourse contexts on a practice-by-practice basis (e.g. first-person vs. third-person voice). As the analysis indicates, such negotiation of discourse elements in the situation of "doing science" is not an external event, which could be conceived of as being unrelated with the science learning situation at hand. Rather, the active confrontation of discourse elements and their practices contribute and constitute the science learning by challenging the norm by means for mediating elements, here the discourse practice.

Secondly, the functioning of multiple languages (Grosjean 1982; Ziegler 2011) rather than the simple fact of having two or more languages available in an interaction is relevant for understanding how multiple resources are referred to, transformed and rejected in the process of science learning. By tracing the mutual elaboration and ratification of target language elements (Luxembourgish) in and through the negotiation of non-target language elements, the process of science learning as a joint achievement is accessible in the diversity setting.

The analysis calls for future detailed studies, which trace not only issues of the diversity setting (such as multilingualism among the classroom population) or home-culture elements brought into play by the children, but should address the discourse elements in terms of topics or reference elements of discourse contexts (such as comic discourse, religious discourse, video-game discourse) together with a critical stance towards multiple linguistic resources and the negotiation of these resources, leading the children towards the target practice of "doing science" and the target language.

References

Cole, M. (1996). *Cultural psychology: A once and future discipline.* Cambridge, MA: The Bellknap Press of Harvard University Press.

Daniels, H. (2008). *Vygotsky and research.* New York: Routledge.

Daniels, H., Edwards, A., Engeström, Y., Gallagher, T., & Ludvigsen, S. R. (2009). *Activity theory in practice: Promoting learning across boundaries and agencies.* London: Routledge.

Edwards, R. (2005). *Literacies for learning in further education in the UK: Theoretical and methodological challenges.* Paper presented at the Australian Association for Research in Education Annual Conference, University of Western Sydney.

Engeström, Y., & Sannino, A. (2010). Studies of expansive learning: Foundations, findings and future challenges. *Educational Research Review, 5,* 1–24.

Engeström, Y., Miettinen, R., & Punamäki, R.-L. (Eds.). (1999). *Perspectives on activity theory.* Cambridge: Cambridge University Press.

Engeström, Y., & Miettinen, R. (1999). Introduction. In Y. Engeström, R. Miettinen, & R.-L. Punamäki (Eds.), *Perspectives on activity theory* (pp. 1–18). Cambridge: Cambridge University Press.

Gabo, L., & Mondada, L. (2000). *Interactions et acquisitions en contexte.* Fribourg: Editions Universitaires.

Gallas, K. (1995). *Talking their way into science.* New York: Teacher's College Press.

Gee, J. P. (2004). *Situated language and learning: A critique of traditional schooling.* London: Routledge.

Grosjean, F. (1982). *Life with two languages.* Cambridge: Harvard University Press.

Gutiérrez, K., Baquedano-Lopez, P., & Tejeda, C. (1999). Rethinking diversity: Hybridity and hybrid language practices in the third space. *Mind, Culture, & Activity, 6*(4), 286–303.

Hedegard, M. (2001). *Learning in classrooms. A cultural-historical approach*. Aarhus: Aarhus University Press.

Inan, H. Z., Trundle, K. C., & Kantor, R. (2010). Understanding natural sciences education in a Reggio Emilia-inspired preschool. *Journal of Research in Science Teaching, 47*(10), 1186–1208.

Kostogriz, A. (2000, December 4–7). *Activity theory and the new literacy studies: Modelling the literacy learning activity system*. Paper presented at the meeting of the Australian Association for Research in Education, Sydney.

Lave, J., & Wenger, E. (1991). *Situated learning. Legitimate peripheral participation*. Cambridge: University of Cambridge Press.

Lemke, J. (1990). *Talking science: Language, learning, and values*. Norwood: Ablex.

Matusov, E. (2007). In search of 'the Appropriate' unit of analysis for sociocultural research. *Culture & Psychology, 13*(3), 307–333.

Max, C., & Ziegler, G. (2009). *Developing discourse-competence in doing science: Insights from multilingual interactions (K1-K2)*. Paper at workshop, ESERA conference 2009, Istanbul.

Melander, H., & Sahlström, F. (2009). In tow of the blue whale. Learning as interactional changes in topical orientation. *Journal of Pragmatics, 41*, 1519–1537.

Pekarek-Doehler, S., & Ziegler, G. (2007). Doing science – Doing language: The sequential organization of science talk and language-talk in immersion classrooms. In Z. Hua, P. Seedhouse, & V. Cook (Eds.), *Language learning and teaching as social interaction*. Basingstoke: Palgrave Macmillan.

Poddiakov, N. (2011). Searching, experimenting and the heuristic structure of a preschool child's experience. *International Journal of Early Years Education, 19*, 55–63. (Special issue: Early childhood education research from a Russian Perspective).

Rogoff, B. (1995). Observing sociocultural activity on three planes: Participatory appropriation, guided participation, and apprenticeships. In J. V. Wertsch, P. Del Rio, & A. Alvarez (Eds.), *Sociocultural studies of mind*. Cambridge: Cambridge University Press.

Rogoff, B. (2003). *The cultural nature of human development*. Cambridge, MA: Harvard University Press.

Roth, W.-M. (2005). *Talking science. Language and learning in science classrooms*. New York: Rowman and Littefield.

Roth, W.-M. (2006). *Learning science: A singular plural perspective*. Rotterdam: Sense Publishers.

Sannino, A., Daniels, H., & Gutiérrez, K. D. (Eds.). (2009). *Learning and expanding with activity theory*. New York: Cambridge University Press.

Sfard, A. (2009). Moving between discourses: From learning-as-acquisition to learning-as-participation. In M. Sabella, C. Henderson, & Ch. Singh (Eds.), *Proceedings of the physics education research conference*. Ann Arbor, Michigan: American Institute of Physics.

Siry, C., Ziegler, G., & Max, C. (2012). 'Doing science' through discourse-in-interaction: Young children's science investigations at the early childhood level. *Science Education, 96*(2), 311–336.

Tomasello, M. (2008). *Origins of human communication*. Cambridge: MIT Press.

Tuomi-Gröhn, T. (2003). Developmental transfer as a goal of internship in practical nursing. In T. Tuomi-Gröhn & Y. Engeström (Eds.), *Between school and work: new perspectives on transfer and boundary crossing*. Amsterdam: Pergamon.

Tuomi-Gröhn, T. (2005). Studying learning, transfer and context: A comparison of current approach to learning. In Y. Engeström, J. Lompscher, & G. Rückriem (Eds.), *Putting activity theory to work: Contributions from developmental work research*. Berlin: Lehmanns Media.

Valsiner, J. (2002). Mutualities under scrutiny: Dissecting the complex whole of development. *Social Development, 11*(2), 296–301.

Van Lier, L. (1996). *Interaction in the language curriculum: Awareness, autonomy, and authenticity*. London: Longman.

Van Oers, B., Elbers, E., Van der Veer, R., & Wardekker, W. (Eds.). (2009). *The transformation of learning: Advances in cultural-historical activity theory*. Cambridge: Cambridge University Press.

Wegerif, R., McLaren, B. M., Chamrada, M., Scheuer, O., Mansour, N., Mikšàtko, J., & Williams, M. (2010). Exploring creative thinking in graphically mediated synchronous dialogues. *Computers in Education, 54*, 613–621.

Wertsch, J. V. (Trans. & Ed.). (1981). *The concept of activity in Soviet psychology*. New York: M. E. Sharpe.

Woods, D. K. (2007). *Transana 2.21* (Mac) [Computer software and tutorial]. Madison: Wisconsin Center for Education Research.

Ziegler, G. (2006). Determination und Referenz im frühen L2 Erwerb Französisch. *Zeitschrift für Literaturwissenschaft und Linguistik, 43*, 25–69.

Ziegler, G. (2011). Innovation in learning and development in multilingual and multicultural contexts. *International Review of Education, 57*, 1–21 (Special issue: Principles and innovations in multilingual education).

Conceptual Frameworks, Metaphysical Commitments and Worldviews: The Challenge of Reflecting the Relationships Between Science and Religion in Science Education

Keith S. Taber

Introduction

Learning is mediated by social interactions, and education involves the induction of learners into facets of culture that are represented, explicitly or implicitly, in curriculum. Yet for many young people growing up in technologically advanced, multicultural societies, learning occurs in a range of contexts and is mediated by diverse groups of others: the home and its extended family and social community, playtime peer groups, indirectly through interaction with the media-rich environment (newspapers and magazines, television programmes, the Internet, etc.) *as well as* in the formal learning context of the school class. Indeed, whilst a school may have an 'ethos' and represent certain values and norms (which may match those of the home to differing degrees), it also offers mediated access to those elite features of culture: the academic disciplines – *each* having their own norms and privileged ways of behaving, thinking, communicating and so forth. Science is one such way of knowing and acting in the world, and Alsop and Bowen (2009, p. 53) have argued that in science education 'an overwhelming emphasis (in research and practice) is put on induction and initiation into a subculture and its associated epistemology - the language, culture and tradition of science'.

Entering the science classroom has been compared to making a 'border crossing' for students (Aikenhead 1996), as the world of science (as represented in school science) may seem quite foreign to many pupils. School science is a form of mediation into a particular way of using language (Lemke 1990), a specific set of customs for how one should think and come to knowledge. The privileged concerns, the ways of doing things and especially the ways of communicating may be quite at odds with the learner's life outside the science classroom (Solomon

K.S. Taber (✉)
Faculty of Education, University of Cambridge, Cambridge, UK
e-mail: kst24@cam.ac.uk

N. Mansour and R. Wegerif (eds.), *Science Education for Diversity: Theory and Practice*,
Cultural Studies of Science Education 8, DOI 10.1007/978-94-007-4563-6_8,
© Springer Science+Business Media Dordrecht 2013

1992). Regardless of whether this is something welcomed by, or alienating to, students, it certainly adds to the 'learning demand' (Leach and Scott 2002) of the subject. It has been argued that many of the common 'alternative' (i.e. contrary to scientific thinking) conceptions exhibited by science learners can be considered as the application of lifeworld knowledge (Schutz and Luckman 1973) that functions effectively in everyday social exchanges, but is inadmissible as part of formal scientific discourse (Claxton 1993; Solomon 1993; Taber 2009). Students may find that something that works well in a more familiar language community seems to be judged as inadequate in the particular context of the science lesson.

This chapter considers one particular aspect of cultural mediation of learning in the science classroom: that of the relationship between science and religion. This issue has attracted much public attention because of the question of teaching scientific theories of origins (the 'big bang', and in particular evolution by natural selection) to those for whom such ideas are perceived as contradicting their own faith commitments (Antolin and Herbers 2001; Poole 2008). In some contexts, such as the UK, it is sometimes perceived as a 'minority' (i.e. fundamentalist) issue, as the mainstream Christian churches have long been happy to accommodate scientific theories. In such a context, suggestions that teachers need to engage in dialogue with pupils on such issues (Reiss 2008) – despite being supported by research evidence (Verhey 2005) – have been criticised (Vallely 2008), for example, as a 'slippery slope' towards relativism.

However, such simplistic responses ignore the complexity of the issue in practice. Many pupils from faith backgrounds (not just those who identify with 'fundamentalist' groups) hold to worldviews that encompasses the supernatural as an *integral* part of their world in which they live. Science does not inherently exclude the existence of a supernatural realm (although *some* scientists, including some high-profile science 'media stars' do vehemently claim otherwise), but does in a sense require it to be put aside when doing scientific work. Moreover, there are a range of positions that can be taken about such matters as: the extent of the magisterium of science, the absolute nature of scientific laws, the potential of science to offer explanations, the ability of human minds to understand the nature of the world and so forth. These are largely metaphysical matters: they are not (and cannot be) determined by empirical work in science, but underpin the very values that inform the enactment of science itself.

School science offers a representation of the nature of science (Millar 1989), and science teachers portray messages about such matters (intentionally or otherwise). However, in many educational contexts, teachers are often not well informed about the nature of science and in particular its philosophical underpinnings (Hodson 2009). The preparation of science teachers often offers them limited support for developing an appreciation of the range of respectable scholarly positions about the relationship of science and religion and the range of views about the nature and limits of science that in part underpin such positions. That is, teachers may not appreciate how the culturally constituted set of 'scientific values' which are shared commitments within the community of science, and into which scientific training inducts new community members (Kuhn 1996), can obscure a diversity of underpinning ontological, epistemological and axiological frameworks informing different individual scientists' work. Teachers are therefore often not well placed to mediate a balanced view about such issues through their own interactions with pupils in the classroom.

Consequently there is much potential for the image of science offered to pupils to be scientistic: one of an all-encompassing and exhaustive approach to understanding the world. Often, the view communicated in school science (i.e. the message as perceived by many students, regardless of whether intended) is that the natural world is all there is and that it can in principle be fully understood by science (Francis et al. 1990; Fulljames et al. 1991; Hansson and Redfors 2007; Taber et al. 2011a, b). Such a world is very much at odds with the one inhabited by many scientists of faith (Berry 2009) and certainly is an alien world for many school pupils.

The tradition in Western science (with its tendencies towards an analytical and reductionist approach) to precede as though the existence and potential role of God in nature is irrelevant to answering scientific questions, if not explicitly explained to students, may well give the impression that because *science* (as a sociocultural activity) does not need to adopt the hypothesis of the divine, *scientists themselves* (as individuals sharing membership of various social groups with their identities as scientists) eschew such an idea. This is likely to be especially the case for learners who have been brought up in a faith tradition that considers God to be immanent in all things and which teaches that the believer should put God at the core of their entire life. A theist who considers God to work through nature might well take the methodological stance that she or he is likely to come to a better understanding of God's work by proceeding in scientific work *as though* God is irrelevant, but this may not be inherently obvious to school students. Indeed, if ones whole life is grounded in a belief that God works through and maintains all aspects of nature, then this may operate as a taken-for-granted commitment that no more needs to be made explicit in accounts of scientific work than the taken-for-granted assumption that the scientist breathed air, or stopped for meal breaks, during scientific work. Yet the potential for alienating many pupils from science (and so advanced scientific study and scientific careers) is clear.

This chapter sets out a discussion of this major problem in science education, by considering some of the range of metaphysical commitments that inform different understandings of the nature of science (in particular, in its relationship to religions) and considering how these can contribute to making pupils' 'border crossings' (Aikenhead 1996) into science more or less problematic. Such an analysis is needed as the first step to supporting teacher education in the issue, something that is essential if we wish to ensure that the image of science mediated by teaching is not more alien to many young people than it needs to be.

Cultural Border Crossing in Science Education

To understand how young people respond to school science, we have to acknowledge its cultural dimension, in the sense that school science offers certain norms (e.g. ways of talking and so legitimised ways of thinking) that may seem strange to students such that they experience science lessons as somewhat 'alien' or 'other'.

For some scientists, science may be an important *part of* culture yet also considered cross-cultural in the sense of science being seen as 'universal'. After all, natural science is intended to be about the way the world is *independently* of human contingencies (Bhaskar 1975/2008). Yet science has been subjected to various critiques that consider that *the practice of science* has reflected norms of particular social or cultural groups rather more than others, for example, feminist critiques suggesting that science has traditionally reflected ways of thinking predominantly associated with what it is to be masculine (Bentley and Watts 1987). Similarly, it has been argued that it is far from clear that science today can be considered pan-cultural and 'universal' (Harding 1994).

Science as Culture

Indeed the notion that any person can completely rise above their sociocultural context (or even fully recognise its influence) is questionable; it is important not to underestimate the significance of culture on each one of us. Indeed Geertz has gone as far as to suggest that

> Whatever else modem anthropology asserts - and it seems to have asserted almost everything at one time or another - it is firm in the conviction that [people] unmodified by the customs of particular places do not in fact exist, have never existed, and most important, could not in the very nature of the case exist.
>
> (Geertz 1973/2000, p. 35)

Science is practised by people from specific cultural contexts, who are products of those contexts. As such, they will bring with them particular ways of thinking and understanding the world and particular value systems, at least parts of which will have in effect become 'second nature' (or perhaps, in view of Geertz's comment, just, 'their nature') during their upbringing and so will in effect be invisible and therefore influence them in an insidious way.

Kuhn's (1996) famous work on the structure of scientific 'revolutions' acknowledged this. Whilst some saw his essay as the justification for taking a relativistic view of science, a weaker version of the thesis (more akin to Kuhn's own position) would be that scientists are never going to be able to be completely immune to biases deriving from extra-scientific background issues. This is, in effect, little more than acknowledging that any person's current thinking is inevitably contingent upon and so influenced by the cognitive resources available (what might variously be described as ideas, knowledge, beliefs, expectations, habits of mind, etc.) This is recognised in science education in how learners very commonly come to alternative understandings of many scientific ideas because their existing ideas provide the interpretative frameworks for making sense of science teaching (Taber 2009).

Indeed, research in student learning in science finds not only some very common alternative conceptions which appear to develop in a range of educational contexts (and so may reflect 'genetically directed' aspects of how the human cognitive

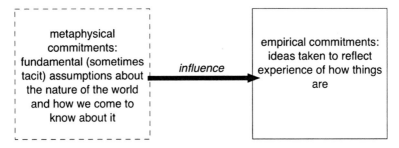

Fig. 1 The ideas we have about the ways things are, which we consider to be based upon our experience in the world, are influenced by background assumptions that we may not always be explicitly aware of

apparatus tends to process information about its environment) but also some differences between populations in different educational contexts (e.g., Brewer 2008), suggesting that different cultural backgrounds channel students' understanding of scientific ideas in different directions.

Much the same will happen with scientists themselves. Professional scientists may be in a better position to recognise and overcome sources of 'bias', and science itself is set up to do just this, but if science is dominated by some cultural groups (e.g. mostly men or mostly those educated in a European/North American tradition) then there are likely to be consequences.

Worldviews

One notion which has come to be increasingly used in considering such issues is that of a person's 'worldview', that is, a set of 'assumptions held by individuals and cultures about the physical and social universe . . . [including] the purpose or meaning of life' (Koltko-Rivera 2006, pp. 309–310). These assumptions may be held implicitly, but concern fundamental commitments:

> Thus, worldview is about metaphysical levels *antecedent to* specific views that a person holds about natural phenomena, whether one calls those views commonsense theories, alternative frameworks, misconceptions, or valid science. A worldview is the set of fundamental non rational presuppositions on which these conceptions of reality are grounded
> (Cobern 1994, p. 6, italics added)

When Cobern refers to such assumptions as 'non-rational', this is not intended in a pejorative sense, but rather reflects their nature as starting points for coming to make sense of the world. Whilst metaphysics *as a topic of discussion* in its own right might be seen as the business of philosophers, fundamental intellectual commitments in terms of how we understand the nature of the world we experience are essential to sense making for all of us (see Fig. 1).

Arguably, there have to be some such commitments as starting points for making sense of the world that in themselves are not open to logical demonstration. The modern ('Western') scientific tradition has been built upon the two foundations of empirical evidence and rational thought, but the claims relating to these grounds have been subject to ongoing discussion by philosophers (Losee 1993). It is worth pausing to consider how one would try to persuade another individual of any scientific claim if they did not accept (a) that evidence from experience had any relevance to the true nature of the world and/or (b) that logic could be relied upon as the basis for sound thinking. Perhaps it seems unlikely that anyone should take such a stand, but if we accept the possibility for argument's sake, then finding grounds for accepting these starting points, *without actually drawing upon them to make the case*, might seem a rather forlorn project. Without an alternative foundation, the whole of science (and much else) could be considered somewhat tautologous.

An individual's worldview may in particular include religious beliefs that are basic assumptions about the nature of the world in which we live. Hodson comments that 'because a worldview includes fundamental beliefs about causality and about humanity's place in the world, it is fairly easy to see how it could be incompatible with the fundamental metaphysical underpinnings of science' (2009, p. 120). Thagard (2008, pp. 385–386) offers an example from considering the history of medicine, where 'popular concepts of life, mind, and disease are tightly intertwined: God created both life and mind, and can be called on to alleviate disease [and so] conceptual change can require not just rejection of a single theory in biology, psychology, and medicine, but rather replacement of a theological world-view by a scientific, mechanist one'. However, this need not always be the case of course. For example, Newton considered that the world was created by God and that activity in the world was primarily due to God's will. So although Newton investigated the nature of what we would now call physical forces, he understood those forces in terms of his own theological beliefs (Tamny 1979).

Worldview as a Part of Culture

The notion of worldviews was adopted in anthropology to describe 'the cognitive, existential aspects' of a culture, where a people's 'world-view is their picture of the way things, in sheer actuality are, their concept of nature, of self, of society. It contains their most comprehensive ideas of order' (Geertz 1957, pp. 421–422). As Matthews (2009, p. 707) points out, one aspect of worldview concerns ontology, 'ideas of what entities exist in the world—matter? spirits? minds? Angels?' So, for example, the Yupiaq people of Alaska view the world as being composed of five elements: earth, air, fire, water and spirit (Kawagley et al. 1998, p. 138), positioning spirit alongside (what would be considered in the scientific tradition) 'material' elements in a way that would seem quite incongruous from a modern scientific perspective.

Hewitt (2000, p. 111) notes that 'worldviews do not arise spontaneously' but are 'shaped in part by the cultural imprint of socialization'. He describes how Australian aboriginal peoples' worldview has developed over many generations in a difficult environment where 'survival depends on cooperation and coexistence with the forces of nature rather than expecting to manipulate and control them' (p. 112), a view rather at odds with the technical-scientific-industrial mentality dominating much of 'Western' culture. Similarly, in working with Kickapoo students in Alaska, Allen and Crawley (1998, p. 126) reported how the young people generally expressed 'a harmonious relationship with nature, recognizing kinship without seeking control'. From the perspective of students from this culture, it was not appropriate for animals to be kept caged in school laboratories and used in investigations. Part of the Kickapoo worldview was to consider (non-human) animals as 'brothers' to people (whereas from some other worldviews the more limited degree of kinship suggested by scientific models of common descent are considered unacceptable).

Worldview encompasses epistemological as well as ontological commitments, and the separation of formal canons of knowledge (e.g. science) as marked apart from what is learned through everyday experience (and indeed the institutions of formal schooling where such knowledge is often decontextualised from its application) makes little sense within the worldviews of many indigenous peoples. So for the Yupiaq, for example, 'their science is interspersed with art, storytelling, hunting, and craftsmanship' (Kawagley et al. 1998, p. 137) and 'Western methods of teaching science often run counter to the students' own cultural experiences' (p. 141).

School Science as a Representation of the Culture of Science

By contrast, a simplistic view of school subjects in Western education might see the curriculum in terms of partitioned portions of content knowledge to be 'delivered' by different subject teachers. There has long been a notion of curriculum as providing access to those aspects of a society's culture that are judged to be of importance and value to the young. From such a perspective science is a 'form of knowledge', that is, a 'complex way of understanding experience which [humanity] has achieved' (Hirst 1974, p. 38), and science lessons are not just about being told some science, but acquiring something more profound: 'the development of creative imagination, judgment, thinking, communicative skills etc., in ways that are peculiar to [that form of knowledge, so here science] as a way of understanding experience'. So part of the role of science education in a 'liberal' curriculum might be to provide access to a scientific way of thinking or a scientific attitude or perspective.

This might be seen to encompass, inter alia, thinking for oneself, questioning and not accepting the views of authority without supporting grounds, being sceptical and so forth, reflecting the famous motto of the London Royal Society to 'take no one's word for it'. Such an attitude might seem to be at odds with other values that could be encouraged in some cultural contexts, such as respecting elders, knowing and having to earn one's place and having faith in 'the Word'.

This might increasingly be the case as the nature of science education itself has shifted. There is increasingly a view that science for citizenship in the modern world encompasses a kind of scientific 'literacy' that goes beyond learning a corpus of presented examples of the products of science, to appreciating the processes of science (Millar and Osborne 1998). Learning about the nature of science is seen as more than learning about a bowdlerised notion of 'the' scientific method (Taber 2008). Increasingly, science classes have come to be seen as about enculturation into the practices and norms of science themselves (e.g., Roth and Bowen 1995). When science is seen in this way, it is increasingly clear that success in school science is likely in part to depend upon how readily a learner can recognise and adapt to the culture being represented in science lessons, something that will surely be influenced by the extent to which that culture fits or challenges her or his own worldview.

Border Crossing into School Science

The metaphor of border crossing has therefore been used to describe the process of entering into the culture of the science classroom. Aikenhead and Jegede (1999, p. 269) note that whilst barriers to border crossing may be most severe among students from developing countries who find 'that school science is like a foreign culture to them' due to 'fundamental differences between the culture of Western science and their indigenous cultures'. nonetheless 'many students in industrialized countries share this feeling of foreignness as well'.

Difficulties for science education are to be expected among 'students whose worldviews conflict with mainstream schooling and Eurocentric science' (Brandt and Kosko 2009, p. 398). So Carambo (2009) reports a study carried out in an urban setting in the USA that explored 'students' development of secondary discursive practices of the scientific community' through a project where youngsters designed and built model racing cars. It was intended that 'the analysis and redesign processes would create a field where students' primary discourse would reflect practices associated with scientific discourse' which would provide a 'border crossing' (p. 477). However, although students were engaged by the premise of building and racing their model cars, they did not adopt the hoped-for scientific approach to analysing the performance of different models as a means to look to improve their designs. Or as Carambo concluded, 'students failed to adopt secondary discursive practices as they refused to engage in the analysis and redesign of their model cars' (p. 477).

Carambo's study suggests that the mentality of science may be at odds with the thinking of many young people, *even* in an area where there would seem to be little sense of conflicting with fundamental personal values of the type that may be related to strong cultural beliefs. Religion may offer very strong commitments that are adopted into an individual's Worldview at a young age, often with strong support from family and the most respected members of the community. This potentially offers a basis for very significant barriers to science learning if science is perceived as in conflict with the learner's own worldview.

'Science and Religion' Is an Issue for Science Educators

This potential for students' worldviews to appear to clash with a scientific perspective on religious grounds has been recognised, for example, by Martin-Hansen:

> When we consider the way we teach science or how the general populous thinks science is conducted, not only are there very naïve views of nature of science concepts, but also different worldviews are coming into conflict. Science teachers are asked to help students understand the way science works, but some teachers as well as many of our students hold rigid theistic worldviews that threaten their understanding of science concepts.
>
> (Martin-Hansen 2008, p. 318)

Clearly different religions have different tenets and so inform worldviews in different ways, in turn leading to different degrees and points of potential contact with scientific principles and ideas. Interestingly, for example, developments in some areas of twentieth-century physics saw some scientists and commentators seeking to make sense of areas such as quantum theory (where mechanistic notions of causality and 'common-sense' thinking about how the world is structured seemed at odds with scientific evidence) by drawing upon religious and philosophical ideas from Eastern cultures (Capra 1983). This chapter will focus in particular on examples relating to Christianity and Islam, where there has been recent concern about the potential for student worldviews to lead to conflict with science teaching, especially regarding evolution (Hameed 2008; Long 2011; Reiss 2009).

Perspectives on 'Origins'

Martin-Hansen (2008, p. 318) gives the example of 'when a student says that they believe the earth is 6,000 years old [which] is usually due to a conflict between a theistic worldview and a naturalistic worldview'. Issues of origins may often be the contexts for explicit perceived conflicts in science classes, as, for example, in the perspective adopted in the comments of student Brent, reported by Roth and Alexander:

> When I hear you and other people talk about how the Earth was created, by referring to the theories of Big Bang and evolution, I say, well that is wrong. I believe that you are wrong and I am right—I am right because God has taught me so; and you are wrong because God did not bring you up that way, you are misinterpreting what the world actually is.
>
> (Roth and Alexander 1997, p. 142)

Where a student takes such a strong stance as Brent ('you are wrong and I am right'), there seems little scope for common ground between teacher and learner, to allow any kind of dispassionate exploration of ideas. Indeed, the very notion that one *should* seek to explore such matters in a dispassionate manner might itself be seen as an alien cultural norm within some communities.

The themes of the beginnings of 'the world' (a notion that itself may be incongruous from different perspectives, i.e. a vast universe including this planet among a myriad of others or humanity's earthly home in its almost incidental or supporting cosmic environment) and the origins of human beings are well recognised to be problematic topics in science lessons for some learners due to apparent clashes with their own worldviews. It does not help that evolution is counter-intuitive and that understanding natural selection as a 'simple' and yet powerful concept first requires the coordination of a number of different key principles (Taber 2009, pp. 287–288).

These principles, once understood, can be integrated into a coherent conceptual framework that allows one to make sense of a great deal of data about the natural world and which with regular use can come to provide the taken-for-granted basis for interpreting new information about life and living things. Such conceptual frameworks can be very powerful in channelling thinking. Science educators have noted how learners' alternative conceptual frameworks can hold a strong influence on their learning (Duit 1991), but scientists' conceptual frameworks can be just as influential in biasing perception and thinking. This may explain how a popular science book by an influential evolutionary biologist proclaims on its dust jacket that 'no one doubts that Darwin's theory of evolution by natural selection is correct' (Eldredge 1995). Brent certainly doubted Darwin's theory is correct, and he is by no means alone.

Hokayem and BouJaoude (2008, pp. 407–408) report the comments of a university student in Lebanon who begins by asserting he agrees with, and even finds obvious, one tenet of evolution, but then immediately goes on to reject evolution as the origin of species: 'survival of the fittest, I accept 100 %, and I don't think its [sic] such an achievement when Mr. Darwin discovered it, but transitions between monkeys and humans and others between reptiles and birds, that is not very credible'. The student supports this position with various arguments, such as:

- [natural selection as the origin of species] has nothing to do with science, there's no research or something they work in the lab…[rather] its like an artist created a picture.
- If organisms did not exist today in essentially the same way they existed in the past, it doesn't mean they evolved from each other, God created them like that…there's no evidence to say they came from each other, but [just] after each other.
- They haven't scanned the whole earth to see if what they're talking about is true, it's fragmented, they find one thing they make up a theory on it, they find something else, they change their theory…they're basing it on their imagination nothing else…they need to show me transitional species that are really found, not just they found a human being with a bigger jaw, it's normal for the jaw to be bigger because he used to eat other kind of food, if this is what they mean by evolution then fine but not the evolution from monkeys to human, this is another idea.

I have drawn upon Hokayem and BouJaoude's original published data in some detail here, because I find something very interesting about this student's position. This student's arguments remind me of scientists interviewed for a sociological study reported by Gilbert and Mulkay (1984), where they found their scientist interviewees operated with two interpretive repertoires when asked about differences of opinion within science. Put simply, scientists tended to present their own view through an *empiricist* repertoire that suggests it is a neutral view based upon what the evidence shows. However, the different views of some of their colleagues would be explained through a *contingent* repertoire that emphasised subjective aspects of how other scientists' views were influenced by factors outside the true interpretation of clear empirical evidence. (This brief outline cannot do justice to the work reported in Gilbert and Mulkay 1984, to whom the reader is referred for a fuller account.)

Hokayem and BouJaoude's informant here seems to present his views in a similar way, in the sense of suggesting that scientific knowledge needs to be empirically grounded ('there's no research or something they work in the lab'; 'there's no evidence to say they came from each other'; 'They haven't scanned the whole earth to see if what they're talking about is true'; 'they need to show me transitional species that are really found') and that the scientists supporting the views he does *not* accept are basing their view on non-empirical contingent, and so subjective, factors ('has nothing to do with science ... [rather] its like an artist created a picture'; 'they find one thing they make up a theory on it, they find something else, they change their theory ... they're basing it on their imagination nothing else'). Presumably this reflects a widespread aspect of human thinking, whether scientist, student or science teacher; my views are rational and well grounded, whilst yours are arbitrary and contingent on chance factors. Perhaps such a bias in human cognition (Nickerson 1998) has had value for survival during the evolution of our species, but it does not help us come to see the merits of a disparate viewpoint.

Science proceeds through the iterative interaction between evidence and imagination (Taber 2011). Sensory data only becomes perception by being interpreted through an existing cognitive apparatus, and those perceptions only become evidence within the context of some existing conceptual framework, i.e. data is always 'theory-laden' (Kuhn 1996). Imagination is always involved in devising a theoretical scheme within which evidence is interpreted and coordinated – in setting up hypotheses and devising tests for them – but *in retrospect* natural science formally focuses on the context of justification, not the context of discovery (Medawar 1963/1990). That is, the modern scientific literature is based in the empirical repertoire which is used to argue how we know, not the contingent repertoire which can tell us who had the idea, and whether it derived from a serendipitous accident in the lab, a chance conversation at a conference, a dream or a flash of inspiration 'popping into' consciousness whilst bathing. So although both imagination and evidence have essential roles to play in the processes of science (Taber 2011), we can understand why it may be rhetorically convenient to emphasise how *our* views are based on evidence, whereas *their* different views derive from their imagination.

Is There a Scientific Worldview?

If individuals are considered to have a set of assumptions about the world making up a 'Worldview' which can sometimes conflict with the science presented in the classroom, then this leads to the question of whether science itself reflects a worldview which would suggest that full admission to the scientific community is only possible to individuals who adopt that worldview. The answer offered here is a clear 'no': that what we might term the scientific perspective or the scientific attitude does involve some features that could be considered constituent of a worldview, but is not a fully encompassing worldview in its own right; 'religions and science answer different questions about the world. Whether there is a purpose to the universe or a purpose for human existence are not questions for science' (National Academy of Sciences Working Group on Teaching Evolution 1998, p. 58).

That is, the nature of science, as currently understood, is informed by a common set of shared metaphysical assumptions about the nature of the material world and how we can come to knowledge of it (but not whether it reflects a purpose). These fundamental assumptions are therefore at the 'level' of a worldview and so may be considered potential components or facets of a worldview, but do not in themselves constitute a worldview. There is therefore scope for a range of different worldviews that may encompass these assumptions.

Consensus Scientific Values

There is unlikely to be a simple consensus on the precise nature of such a list (or indeed on who exactly might be considered as a member of the scientific community). Given this proviso, I would suggest that the following candidates for metaphysical commitments to underpin science (as it is generally currently understood within the scientific community):

O1: There exists a (natural) physical world.
O2: The physical world has a degree of permanence and underlying order.
E3: Experience offers a meaningful guide to the nature of that world.
E4: It is possible to construct useful 'knowledge' of the world.
E5: It is possible to develop knowledge of the world, which is objective in the sense that it is independent of the standpoint of the particular observer.

I suspect it is very likely that any reader will find things to quibble with on this list, especially in the choice of phrasing. The term 'knowledge' here certainly will not match the philosophical notion of 'justified, true belief' (Matthews 2002). Moreover, scientists will show a range of views on the extent to which the natural world is 'knowable' to human minds: (E3, E4) varying from those who are strongly realist to those who take a much more instrumentalist approach, i.e. whether our theories and models are good approximations to reality or best just considered useful tools that often do a good enough job for us (Taber 2010). Scientists will also take a range of views on quite how much the objectivity of science (E5) is best seen as an ideal and aspiration (Springer 2010), rather than something that is regularly achieved.

Whilst the diversity in such matters might be quite significant (and probably in part varies from field to field within the sciences), it would seem these assumptions, or at least something quite similar to them, are essential for what we *currently* understand as science. There would seem little point in doing science if one thought the world was an illusion, or that it was continually and fundamentally changing its nature in unpredictable ways, or that it was completely beyond human comprehension, or that at the most basic level, it was really different for different observers.

My argument would be that:

(a) These ontological (O1–2) and epistemological (E3–5) commitments, or a set quite similar, are necessary to what we understand as science.
(b) Moreover, that these commitments are also sufficient as a starting point for doing science.

That is, that what is excluded here is *not essential* to science as currently (but see below) understood. What is excluded includes both greater specification of the statements above and what is not mentioned. So, for example, O2 refers to the physical world having *a degree of* permanence and underlying order. This does not specify total permanence and order (although *some* individual scientists might certainly assume something at least approaching that), but rather implies enough permanence and order to make systematic observation meaningful and worthwhile.

Not referred to above is any sense of extra-scientific values. So, for example, *many* scientists might share an axiological commitment along the lines:

A6: Scientific work should be carried out for the benefit of all the peoples of the world and taking care not to damage the ecosystem.

We might like to see all science informed by such a principle, but it is *not* part of science as currently understood, and there has been a great deal of science that has been motivated quite differently. Perhaps, in some more perfect future, such a commitment would be shared by all scientists, but it would still be 'prior' to science itself, in the sense that like the ontological and epistemological assumptions O1–E5, it informs science rather than making up part of its content.

Relating Worldviews to Science

Space limitations here do not allow a detailed exposition of the idea that different worldviews may be consistent with science, but some exemplification is possible. I will here simply illustrate the general point with brief consideration of a small range of examples.

Natural Philosophy and Belief in God

Many of those considered as leaders of the first generations of scientists (in a modern sense), or natural philosophers as they would have seen themselves at the time, would have seen no problem with committing to something like my list of metaphysical commitments for science, without seeing any conflict with deeply held religious beliefs. Indeed, it has long been argued that religion was among the factors contributing to the social nexus which was 'favourable to scientific interests' in England in the seventeenth century (Merton 1938). These early scientists commonly shared a worldview that included commitments to God as the creator of the universe. Indeed to suggest that such luminaries did *not see their scientific activities in conflict* with their faith would not do justice to the motivation of someone like Isaac Newton. It seems clear Newton was a devout believer, who saw his science as finding out *how* God worked through His creation, for 'if you believe, as Newton did, that God has created our world and all of its operations, then you cannot invoke God to function as an explanation for the cause of any particular effect. You must assume that God provided a natural cause for that effect, and it is the task of the natural philosopher to discover it' (Grant 2000, p. 290).

Newton's work was clearly influenced by metaphysical commitments. He recognised that the spectrum he obtained by passing white light through a prism did not give distinct colours with sharp divisions, but rather included 'an indefinite variety of intermediate graduations' (quoted in Wörne 2008, p. 19). Yet he revised his early view that the spectrum should be described in terms of five colours to the now canonical seven. Scholarship suggests that in this Newton was strongly influenced by an analogy with the musical scale that derived from his commitment to certain ideas we might now describe as numerology (Pesic 2006), that is, metaphysical commitments that would now be considered external to science. How such a cognitive 'bias' compares to contemporary physicists expecting to find symmetries in nature is an interesting theme, but arguably this is an example of another extra-scientific commitment that was part of Newton's worldview, but is *not* a core scientific value. This might be considered simply part of the variety encompassed by the less specific commitment that the physical world has a degree of permanence and underlying order (O2).

It is suggested here that it is useful to consider the views of an individual, such as Newton in terms of his metaphysical and scientific commitments, and how these match to the scientific consensus (which of course is historically labile). Clearly many of Newton's scientific ideas are still influential and accepted as useful within science today. Some of his work, however, such as his alchemical ideas, would not today be considered as scientifically acceptable, and in such cases it is easy to suggest how he was led to misinterpret nature due to the influence of metaphysical ideas external to science (Dobbs 1982), but it is just as much the case that aspects of Newton's thinking that are now established as canonical parts of science and the school science curriculum were also influenced by his metaphysics. This is reflected schematically in Fig. 2, which includes some examples of ideas associated with Newton.

In this regard, although Newton's cosmology is sometimes described as a 'clockwork' universe, implying that God had set it up (and metaphorically wound it up) and left it to play out, Newton seemed to consider that it was necessary for God to occasionally intervene to make fine adjustments (Cooper 1980); even

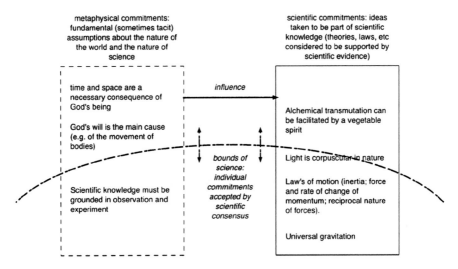

Fig. 2 An individual (such as Newton)'s thinking is informed by metaphysical commitments and includes various ideas about the material world, which may fit to the current scientific consensus to different degrees

the Omni-powerful was apparently unable to set up a celestial mechanics that did not need some occasional tinkering! It may seem arrogant of Newton (and he has been accused of that) to assume that if he could not calculate stable orbits for the planets, then God could not set them up: yet perhaps if one holds an epistemological commitment that the world should be comprehendible because that is God's intention, then this seems less arrogant. The point, however, is that even Newton with his establishment of universal principles and mathematically described laws did not assume such a high inherent order to the world (re-O2) to exclude supernatural intervention in natural laws.

Scientific Creationism

That many of the early pioneers of science were theists, who considered God created the World, is a point that was commonly made by Henry Morris, who was a leading advocate of young earth creationism in the USA. Morris considered himself to work within science (being trained as an engineer, and having taught at various universities), and I suspect would have no problems ascribing to my list of core scientific commitments above. However, his worldview also included not only a commitment to a belief in a creator God but also a commitment to the mode of creation being *as* described in Christian scripture. Now the Christian Bible includes two accounts of God creating the World in 6 days, by a series of discrete acts of special creation for each kind of living thing. The Bible also includes a good deal of genealogical detail, which if assumed to be complete and accurate, allows scholars to date the life of Christ relative to the creation, leading to the conclusion that the creation of world was a historical event that occurred at most about 10,000 years ago. Within this perspective, all living things alive today are the descendents of, and are of the same kinds as, those created directly by God. From this perspective, it is generally accepted that the original stock has given rise to variations, but only within the basic types of living thing created by God (just as suggested by Hokayem & BouJaoude's interviewee, reported above).

As the geological sciences suggest the world is billions of years old, and astronomy that the universe is further billions of years older than the earth, and as modern biology considers all life on earth to have evolved from a common ancestral stock that was single-celled, there are some clear contradictions between the currently accepted conceptual frameworks of science and some of the metaphysical assumptions incorporated in the worldview of Morris and others who share his stance. So Morris (2000, p. 18) describes evolution as 'completely anti-biblical and even anti-theistic'.

Yet, Morris, just as Hokayem and BouJaoude's (2008, pp. 407–408) informant above, is able to consider that his worldview is quite consistent with science, as he is able to support his view, with *his* interpretation of the available scientific evidence:

> creationists do not reject the actual, factual data of any of these sciences. They are all legitimate sciences (the founding fathers of which, incidentally, were almost all creationists!), and they have contributed immeasurably to our knowledge about God's created world and our ability to use its resources for man's benefit. All of the real [sic] data of these sciences can be understood much better in the context of creationism.
>
> (Morris 2000, p. 32)

His rhetoric again reflects the work of Gilbert and Mulkay (1984), i.e. that his own position (see Fig. 3) is empirically supported, whilst it is others who are misinterpreting the available data:

> The fact is, however, that although the natural sciences are commonly interpreted in an evolutionary framework, no one has ever observed real [sic] evolution to take place, not even in any of the life sciences, let alone the earth sciences or the physical sources. True science is supposed to be observable, measurable, and repeatable. Evolution, however, even if it were true, is too slow to observe or measure and has consisted of unique, non-repeatable events of the past. It is therefore outside the scope of genuine [sic] science and has certainly not been proven by science.
>
> (Morris 2000, p. 23)

So according to Morris (2000):

- The most significant feature about the fossil record is the utter absence of any true [sic] evolutionary transitional forms (p. 27).
- ...the real scientific evidence in both domains of science [i.e. the earth sciences and the life sciences] is firmly opposed to evolution (p. 28).
- As long as people have been observing the stars, no one has ever seen a star evolve from anything (p. 30).
- ...evolution is quite false and is utterly devoid of any scientific evidence... (p. 91).

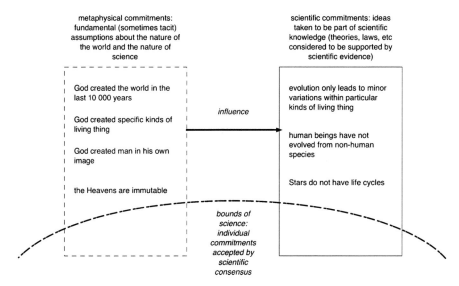

Fig. 3 Young earth creationism: Empirical evidence is interpreted differently when thinking is channelled by prior commitments to how the world must necessarily be

Morris refers to the empirical support for the second law of thermodynamics as excluding the possibility of evolution (p. 31) – a rather surprising misconception of the law from an engineer – and refers to claims of 'the great age and evolution of the cosmos' as 'arbitrary' (p. 124). Nothing in Morris's claims would put him outside of science in terms of his espoused commitments to the fundamental ontological and epistemological commitments underpinning science as listed above (O1–2, E3–5), yet he was able to *interpret* scientific evidence generally considered highly stacked in favour of evolution, as consistent with, and indeed supportive of, the antievolutionary commitments in his worldview. That he could interpret scientific evidence in this way must seem just as puzzling and bizarre to evolutionary biologists such as Niles Eldredge, as Eldredge's publisher's claim that 'no one doubts that Darwin's theory of evolution by natural selection is correct' would have seemed to Morris.

Scientific Materialism

Where fundamentalist Christians such as Morris are able to see science as fitting with a theistic worldview, some scientists who are committed atheists have not only managed to see science as consistent with their own worldview but have argued that science itself is inherently atheistic. For these individuals, the natural world is all there is and is not only open to scientific investigation but ultimately only capable of being meaningfully understood in scientific terms. For such extreme materialists,

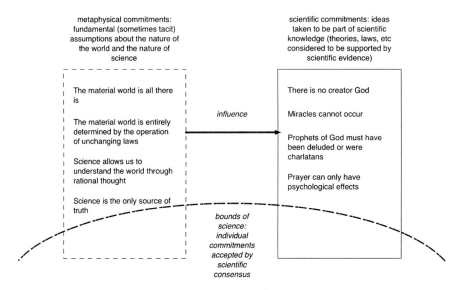

Fig. 4 Materialistic metaphysical assumptions lead to interpretations of phenomena for which there is no empirical evidence or viable mechanism being assumed to be imaginary

i.e. 'people who believe that because there is no evidence of God in nature, God must play no role in the development of the cosmos or of life on earth', their own atheistic worldview is considered as a (the) scientific worldview, leading to 'the belief that science and religion must inevitably conflict' (Brickhouse et al. 2000, p. 349).

In view of the model presented in this chapter, these individuals argue for the adoption among the fundamental assumptions shared by scientists, of additional (materialistic, atheistic, scientistic) commitments from their own worldviews (see Fig. 4), and 'hope science, beyond being a measure, can replace religion as a worldview and a touchstone' (Cray et al. 2006). So, for example, whereas many scientists would not exclude miracles from occurring (because their worldviews encompass a God capable of acting in the World), Richard Dawkins would argue that 'any belief in miracles is flat contradictory not just to the facts of science but to *the spirit of* science' (Cray et al. 2006, emphasis added).

That is, that (i) if one assumes that the material realm is all there is, and therefore all there is will be open to scientific investigation and explanation, then these (metaphysical) assumptions exclude the possibility of miracles; and (ii) if one considers such commitments as necessary for and inherent in science, then it follows that *science itself excludes* the possibility of miracles. This is of course tautological, because the metaphysical sets bounds on the scientific interpretations possible, so that what is scientifically accepted *necessarily* fits with those original assumptions: just as Morris assumed evolution could not be the case because he saw it contrary to scripture and then found that all the evidence seemed to him to fit his prior assumption.

Natural Theology

In both the cases of Morris and Dawkins, it is possible for *a form of science* to fit their worldviews. However, in both cases they understand science in terms of fundamental commitments, some of which fall well outside the common ground of the current scientific community (Figs. 3 and 4). To the extent that science is part of culture, it can change. At one time the scientific consensus would have reflected theologically based metaphysical assumptions that are now no longer part of the common commitments of science (i.e. that science is the study of God's work). Indeed there developed a whole tradition of 'natural theology' where the 'book of nature' was to be 'read' and considered to offer insights into God's work and his nature (Grumett 2009; Sagan 1985/2006), alongside but independent from the revealed Word in the book of scripture. From this perspective, it was easy to adopt a commitment that 'the physical world has a degree of permanence and underlying order' (O2), because it was ordered by God, and a commitment that 'experience offers a meaningful guide to the nature of that world' (E3) was more than just an 'article of faith' in science, but actually derived support from a worldview commitment to God having set up the world with humans in mind, humans who would come to know Him and appreciate the glory of His work, perhaps something along the lines:

> E7: The World can be understood by man, because God has created man in the image of God, to appreciate His works

However, this is no longer a shared commitment of the scientific community (whilst being retained as a commitment by some members of that community). Other such shifts may occur in the future. So, for example, if the scientific community were over time to come to adopt scientistic metaphysical commitments as part of the common core of fundamental assumptions underpinning science, then it would in principle be possible that science may become secularised and Dawkin's prescription for what science should be would indeed have become a descriptive account.

Islam and Science

However, such a shift to adopting atheistic and materialist commitments as common values in science is certainly not imminent. Indeed, in much of the Islamic world, the Worldview of Muslims includes metaphysical commitments to the existence of God at work in the world whilst sharing commitments necessary for empirical science:

> From an Islamic perspective, science is the study of the material processes and forces of the natural world. Science is not about belief; it is about how things work. Science is about the exploration of natural causes to explain natural phenomena. Science is empirical, which

means that questions of truth are established through experimenting and testing. There are
no absolutes in science; all issues are open to retesting and reconsideration. In contrast,
religion is about belief, meaning, and purpose. Religious truths are evaluated by an appeal
to authority, by contextualization in history, by their philosophical coherence.

(Mansour 2009, p. 109)

This is in contrast to the materialist position that would not grant epistemological
power to authority, history or philosophy (or indeed anything other than objective,
reproducible empirical evidence) and so would consider religious truth as something
of an oxymoron. Islamic scientists traditionally, as in Christian natural theology,
saw nature as reflecting God and the study of nature as a way of coming to know
God better. This type of thinking is still reflected in science curricula in some Arab
states. So in Jordan, one goal of the science curriculum is to enable students to
better understand the universe, as this should strengthen their faith in its creator
(Dagher 2009). Similarly, there are explicit references to Islam in the science
curriculum in Oman. For example, the biology curriculum aims to help students
strengthen their Islamic beliefs through learning about the cell (Ambusaidi and Al-
Shuaili 2009). From the prior assumption that God created the world, the cell as
a building block of all livings things becomes interpreted as an aspect of God's
way of creating complex organisms. So science education in these states reflects
shared metaphysical commitments of the culture, which become adopted as part of
the metaphysical underpinning of science itself. In terms of the common core of
assumptions shared by the *international* scientific community, these commitments
to the World as God's handiwork are just as much local adjuncts as the materialist's
commitment to excluding such notions.

Evolutionary Creationism

Something of the mentality of natural theology led Charles Darwin to ask what
kind of a God would have set up the wasteful world of excessive suffering that his
scientific work (as well as his personal experience of the loss of loved children to
disease) seemed to imply, a question that led *him* to reject a personal God that loved
and cared about each of his subjects (Phipps 2002). A caricature of the science and
religion debate often suggests that Darwin's 'discovery' of evolution challenged
the established Christian Church's model of creation and the origins of man, by
providing scientific evidence that Biblical accounts were false. This is far from an
accurate account, both as evolution was not a new idea with Darwin and because
there had long been a tradition in Christian thinking that where scientific evidence
seemed to contradict scripture, then the interpretation of scripture needed to be
revisited – a point raised, for example, by Galileo in his sometimes troubled dealings
with the Church (Johnston 1993). The Darwin-Wallace notion of natural selection
(Darwin and Wallace 1858) certainly did not rule out a creator God, although it did
for Darwin and others raise issues of what kind of God would go about His work in
such a way.

However, while Darwin's faith was challenged by his scientific discoveries, a great many religious people (in accord with the natural theology tradition) accepted that science had made progress finding out more about how God had gone about his work. From the perspective of many believers, i.e. those people who had a theistic worldview, evidence that strongly implied that certain traditional ideas (such as a 6 day creation of the world; the special creation of distinct species; a worldwide flood leaving four couples to repopulate the world) were not accurate historical accounts did not count as evidence against a creator God – only evidence that scripture needed to be understood figuratively as offering narratives with moral truth rather than scientific truth (Alexander 2008).

From such a perspective, evolution can be seen as part of the mechanism of God's ongoing creation (after all, although according to scripture the world itself was created ex nihilo, Genesis suggests that Adam and Eve both derived from materials that God had already created as part of the World). Today, as in the more immediate aftermath of Darwin's publication of his *Origin of Species* (1859/1968) and *Descent of Man* (1871/2006), there are a great many scientists able to commit to the core scientific values (e.g. as represented in my list O1–E5 above) without finding any conflict with their theistic worldview. Indeed it has been argued that 'a more systematic integration can occur if both science and religion contribute to a coherent worldview elaborated in a comprehensive metaphysics' (Barbour 2000, p. 34).

Coda

So here I have just sketched a few of the positions taken by people with different worldviews, who understand science in accordance with their own metaphysical commitments. Slezak (2008) has argued that 'the Gospels only support Christianity if you already believe it. If that's the best that philosophers can offer, it's hard to see how Christian theism could provide a 'metaphysical' alternative to the naturalism of our best science'. Yet any deeply held metaphysical commitments (theism, atheism, young earth creationism, etc.) will necessarily inform the interpretation of empirical evidence to construct conceptual frameworks about the world which are consistent with, and so can readily seem to support, those particular metaphysical underpinnings.

We each live our lives as a personal version of a scientific research programme (Taber 2009, pp. 92–110) in the sense of Lakatos (1970), building up and revising our model of the world in view of ongoing experience (von Glasersfeld 1989) and sacrificing auxiliary ideas in order to maintain conceptual frameworks consistent with our 'hard-core' assumptions. This is why Lakatos referred to these auxiliary ideas as making up a 'protective belt'. For most theists in the Christian tradition, the place of the earth at the centre of the universe, the recent creation, the global nature of Noah's flood, special creation of species and so forth are (in Lakatosian terms) 'refutable variants' that have been allowed to fall to maintain congruence between empirical evidence and the core notions of theistic creation. What seems from outside that programme as desperate patching up of a faith position makes perfect sense from within the programme as indications of its progressive nature.

We each start from a metaphysical position that will seem to be supported by empirical evidence (because of its role in influencing our interpretation of that evidence). Such commitments will inform our science and also our teaching of science. Hodson suggests that 'border crossing' into science classrooms

> is inhibited not so much by the cognitive demand of the learning task as by the discomfort caused by some of the distinctive features of science, features that are often exaggerated and distorted by school science curricula into a scientistic cocktail of naïve realism, blissful empiricism, credulous experimentation, excessive rationalism and blind idealism.
>
> (Hodson 2009, p. 121)

That is, inappropriate representations of the nature of science put up barriers for some students. This is surely going to be the case when science is presented as, for example, inherently about studying God's work, or as inherently excluding the possibility of God being at work in the world, when such assumptions (themselves external to science itself) are contrary to the strong personal convictions, and community commitments, of learners.

Ways Forward?

Cobern has argued that

> it is important for science educators to understand the fundamental, culturally based beliefs about the world that students bring to class, and how these beliefs are supported by students' cultures; because, science education is successful only to the extent that science can find a niche in the cognitive and socio-cultural milieu of students.
>
> (Cobern 1994, p. 22)

Of course, this does not mean compromising on scientific values. Logical analysis of empirical evidence remains at the core of science. But as science educators we must be very careful to ensure that the nature of science we present reflects the shared commitments of the scientific community and is not an amalgam including extra-scientific features imported from our own individual worldviews. As Cobern suggests,

> teachers and curriculum developers need to examine and then come to understand the fundamental, culturally based beliefs about the world that they bring to class through teaching and the curriculum. They likely will find that some of these fundamental beliefs are neither necessary for science nor for the effective teaching of science.
>
> (Cobern 1994, p. 22)

This will allow us to be clear with learners about which metaphysical commitments are inherent in science, and those which are not, but which may be adopted by some individual scientists (Hansson and Redfors 2007). This is especially important given that research suggests that many school-age students adopt 'a stereotyped image of scientists [that makes] no distinction between their personal and professional concerns' (Driver et al. 1996, p. 84).

Martin-Hansen (2008, p. 318) suggests that 'by involving students in explicit nature of science activities which illustrate the boundaries of science, they can begin to see that an acceptance of a scientific theory does not eliminate the existence of a supernatural entity'. I would add that such activities should equally make it clear that the acceptance of a scientific theory should not follow from a belief in the existence of a supernatural entity. Science teachers are generally not in a position to offer informed instruction on religion(s) – but an exploration of the possible metaphysical bases of science, and how these may be congruent or not with different worldviews, could be considered as a key feature of the nature of science.

This is of course going to be a sensitive matter, and rather than directly engage students in consideration of their own worldviews (which may in part be tacit and in any case are by definition going to be beyond question), an alternative may be to consider historical cases. These should of course be both inclusive (e.g. not just male European Christians) and selected to be linked to curriculum topics such that learners can realistically be expected to understand enough of the science to be able to engage with them.

One framework that might be suitable for this is that drawn upon earlier in this chapter, which distinguishes metaphysical ('background') assumptions from scientific ideas and considers:

- Which background assumptions that the scientist brought to bear would (and would not) be considered as agreed scientific values by the scientific community today
- Which of the scientific ideas the scientists adopted are still considered sound today and which would no longer be considered supported by the available evidence

This will illustrate both how background assumptions can lead to conclusions we would not accept today and how the same ideas can sometimes be supported from very different starting points.

Whether such an approach will prove helpful is an empirical question. Research would be needed both to identify teaching schemes and resources that can be effective at helping students tease out scientists' metaphysical assumptions from their scientific ideas – including the development of teaching models that are accessible to school-age learners whilst offering intellectually valid simplifications (Taber 2008) – and then to determine whether time spent exploring such ideas helps students 'cross the borders' when material met in science classes is potentially strange from the perspective of, or indeed antithetical to, their own worldviews.

This is of course only one outline idea for tackling this issue. However, if as science educators we are not able to disentangle scientific from extra-scientific commitments when we present science to learners, we will both be offering a biased image of the nature of science and risk setting up uninviting border controls that make visits to the scientific landscape seem even less enticing to many learners.

Acknowledgement The author would like to acknowledge useful discussions on the issues considered in this chapter with colleagues working on the Learning about Science and Religion project, in particular Dr Berry Billinglsey (University of Reading).

References

Aikenhead, G. S. (1996). Science education: Border crossing into the sub-culture of science. *Studies in Science Education, 27*, 1–52.

Aikenhead, G. S., & Jegede, O. J. (1999). Cross-cultural science education: A cognitive explanation of a cultural phenomenon. *Journal of Research in Science Teaching, 36*(3), 269–287.

Alexander, D. R. (2008). *Creation or evolution: Do we have to choose?* Oxford: Monarch Books.

Allen, N. J., & Crawley, F. E. (1998). Voices from the bridge: Worldview conflicts of Kickapoo students of science. *Journal of Research in Science Teaching, 35*(2), 111–132.

Alsop, S., & Bowen, M. G. (2009). Inquiry science as a language of possibility in troubled times. In W.-M. Roth & K. Tobin (Eds.), *Handbook of research in North America* (pp. 49–60). Rotterdam: Sense Publishers.

Ambusaidi, A., & Al-Shuaili, A. (2009). Science education development in the Sultanate of Oman. In S. BouJaoude & Z. R. Dagher (Eds.), *Arab States* (Vol. 3, pp. 205–219). Rotterdam: Sense Publishers.

Antolin, M. F., & Herbers, J. M. (2001). Evolution's struggle for existence in America's public schools. *Evolution, 55*(12), 2379–2388.

Barbour, I. G. (2000). *When science meets religion: Enemies, strangers or partners?* San Francisco, CA: HarperCollins.

Bentley, D., & Watts, D. M. (1987). Courting the positive virtues: A case for feminist science. In A. Kelly (Ed.), *Science for girls?* (pp. 89–98). Milton Keynes: Open University Press.

Berry, R. J. (Ed.). (2009). *Real scientists real faith.* Oxford: Monarch Books.

Bhaskar, R. (1975/2008). *A realist theory of science.* Abingdon: Routledge.

Brandt, C. B., & Kosko, K. (2009). The power of the earth is a circle: Indigenous science education in North America. In W.-M. Roth & K. Tobin (Eds.), *Handbook of research in North America* (Vol. 1, pp. 389–407). Rotterdam: Sense Publishers.

Brewer, W. F. (2008). Naïve theories of observational astronomy: Review, analysis, and theoretical implications. In S. Vosniadou (Ed.), *International handbook of research on conceptual change* (pp. 155–204). New York: Routledge.

Brickhouse, N. W., Dagher, Z. R., Letts, W. J., & Shipman, H. L. (2000). Diversity of students' views about evidence, theory, and the interface between science and religion in an astronomy course. *Journal of Research in Science Teaching, 37*(4), 340–362.

Capra, F. (1983). *The Tao of Physics: An exploration of the parallels between modern physics and eastern mysticism* (Revised edition ed.). London: Fontana.

Carambo, C. (2009). Evolution of an urban research project: The Philadelphia project. In W.-M. Roth & K. Tobin (Eds.), *Handbook of research in North America* (Vol. 1, pp. 473–489). Rotterdam: Sense Publishers.

Claxton, G. (1993). Minitheories: A preliminary model for learning science. In P. J. Black & A. M. Lucas (Eds.), *Children's informal ideas in science* (pp. 45–61). London: Routledge.

Cobern, W. W. (1994). *Worldview theory and conceptual change in science education.* Paper presented at the National Association for Research in Science Teaching. Anaheim, CA.

Cooper, L. N. (1980). Source and limits of human intellect. *Daedalus, 109*(2), 1–17.

Cray, D., Dawkins, R., & Collins, F. (2006, November 5). God vs. Science. *Time.* Retrieved from http://www.time.com/time/printout/0,8816,1555132,00.html

Dagher, Z. R. (2009). Epistemology of science in curriculum standards of four Arab countries. In S. BouJaoude & Z. R. Dagher (Eds.), *Arab States* (Vol. 3, pp. 41–60). Rotterdam: Sense Publishers.

Darwin, C. (1859/1968). *The origin of species by means of natural selection, or the preservation of favoured races in the struggle for life*. Harmondsworth: Penguin.

Darwin, C. (1871/2006). The descent of man, and selection in relation to sex. In E. O. Wilson (Ed.), *From so simple a beginning: The four great books of Charles Darwin* (pp. 767–1248). New York: W W Norton & Company.

Darwin, C., & Wallace, A. (1858). On the tendency of species to form varieties; and on the perpetuation of varieties and species by natural means of selection. *Proceedings of the Linnean Society, 3*, 45–62.

Dobbs, B. J. T. (1982). Newton's alchemy and his theory of matter. *Isis, 73*(4), 511–528.

Driver, R., Leach, J., Millar, R., & Scott, P. (1996). *Young people's images of science*. Buckingham: Open University Press.

Duit, R. (1991). Students' conceptual frameworks: Consequences for learning science. In S. M. Glynn, R. H. Yeany, & B. K. Britton (Eds.), *The psychology of learning science* (pp. 65–85). Hillsdale: Lawrence Erlbaum Associates.

Eldredge, N. (1995). *Reinventing Darwin: The great evolutionary debate*. Lonon: Weidenfeld and Nicolson.

Francis, L. J., Gibson, H. M., & Fulljames, P. (1990). Attitude towards Christianity, creationism, scientism and interest in science among 11–15 year olds. *British Journal of Religious Education, 13*(1), 4–17.

Fulljames, P., Gibson, H. M., & Francis, L. J. (1991). Creationism, scientism, christianity and science: A study in adolescent attitudes. *British Educational Research Journal, 17*(2), hbox171–190.

Geertz, C. (1957). Ethos, world-view and the analysis of sacred symbols. *The Antioch Review, 17*(4), 421–437.

Geertz, C. (1973/2000). The impact of the concept of culture on the concept of man. In *The interpretation of cultures: Selected essays* (pp. 33–54). New York: Basic Books.

Gilbert, G. N., & Mulkay, M. (1984). *Opening Pandora's box: A sociological analysis of scientists' discourse*. Cambridge: Cambridge University Press.

Grant, E. (2000). God and natural philosophy: The late middle ages and Sir Isaac Newton. *Early Science and Medicine, 5*(3), 279–298.

Grumett, D. (2009). Naturla theology after Darwin: Contemplating the vortex. In M. S. Northcott & R. J. Berry (Eds.), *Theology after Darwin* (pp. 155–170). Milton Keynes: Paternoster.

Hameed, S. (2008). Bracing for Islamic Creationism. *Science, 322*(5908), 637–1638.

Hansson, L., & Redfors, A. (2007). Physics and the possibility of a religious view of the universe: Swedish upper secondary students' views. *Science & Education, 16*(3–5), 461–478.

Harding, S. (1994). Is science multicultural? Challenges, resources, opportunities, uncertainties. *Configurations, 2*(2), 301–330.

Hewitt, D. (2000). A clash of worldviews: Experiences from teaching aboriginal students. *Theory Into Practice, 39*(2), 111–117.

Hirst, P. H. (1974). Liberal education and the nature of knowledge. In P. H. Hirst (Ed.), *Knowledge and the curriculum: A collections of philosophical papers* (pp. 30–53). London: Routledge & Kegan Paul.

Hodson, D. (2009). *Teaching and learning about science: Language, theories, methods, history, traditions and values*. Rotterdam: Sense Publishers.

Hokayem, H., & BouJaoude, S. (2008). College students' perceptions of the theory of evolution. *Journal of Research in Science Teaching, 45*(4), 395–419.

Johnston, G. S. (1993). The Galileo Affair. *Lay Witness Magazine*. Retrieved from https://www.scepterpublishers.org/product/samples/9716.pdf

Kawagley, A. O., Norris-Tull, D., & Norris-Tull, R. A. (1998). The indigenous worldview of Yupiaq culture: Its scientific nature and relevance to the practice and teaching of science. *Journal of Research in Science Teaching, 35*(2), 133–144.

Koltko-Rivera, M. E. (2006). Rediscovering the later version of Maslow's hierarchy of needs: Self-transcendence and opportunities for theory, research, and unification. *Review of General Psychology, 10*(4), 302–317.

Kuhn, T. S. (1996). *The structure of scientific revolutions* (3rd ed.). Chicago: University of Chicago.

Lakatos, I. (1970). Falsification and the methodology of scientific research programmes. In I. Lakatos & A. Musgrove (Eds.), *Criticism and the growth of knowledge* (pp. 91–196). Cambridge: Cambridge University Press.

Leach, J., & Scott, P. (2002). Designing and evaluating science teaching sequences: An approach drawing upon the concept of learning demand and a social constructivist perspective on learning. *Studies in Science Education, 38*, 115–142.

Lemke, J. L. (1990). *Talking science: Language, learning, and values*. Norwood: Ablex Publishing Corporation.

Long, D. E. (2011). *Evolution and religion in American Education: An ethnography*. Dordrecht: Springer.

Losee, J. (1993). *A historical introduction to the Philosophy of Science* (3rd ed.). Oxford: Oxford University Press.

Mansour, N. (2009). Religion and science education: An Egyptian perspective. In S. BouJaoude & Z. R. Dagher (Eds.), *Arab States* (Vol. 3, pp. 107–131). Rotterdam: Sense Publishers.

Martin-Hansen, L. M. (2008). First-year college students' conflict with religion and science. *Science Education, 17*, 317–357.

Matthews, M. R. (2002). Constructivism and science education: A further appraisal. *Journal of Science Education and Technology, 11*(2), 121–134.

Matthews, M. R. (2009). Teaching the philosophical and worldview components of science. *Science Education, 18*(6), 697–728.

Medawar, P. B. (1963/1990). Is the scientific paper a fraud? In P. B. Medawar (Ed.), *The threat and the glory* (pp. 228–233). New York: Harper Collins.

Merton, R. K. (1938). Science, technology and society in seventeenth century England. *Osiris, 4*, 360–632 (ArticleType: research-article / Full publication date: 1938 / Copyright © 1938 Saint Catherines Press).

Millar, R. (Ed.). (1989). *Doing science: Images of science in science education*. London: The Falmer Press.

Millar, R., & Osborne, J. (1998). *Beyond 2000: Science education for the future*. London: King's College.

Morris, H. (2000). *The long war on god: The history and impact of the creation/evolution conflict*. Green Forest: Master Books.

National Academy of Sciences Working Group on Teaching Evolution. (1998). *Teaching about evolution and the nature of science*. Washington, DC: National Academy Press.

Nickerson, R. S. (1998). Confirmation bias: A ubiquitous phenomenon in many guises. *Review of General Psychology, 2*(2), 175–220.

Pesic, P. (2006). Isaac Newton and the mystery of the major sixth: a transcription of his manuscript 'Of Musick' with commentary. *Interdisciplinary Science Reviews, 31*, 291–306.

Phipps, W. E. (2002). *Darwin's religious odyssey*. Harrisburg: Trinity Press International.

Poole, M. (2008). Creationism, intelligent design and science education. *School Science Review, 90*(330), 123–129.

Reiss, M. J. (2008). Should science educators deal with the science/religion issue? *Studies in Science Education, 44*(2), 157–186.

Reiss, M. J. (2009). Imagining the world: The significance of religious worldviews for science education. *Science Education, 18*(6), 783–796.

Roth, W.-M., & Alexander, T. (1997). The interaction of students' scientific and religious discourses: Two case studies. *International Journal of Science Education, 19*(2), 125–146.

Roth, W.-M., & Bowen, G. M. (1995). Knowing and interacting: A study of culture, practices, and resources in a grade 8 open-inquiry science classroom guided by a cognitive apprenticeship metaphor. *Cognition and Instruction, 13*(1), 73–128.

Sagan, C. (1985/2006). *The varieties of scientific experience: A personal view of the search for God*. New York: Penguin.

Schutz, A., & Luckmann, T. (1973). *The structures of the life-world* (R. M. Zaner & H. T. Engelhardt, Trans.). Evanston: Northwest University Press.

Slezak, P. (2008, November). Opinion: Theism vs naturalism. *International History, Philosophy and Science Teaching Group Newsletter,* 8–9.

Solomon, J. (1992). *Getting to know about energy – In school and society.* London: Falmer Press.

Solomon, J. (1993). The social construction of children's scientific knowledge. In P. Black & A. M. Lucas (Eds.), *Children's Informal ideas in science* (pp. 85–101). London: Routledge.

Springer, K. (2010). *Educational research: A contextual approach.* Hoboken: Wiley.

Taber, K. S. (2008). Towards a curricular model of the nature of science. *Science Education, 17*(2–3), 179–218.

Taber, K. S. (2009). *Progressing science education: Constructing the scientific research programme into the contingent nature of learning science.* Dordrecht: Springer.

Taber, K. S. (2010). Straw men and false dichotomies: Overcoming philosophical confusion in chemical education. *Journal of Chemical Education, 87*(5), 552–558.

Taber, K. S. (2011). The natures of scientific thinking: Creativity as the handmaiden to logic in the development of public and personal knowledge. In M. S. Khine (Ed.), *Advances in the nature of science research – Concepts and methodologies* (pp. 51–74). Dordrecht: Springer.

Taber, K. S., Billingsley, B., Riga, F., & Newdick, H. (2011a). Secondary students' responses to perceptions of the relationship between science and religion: Stances identified from an interview study. *Science Education, 95*(6), 1000–1025.

Taber, K. S., Billingsley, B., Riga, F., & Newdick, H. (2011b). To what extent do pupils perceive science to be inconsistent with religious faith? An exploratory survey of 13–14 year-old English pupils. *Science Education International, 22*(2), 99–118.

Tamny, M. (1979). Newton, creation, and perception. *Isis, 70*(1), 48–58.

Thagard, P. (2008). Conceptual change in the history of science: life, mind, and disease. In S. Vosniadou (Ed.), *International handbook of research on conceptual change* (pp. 374–387). New York: Routledge.

Vallely, P. (2008, October 11). Religion vs science: can the divide between God and rationality be reconciled? *The Independent.* Retrieved from http://www.independent.co.uk/news/science/religion-vs-science-can-the-divide-between-god-and-rationality-be-reconciled-955321.html

Verhey, S. D. (2005). The effect of engaging prior learning on student attitudes toward creationism and evolution. *BioScience, 55*(11), 996–1003.

von Glasersfeld, E. (1989). Cognition, construction of knowledge, and teaching. *Synthese, 80*(1), 121–140.

Wörne, C. H. (2008). Some physics teaching whispered fallacies. *Latin-American Journal of Physics Education, 2*(1), 18–20.

Science Curriculum Reform on 'Scientific Literacy for All' Across National Contexts: Case Studies of Curricula from England & Wales and Hong Kong

Sibel Erduran and Siu Ling Wong

Introduction

Internationally science education research (e.g. Brown et al. 2005; Gott and Roberts 2004; Holbrook and Rannikmae 2009; Laugksch 2000; Lemke 2004; Norris and Phillips 2003) and policy (National Research Council 1996; OECD 1999) have paid considerable attention to the notion of 'scientific literacy' in the past 30 years. In the policy rhetoric of the 1990s, scientific literacy has emerged as a key theme across the world. This is exemplified in the definition of scientific literacy provided by the National Research Council (1996, p. 22) in the United States:

> Scientific literacy means that a person can ask, find, or determine answers to questions derived from curiosity about everyday experiences. It means that such a person has the ability to describe, explain, and predict natural phenomena.

In 1998 the OECD set up the Programme for International Student Assessment (PISA) to assess student knowledge and skills in a way that transcended cultural and regional boundaries. Their science tests are set in real-world contexts highlighting the need for scientific literacy in everyday life and signal the following components as being of most importance (OECD 1999):

Recognising scientifically investigable questions
Identifying evidence needed in science investigations
Drawing and evaluating conclusions
Communicating valid conclusions
Demonstrating understanding of scientific concepts

S. Erduran (✉)
Graduate School of Education, University of Bristol, Bristol, UK
e-mail: sibel.erduran@bristol.ac.uk

S.L. Wong
Faculty of Education, The University of Hong Kong, Hong Kong

N. Mansour and R. Wegerif (eds.), *Science Education for Diversity: Theory and Practice*, 179
Cultural Studies of Science Education 8, DOI 10.1007/978-94-007-4563-6_9,
© Springer Science+Business Media Dordrecht 2013

A central contributor to the conceptualisation of scientific literacy is Roberts (2007) who has outlined two visions for scientific literacy. Vision 1 describes an understanding of the enterprise and epistemology of science and could be considered as what the public should know about the science used by society. This vision might also foster the development of positive attitudes towards science and scientists. Vision 2 involves understanding the world as a scientist would, i.e. being able to offer explanations and hypotheses about the world. Generating theories and knowledge claims are seen as the key activities of science, above the processes of investigating and gathering experimental data. This is exemplified in the definition of scientific literacy provided by the National Research Council (1996, p. 22). Vision I surfaced in 1985 with the beginning of AAAS *Project 2061*. The project's *Benchmarks for Science Literacy* and *Atlas of Science Literacy* have influenced the thinking of educationalists in the USA and worldwide. The expression 'science literacy' in the Project 2061 materials is used consistently and deliberately instead of the more widespread expression of 'scientific literacy' (Bybee 1997).

Scientific literacy has further been considered in fundamental and derived senses (Norris and Phillips 2003). In the fundamental sense, literacy refers to the use of language as in reading and writing. In the derived sense, literacy is more broadly construed to denote knowledgeability, learning and education. In terms of scientific literacy, the fundamental sense refers to the use of language in science contexts, whereas the derived sense deals with understandings or abilities relative to science. Norris and Philips further distinguish between the simple and expanded views of fundamental literacy. The simple view transcends the boundaries of science education. The expanded view of literacy positions reading as inferring meaning from text. Brickhouse (2007) outlined at least four dimensions related to scientific literacy: civic, personal, cultural and critical. A key premise of these various conceptualisations of scientific literacy is that *all* students are entitled to learn science and learning it well. In other words, science is not for a selected elite group of students as has traditionally been conceived but rather that it is within the reach of a diverse range of students if the curriculum is effectively structured and the learning goals are set to embrace inclusion rather than exclusion.

In the rest of this chapter, we will review the curriculum contexts of England & Wales and Hong Kong to provide an illustration of how scientific literacy and related themes (e.g. nature of science and scientific inquiry) have been situated within these curricula. Our rationale for the choice of these curricula includes the existence of many similarities between the two educational systems as a result of the British colonial governance of Hong Kong over a century before its handover to China in 1997. The contrast is also useful to illustrate the many distinctive and unique features that have arisen from the very different cultural values and educational practices typical of the East and the West. In each national context, we will first give an overview of the national science curriculum context including the recent history in the development of scientific literacy as a curricular goal. Our approach is based on qualitative case study methodology drawing out a set of key features that surround the content, aims and projected outcomes of the science curricula in England & Wales and Hong Kong, based on our reading and analysis of each curriculum document. We also provide the broad professional development agenda in each national context to illustrate how the implementation of curricula is put into practice in teacher training. We will then use example projects to illustrate efforts in incorporating 'Scientific Literacy for All' at the level of teaching and learning in secondary science lessons. We will conclude with a synthesis of the contrast between the two curricular cases offering some recommendations for future science curriculum design and implementation.

Developments on Scientific Literacy in the Science Curriculum in England & Wales

The National Curriculum in Science (NC) first appeared in 1988 (Department for Education and Science 1988) and represented the first attempt in England and Wales to standardise the provision of science education across the country. There were two overarching principles: (a) that science would be taught from the age of 5 to 16 and (b) that the curriculum would emphasise scientific content knowledge in the broad context of the scientific endeavour. The first NC document incorporated such areas as microelectronics, weather, information technologies and also the nature of science in its attainment targets. The National Curriculum Council, then the curriculum authority, conducted a series of surveys to gauge the effect of the science NC, and it was realised that the focus on content was excessive. Following a consultation (National Curriculum Council 1991), the NC was revised in 1992 and the number of attainment targets was dramatically reduced. Some skills of scientific enquiry were included, for example:

- Exploration and Investigation: Doing (p. 52)
- Exploration and Investigation: Working in Groups (p. 56)
- Communication: Reporting and Responding (p. 60)
- Communication: Using Secondary Sources (p. 64)
- Science in Action: Technological and Social Aspects (p. 68)
- Science in Action: The Nature of Science (p. 70)

Yet the emphasis was still very firmly on science knowledge content. The emphasis on 'processes' of science including not only the practical experimentation but also the development of explanations, models and theories was gradually lost. Likewise, related skills associated with scientific processes, such as discussion and literacy, were ignored.

The subsequent revisions of the national curricula (Department for Education 1995; Department for Education and Employment and Qualifications and Curriculum Authority 1999) outlined a major new area of the curriculum: 'Experimental and Investigative Science', commonly known as Sc1. The focus was now on 'Ideas and Evidence' in science, a theme that emphasised the nature of science and scientific inquiry. The subsequent 2004 revision built on the 'How Science Works'

Table 1 How Science Works in the Science National Curriculum and potential target skills in argument (From La Velle and Erduran 2007)

Curriculum descriptor	Argument skills
Data, evidence, theories, explanations	Understanding the nature of evidence and justifications in scientific knowledge
Practical and enquiry skills	Justifying procedures and choices for experimental design
	Generating and applying criteria for evaluation of evidence
Communication skills	Constructing and presenting a case to an audience either verbally or in writing
Applications and implications of science	Applying argument to everyday situations including active social, economic and political debates

(HSW) framework had further emphasis on inquiry and literacy, particularly in terms of science in its social, cultural and economic contexts.

Argumentation concerns the coordination of evidence and theory to advance an explanation, a model, a prediction or an evaluation (Erduran and Jimenez-Aleixandre 2008, 2012). At its core, argumentation relies on the presence of a range of viewpoints in the context of a debate where multiple claims are advanced, evaluated and refuted in an effort to yield a conclusion that is best supported by evidence. In this sense, it is a strategy that invites the participation of all students in a discussion even if their viewpoints may not be scientifically valid. The extent to which students can refute alternative claims with substantiating evidence is just as valued a skill as the very construction of a valid argument (Erduran et al. 2004). The contemporary context for argumentation in England and Wales is the 'How Science Works' (HSW) component of the Science National Curriculum (DfES/QCA 2004). The HSW agenda suggests the incorporation of evidence-based reasoning and argumentation in various aspects of science teaching and learning (Table 1). For instance, not only should students learn about coordination of evidence and explanation, but they should also be communicating arguments.

In a study of the content of the exam board syllabi, La Velle and Erduran (2007) concluded that the GCSE specifications employ a range of interpretations of HSW. These syllabi are produced by a range of companies that develop the assessment specifications as well as the teaching and learning resources that schools adopt in following a particular approach to science education provision. They can range significantly in the way that they interpret the national curriculum. For example, the syllabi produced by the exam boards in England and Wales make explicit (e.g. WJEC) and implicit (e.g. OCR) reference to themes such as scientific argument, an aspect of scientific literacy, while two (e.g. Edexcel and C21st) make an explicit reference to HSW with no mention of 'argument' (La Velle and Erduran 2007).

An important vehicle in the implementation of the national curriculum is continuous professional development (CPD) of science teachers. In recent years, the establishment of the Science Learning Centres (SLCs) across England has systematised the CPD in a range of innovative areas that are relatively unfamiliar to teachers. SLCs across England provide in-service teacher education in 'How

Science Works' which include themes such as scientific literacy. For example, a course run by Science Learning Centre London is called 'Decision-Making Activities in Science'. It is described as follows:

> This session explores using decision-making activities in science. In this session you and your colleagues will practise using some decision-making activities that have been developed for small group discussion. The activities have been designed to reveal differences in opinion so that children can explore their reasoning and expose their thinking. There are four activities presented in different formats and in the session you will discuss the merits and limitations of the resources for your classes. You will have the opportunity to work in groups to adapt the activities for a science lesson of your choice. The session is designed to enable you to work towards enhancing your children's decision-making and encourage group discussion and argumentation in your lessons. (https://www.sciencelearningcentres. org.uk/centres/london)

Apart from the relatively new CPD initiatives captured in the Science Learning Centres, schools tend to provide in-service training days for teachers. These 'INSET Days' tend to focus on particular themes (for instance 'Assessment for Learning') and are often coordinated by outside experts in the related theme. Professional development literature has long advocated that the development of expertise in teaching is a long-term endeavour (Berliner 2001). Based on novice-expert work, Berliner (1994) describes five levels of skill development: novice, advanced beginner, competent, proficient and expert. Berliner's stages are somewhat similar to preservice, induction, mid-career and advanced-career years that are often used to describe teachers' experience even though necessary transition across these stages is not a foregone conclusion. The Training and Development Agency for Schools, the key government body that oversaw teacher training provision in England, developed a set of career development stages that provided nationwide specification to guide teachers' professional development.

The preceding context of curriculum reform and teacher training provision in England illustrates that 'Scientific Literacy for All' has been contextualised recently in the 'How Science Works' agenda. In particular, strategies such as argumentation which solicit a diversity of opinion and participation of a range of students have gained prominence in the curriculum. Teachers' professional development efforts too have embraced those pedagogical strategies such as group discussions to develop teachers' skills in managing diversity of ideas. In the next section, we will illustrate some examples of funded projects that have aimed to promote diversity through inquiry-based approaches in science teaching and learning with an emphasis on supporting the development of scientific literacy skills.

Infusing Scientific Literacy in Teaching and Learning: Examples from England

In this section, we will describe some projects funded by the European Union that aimed to develop science teachers' skills in scientific literacy. The *Mind the Gap* project ran from 2007 to 2009 and involved 9 European institutions (Jorde

2009). The *S-TEAM* (Science Teaching Advanced Methods) project was conducted from 2009 to 2012 as an extension of the *Mind the Gap* project as well as other projects into a larger-scale consortium of teacher professional development work involving 25 European institutions all focusing on various aspects of inquiry-based science teaching (https://www.ntnu.no/wiki/display/steam/SCIENCE-TEACHER+ EDUCATION+ADVANCED+METHODS).

The main focus of the *Mind the Gap* project was to promote inquiry-based science teaching. The project targeted particular themes such as scientific literacy and argumentation, and researchers worked with teachers to develop and disseminate professional development strategies with the aim of bridging gaps between the curriculum, teaching and learning (e.g. Erduran and Yan 2009, 2010). The ethos of the project was that the rapid pace of technological change and globalisation has replaced the former focus on content knowledge with an emphasis on broad and general science education. At the same time, the growing importance of scientific issues in our daily lives demands an insight in science and a willingness to engage in the socio-scientific debates in an informed manner. The ability to do this is often captured by the concept of scientific literacy.

The traditional transmissive teaching style is replaced by more demanding pedagogical models which rely on perspectives on cognitive sciences (e.g. Duschl and Erduran 1996). The rationale for this switch is that students are now expected to use and communicate their knowledge also outside the school setting and that they need to be prepared for active citizenship and lifelong learning. In this context, inquiry-based science teaching (IBST) is seen as a relevant teaching approach. By focusing on the students' own questions and their abilities to answer them, IBST is an efficient way to obtain scientific literacy. However, many science teachers are challenged by the new trends related to IBST and scientific literacy, because they by and large have been enculturated into the academic scientific society with its focus on science content and traditional pedagogy. New curricula trends like IBST and scientific literacy are often introduced by policymakers and transnational agencies (i.e. OECD), and consequently science teachers often understand curriculum changes as coming top-down and from another world (Gitlin and Margonis 1995). This often causes problems and tensions in the implementation of new curricula.

In order to bridge the gap between teachers' and policymakers' ideas about intentions, challenges and possibilities in new curricula, a strand of the *Mind the Gap* project focused on scientific literacy and had partners from Denmark, Hungary as well as England & Wales. The strand brought together a range of stakeholders involved in the curriculum development and implementation process, i.e. policymakers, teachers, teacher trainers and researchers. There were national as well as cross-national exchanges between colleagues in an effort to understand how gaps between policy, research and practice can be bridged in the context of teaching scientific literacy.

Maps have been produced for central curriculum texts from Denmark, England & Wales and Hungary, and by a special overlay technique, these maps have been compared to each other and to the PISA definition. These representations make it possible to separate characteristic features of the different countries' understanding of scientific literacy. The methodological challenge was to find a way of comparing how different curriculum texts conceptualise scientific literacy (http://www1.ind. ku.dk/mtg/wp3/scientificliteracy/maps/3). Rather than just adding to the textual analyses of scientific literacy available from each country and cross-nationally, maps were created in order to make analyses and comparisons more precise and informative.

These maps have been produced in the PAJEC software based on complex network theory and were led by the team of researchers in Denmark including Jesper Bruun, Robert Evans and Jens Dolin from University of Copenhagen. Monika Reti from Hungarian Science Teachers' Association and Xiaomei Yan and Sibel Erduran from University of Bristol in England worked on the generation of maps for their respective countries. As a special feature, these maps reveal strings of defining elements (concepts, actions, contexts, levels, etc.) showing the relative importance of the connections between the elements. The maps thus make up a visual representation of often complex texts with an integrated quantitative approach. In the case of the English curriculum, the emphasis is on understanding of the nature of scientific inquiry and scientific knowledge in its social context (Erduran 2012). Students are expected to be able to conduct, communicate and evaluate evidence gathered through scientific inquiries.

The Scientific Literacy Map (see Fig. 1) used the visual tool to highlight the above features. It can be seen that the thick connection between the node 'students' and 'use' represents the practical purpose emphasised in the curriculum. If the readers visit the active website, they can use the mouse to activate links and see the strings, for instance, the link 'The students have been expected to be able to use tools to learn science, to use collected data and evidence to draw and evaluate scientific conclusions, to use various ways to represent scientific language'. There are other verbs that have been highlighted in the map, such as 'understand', 'present', 'draw (conclusions)' and 'collect (data)'. These reflect the different areas of scientific literacy that have been addressed in the curriculum. In the Scientific Literacy Map, we can also see the several concept nodes such as the 'symbol', 'convention', 'language', 'tool' and 'information' which represent the communication and representation aspects of scientific literacy. The adjectives of 'technical', 'social', 'environmental' and 'contemporary' highlight the link between scientific knowledge and its influence and connection to its social context.

The Scientific Literacy Map of England & Wales compared to other countries' maps is quite general as a guideline not providing specific details on each curricular goal. This might be due to the function of the curriculum in the English school system in the sense that the national curriculum is intended to be subsequently interpreted by examination boards which then produce syllabi in more detail. Apart from these features of the map, the *Mind the Gap* project researchers built in extra resources to enable effective use of the maps by teachers and teacher educators. For example, the map also includes examples of teaching practice linked to some example curricular statements supplemented by comments on how to bridge the gap between policy and practice. In other words, there are links to concrete

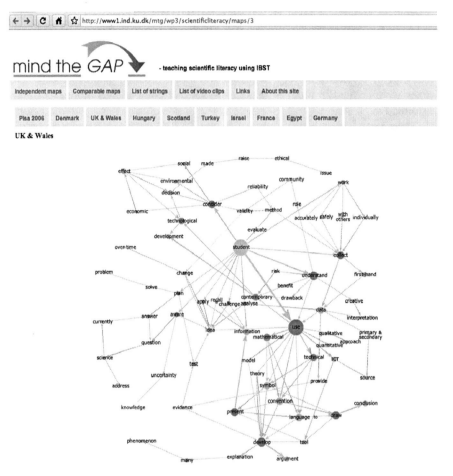

Fig. 1 Scientific Literacy Map for the Science Curriculum for England and Wales 2006 (http://www1.ind.ku.dk/mtg/wp3/scientificliteracy/maps/3)

teaching scenarios where some key curricular descriptors are illustrated in action. Furthermore, in the England map, 'technological' and 'mathematical' have been highlighted. This represents the cross-subject connections of scientific literacy. Scientific literacy is based on the mathematical language, the technological tools and also the representation and communication skills. The emphasis of using small group discussions, group presentations and ICT has also been presented in the map.

Overall, the semantic map analysis of the England and Wales curriculum provides some comparative analyses around the theme of scientific literacy to be conducted across some example curricula in Europe. Within the *Mind the Gap* project, these maps have also been used as professional development tools. A similar approach to the use of video data in exemplifying policy statements has been adopted within the *S-TEAM* project in the context of the University of Bristol work

(Erduran et al. 2012). Various aspects of the 'How Science Works' agenda of the national curriculum have been contextualised in teachers' reflections about their teaching practices captured on video (www.apisa.co.uk).

In summary, our discussion on the instantiation of 'Scientific Literacy for All' in England has included a brief survey of the key curricular developments in recent years: the development of particular strategies such as argumentation that promote the inclusion of a diversity of perspectives in discussions, hence student voice. Furthermore example professional development projects attempted to make specific links between curricular goals and teaching practices by providing teachers and teacher trainers visual tools in coordinating lessons in addressing scientific literacy at the level of the classroom. We will next turn to the curricular and professional development context of Hong Kong to illustrate how scientific literacy has been addressed in this case.

Brief Historical Background About Hong Kong Education System

Basic education for all with junior secondary education (S1–3) was only available in Hong Kong since 1978. While almost 85 % of junior secondary students completing S3 could continue their education to S5, there were very limited places for university education offered by two universities in Hong Kong until 1991. Students needed to go through the highly selective and competitive public examinations at S5 and S7, the Hong Kong Certificate of Education (HKCEE) and the Hong Kong Advanced Level Examination (HKALE), to secure a place in the universities. In the early 1980s, the admission rate of full-time first-degree undergraduate students supported by the University Grants Committee (UGC) was only about 3–4 % of the population of the age group of 17–20 years old. The city transformed from a predominantly manufacture-based economy to a service hub providing financial, professional, logistics and other services from the late 1970s to 1990s. In addition to mass education up to S3, the government started to invest more in higher education to meet the increasing demand for more citizens with higher-order skills such as communication, numeracy and information technology (IT) skills.

By 1990, more than 9 % of the relevant age group could receive UGC-funded tertiary education. There was further expansion of the tertiary education sector from two to eight universities during the 1990s, and admission rate of first-year-first-degree students funded by the UGC increased to 18 % of the age cohort, equivalent to 14,500 students per annum, by 1994 as planned (Morris et al. 1994). This number has maintained at the same level until today as reported by the UGC (2010). Although there has been considerable increase in the opportunity for students to receive publicly funded university education, 18 % is still on the low side as compared to an average of 37 % of 25- to 34-year-olds achieving tertiary education across OECD countries (OECD 2011, p. 40).

What this brief survey illustrates is that access to education has been diversified broadly in the Hong Kong education system. Yet, examination pressure on students, parents, teachers and schools remains huge in Hong Kong (Tang et al. 2010). The high expectation on the results of public examinations from all stakeholders has led to an examination-driven learning approach and excessive drilling of students for the high-stake examinations (Bray and Kwok 2003) in particular on the learning outcomes related to content knowledge which is more easily assessed in written format. In fact, the deeply rooted examination culture among Chinese was long established since the sixth century (Sui Dynasty) when the imperial examination system was instituted for selection of elites in taking up high offices in the government and the top scholar was often granted marriage to the royal family. The rewards, respectable social status and glories brought to the scholar and his family upon the success of examination resulted in a lopsided emphasis on the utilitarian values of education (Leung 2008). Such views about examinations and values of education have perpetuated into the present day of Hong Kong and pose challenges in the recent major curriculum reform in the city.

Developments on Scientific Literacy in the Science Curriculum in Hong Kong

As a British colony for over a century before the handover of the sovereignty back to mainland China in 1997, Hong Kong's educational system and curricula of many subjects, including science, has been much influenced by England. Science teachers who have more than 20 years of experience might remember well the discovery approach incorporated in the Ordinary and Advanced Level Nuffield curricula in the 1960s–1970s in England as the syllabuses of the science subjects in Hong Kong then even referred to the experiments in the Nuffield textbooks. However, such an approach had never taken root in Hong Kong science classrooms. Instead, lessons were commonly predominated by teacher laboratory demonstrations or cookbook verification-type experiments. Worse still, some teachers regarded laboratory demonstrations or student experiments as inefficient means to achieve the key objective of transmitting science subject knowledge for preparing students for the highly competitive content-based examinations.

Thus, practical activities were mostly 'talked through' by teachers rather than 'carried out' by students. During the 1970s–1990s, Hong Kong economy underwent a dramatic structural change from labour-intensive manufacturing to skill-intensive service industries which demands school leavers and university graduates to possess generic skills such as problem solving, investigative skills and self-learning ability. Such changes have resulted in the widening of the curriculum goals of science education from a knowledge-focused one to an expanded one covering the development of skills and attitude (Education Commission 2000). In line with international trends, science education in Hong Kong has undergone considerable changes since the implementation of the revised junior secondary science curriculum (grades 7–9)

(Curriculum Development Council [CDC] 1998). The new curriculum encourages teachers to conduct scientific investigations in their classes; advocates scientific investigation as a desired means of learning scientific knowledge; and highlights the development of inquiry practices and generic skills such as collaboration and communication. It is the first local science curriculum that embraced some features of nature of science (NOS), e.g. being 'able to appreciate and understand the evolutionary nature of scientific knowledge' (CDC 1998, p. 3) was stated as one of its broad curriculum aims. In the first topic, 'What is science?', teachers are expected to discuss with students some features about science, e.g. its scope and limitations, and some typical features about scientific investigations, including fair testing, control of variables, predictions, hypothesis, inferences and conclusions.

Such an emphasis on NOS was further reinforced in the revised secondary 4 and 5 (grade 10 and 11) physics, chemistry and biology curricula (CDC 2002). Scientific investigation continued to be an important component, while the scope of NOS was slightly extended to include recognition of the usefulness and limitations of science as well as the interactions between science, technology and society (STS). In preparation for the implementation of a new curriculum structure (from a 7-year secondary education system to a 6-year one) in September 2009, a new set of Curriculum and Assessment Guides was devised for senior secondary level science subjects (CDC-HKEAA 2007). Promotion of scientific literacy is stated as the overarching aim in the physics, the chemistry and the biology curricula and Assessment Guides. For example, the physics curriculum states:

> The overarching aim of the Physics Curriculum is to provide physics-related learning experiences for students to develop scientific literacy, so that they can participate actively in our rapidly changing knowledge-based society, prepare for further studies or careers in fields related to physics, and become lifelong learners in science and technology. (p. 4)

We note a further leap forward along the direction set in the junior science and the S4–5 physics/chemistry/biology curriculum documents. As a key component for achieving scientific literacy, the importance of promoting students' understanding of NOS is explicitly spelt out. To put greater emphasis on environmental issues, students' appreciation of STS is extended to STSE, where 'E' stands for environment. For example, in the physics curriculum, students are expected to 'appreciate and understand the nature of science in physics-related contexts', 'develop skills for making scientific inquiries', 'be aware of the social, ethical, economic, environmental and technological implications of physics, and develop an attitude of responsible citizenship' and 'make informed decisions and judgments on physics-related issues' (CDC-HKEAA 2007, p. 4).

There is a clear intention to develop students' awareness and understanding of issues associated with the interconnections among science, technology, society and the environment. A separate subsection entitled STSE connections is embedded in each science topic of the Curriculum and Assessment Guides. It suggests examples of issues that teachers could make use of in developing students' awareness and understanding of STSE connections. Suggestions of teaching and learning activities in the guides include some science-related social issues, e.g. in the topic mechanics;

one of the issues suggested is a dilemma of choosing between convenience and environmental protection in modern transportation. The Biology Guide is even one step ahead of the Physics and Chemistry Guides for the inclusion of another subsection entitled Nature and History of Biology which reflects the strongest intention to achieve a more comprehensive understanding of NOS in biology-related content.

In sum, to achieve the overarching aim of promoting students' scientific literacy, the former overloaded knowledge component in the science curricula is tailored to free space for greater emphasis on development of generic skills and NOS understanding. The curriculum developers aim to focus on promoting understanding of some relatively basic features of NOS and related learning outcomes at junior science level (S1–3), including the basic understanding of NOS (e.g. evolutionary nature of science and limitation of science), the skills for the planning and conducting simple scientific investigations and basic understanding of the nature of scientific inquiry. At senior secondary levels, more sophisticated NOS concepts are infused, e.g. theory-laden observation, nature of scientific models and the social dimension of the NOS including how scientists interact within and beyond the scientific community. Students are also expected to apply the relevant science content knowledge and NOS understanding to practise making sense of some science-related social issues and subsequently make ethically and morally sound actions (Hodson 2003). The following sections describe the effort of the science educators in the promotion of NOS teaching and learning in the last decade in the junior and senior secondary levels, respectively.

Preparing Hong Kong Science Teachers for Promoting Students' NOS Understanding

Science educators in Hong Kong started to reform initial teacher education programmes and provide professional development training courses for in-service teachers since the beginning of the twenty-first century to strengthen the knowledge and pedagogical skills in teaching NOS – an area of void in their own schooling and teaching experiences.

Use of Historical Science Stories

With the intention to support junior secondary science teachers of the implementation of the revised junior secondary science curriculum, Tao with his co-authors (Tao et al. 2000) wrote a new series of junior science textbooks which included four science stories for introducing science at S1: *Discovery of Penicillin, Development of Cowpox, Newton's Proposition of the Law of Universal Gravitation* and *Treatment of Stomach Ulcers* (Tao 2002). These stories were designed such that teachers could highlight the NOS aspects through an explicit approach (Abd-El-Khalick and

Lederman 2000). However, it was found that students' learning of NOS based on these stories was disappointing (Tao 2003). The key underlying reason was that many junior science teachers only made use of them for arousing student interest. The example of a junior science teacher illustrates how he came to realise his oversight of not having made good use of the stories for teaching NOS after he attended one of the NOS sessions in our restructured teacher training course:

> I found the story on stomach ulcers very interesting... Marshall tested his hypothesis by trialling out himself....Students all enjoyed the story... I only realise now that there are deeper meanings behind the story and other important learning outcomes to be achieved through it and other stories.

This comment revealed that availability of teaching resources would not by itself result in teachers making use of the materials to teach NOS unless teachers had the ability to understand and appreciate the intended learning outcomes of the instructional materials. It is likely that they would overlook the targeted learning objectives (McComas 2008) and cling to the parts which are more appealing to them (e.g. dramatic stories which promote students' interest). We also reckoned that there were some inadequacies of these relatively 'old' stories. Teachers expressed that though these stories aroused their interests, they happened quite a while ago. Those who did not have the historical and cultural backgrounds of the scientific discoveries and inventions would fail to develop an in-depth understanding of, and hence appreciate, the thought processes of the scientists related to what they encountered at their time.

Promoting Nature of Science Through Conducting Scientific Investigation

There are however some promising learning outcomes relating to the skills and understanding of scientific inquiry. The considerable reduction of the factual knowledge in junior science level does encourage teachers to arrange more scientific investigations particularly at S1 level. Some teachers even comment that the content knowledge is now so thin that they could finish the syllabus in less than half of the academic year! The fact that the junior textbooks are compiling well with the revised S1–3 Science Curriculum Guide and they all include a number of scientific investigations, ranging from some with more stepwise procedures to a few with less structured ones, helps a lot in turning the classroom environment from the traditional knowledge transmission one to a more interactive one with greater student involvement.

The social environment in Hong Kong is also evolving as a result of smaller families of one or two children that most of the families' attention and resources are spent on children's learning, both inside and outside school.

Parent–child activities are strongly encouraged by the government through media. There are increasingly more science competitions organised by the Hong Kong

Education Bureau, Hong Kong Science Museum, local universities and other local and international organisations having similar visions to promote students' interest in science and technology. Through participation of these science competitions, students are practising the investigative skills and experiencing what scientific inquiry is like. Such events are welcome by students, teachers, parents and schools for a variety of reasons. Students mostly like these activities as they have autonomy in choosing a topic that they like to work on together with their good friends. Parents in the East have a reputation for their concerns and high expectation on their children's education. They are certainly glad to see their children spending time meaningfully in learning during their preparation for the competition instead of playing video games. Teachers and schools see the competition as a great opportunity for the students to learn by themselves with certain support and encouragement from them. For some teachers and schools, they might also see the pragmatic value in these competitions as a means to promote the schools' reputation.

Due to the extremely low birth rate of Hong Kong (about 8–11 births per 1,000 population[1]) in recent years, it has resulted in closing down of some schools or reducing the number of classes. The concerns about closing down of the schools have led some schools to take these competitions more seriously in encouraging and even putting efforts in training their students to achieve better results for gaining some brownie points in convincing the parents and the Education Bureau that they are good enough to keep the survival of the school. Another important favourable factor is that there are no high-stake examinations for students until they reach S6; thus, the pressure on junior students, teachers and schools is minimal. Teachers can afford carrying out scientific investigations instead of doing examination drilling.

It is pleasant to see some supporting evidence showing improved basic under-standing of NOS by the students as reflected by comparing the TIMSS performance of Hong Kong eighth-grade students before and after the implementation of the revised junior secondary science curriculum in 2000. Figure 2 compares the students' performance of a pair of very similar questions, one in the TIMSS 1995 and the other in the TIMSS 2003. For simplicity only the question of the 2003 one is shown. Both questions assess students' understanding of fair test and basic experimental design with carts of different conditions rolling down an inclined plane. The percentage of students choosing each option is provided. The correct answer of each of the questions is marked with an asterisk. Significant improvement in students' performance in TIMSS 2003 is noted ($p < 0.001$). Figure 3 compares the students' performance of another pair of two questions which assess students' skills of reasoning based on the concept of control experiment in the topic related to growth of plants. Again, one is from TIMSS 1995 and another one is from TIMSS 2003. Significant improvement in students' performance in TIMSS 2003 is also noted ($p < 0.001$).

When such encouraging improvements in students' achievement is conveyed to the junior science teachers, it gives teachers a strong message which is best

---

[1]http://www.gov.hk/en/about/abouthk/factsheets/docs/population.pdf

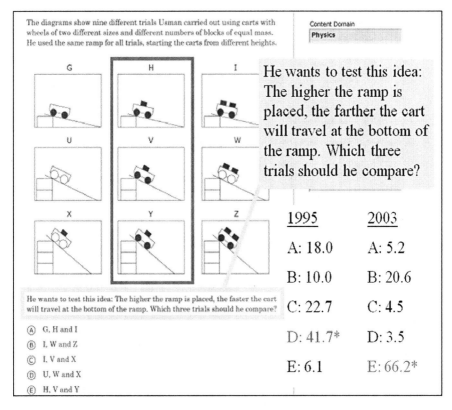

Fig. 2 A question testing on students' experimental design and control of variables

described by a Chinese saying, 種瓜得瓜、種豆得豆, meaning 'what you reap is what you sow'. The TIMSS results serve as convincing evidence to teachers that students could perform better in examination through doing science and learning about science. Many teachers have not realised the improvement of their students as they have not yet included examination questions assessing a boarder set of intended learning outcomes of the revised curriculum. Thus, in addition to training teachers how to implement the new teaching and learning approach, it is equally important to train teachers about the assessment strategies, particularly, given the deeply rooted examination culture.

Use of the SARS Crisis to Teach About NOS

In the summer 2003, when the crisis due to the severe acute respiratory syndrome (SARS) in Hong Kong was coming to an end, my colleague and I, Siu Ling Wong, saw a golden opportunity to turn the crisis into a set of meaningful instructional

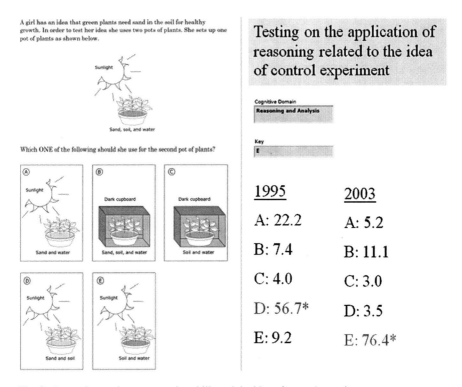

Fig. 3 A question testing on reasoning skills and the idea of control experiment

resources which might help promote good understanding of NOS. The recent history of the story of SARS incident was a unique experience that everyone in Hong Kong had lived through and the memories of which would stay for years to come. At the beginning of the outbreak, the causative agent was not known, the pattern of spread was not identified and the number of infected cases was soaring, yet an effective treatment regimen was uncertain. It attracted the attention of the whole world as scientists worked indefatigably to understand the biology of the disease, develop new diagnostic tests and design new treatments. Extensive media coverage kept people up to date on the latest development of scientific knowledge generated from the scientific inquiry about the disease. It was believed that the incident would have much to reveal about NOS. As the team anticipated, many interesting aspects of NOS based on the interviews with key scientists who played an active role in the SARS research and analysis of media reports, documentaries and other literature were published during and after the SARS epidemic.

The SARS incident illustrated vividly some NOS features advocated in the school science curriculum. They included the tentative nature of scientific knowledge, theory-laden observation and interpretation, multiplicity of approaches adopted in scientific inquiry, the interrelationship between science and technology and the nexus of science, politics, social and cultural practices. The incident also provided some insights into a number of NOS features less emphasised in the school curriculum. These features included the need to combine and coordinate expertise in a number of scientific fields, the intense competition between research groups, the significance of affective issues relating to intellectual honesty, the courage to challenge authority and the pressure of funding issues on the conduct of research. The details on how we made use of the news reports and documentaries on SARS, together with episodes from the scientists' interviews to explicitly teach the prominent features of NOS, are reported in Wong et al. (2009). Since January 2005, we have been using the SARS story in promoting understanding of NOS of hundreds of preservice and in-service science teachers. The learning outcomes have been encouraging (Wong et al. 2008).

Development of Instructional Resources Integrating Teaching of NOS and Subject Knowledge

While the SARS story has been effective in promoting NOS understanding, there was still a lack of NOS instructional materials, preferably grounded in the local contexts and language, for school students. Thus, in September 2005 Siu Ling Wong and colleagues in Hong Kong embarked on a two-year project which aimed to produce local NOS curriculum resources in both English and Chinese languages while preparing more teachers for NOS teaching. They deliberately involved teachers at the beginning stages of the design process of instructional materials. More than 50 senior secondary science teachers worked together with the university team members to develop 12 sets of teaching resources[2] which integrate NOS knowledge with subject knowledge of the new senior secondary biology, chemistry and physics curricula. Efforts were made to include as many local examples as possible on top of some classic stories of science. The topics included development of the Disneyland in Hong Kong, Consumption of Shark Fins, an abridged version of the SARS Story, Nature of Light, Discovery of Electric Current, etc. In doing so, teachers would be more ready to make use of the materials in their own classrooms. This was important for the project as the team wanted to collect data to refine the teaching materials as well as provide opportunities for the teachers to learn how to teach NOS with the support of the university team members in their own classroom settings.

The review and analysis of the lesson videos facilitated teachers' reflection on which areas needed improvement. It also allowed them to appreciate and value the teaching of NOS when they saw students' ability and interest in learning NOS. The proof of workability of both teaching and learning NOS had prompted teachers who were reserved about teaching NOS to follow suit. Details of this teacher professional development project, the favourable learning outcomes of the teachers and their concerns were reported in Wong et al. (2010). Although the Hong

---

[2]The teaching resources can be accessed through the website http://learningscience.edu.hku.hk

Kong team managed to equip many teachers with the knowledge and pedagogical skills to teach NOS, some concerns are expressed by some teachers. Three major concerns included the following: (1) Teacher members of the project or those who participated in our workshops expressed that they lacked the collegial support in sustaining their practice in NOS teaching; (2) some teachers (particularly those who had not participated in the workshops but just got the instructional resources through the DVDs we disseminated to schools or our website) expressed that the richness of the instructional resources gave them pressure as they had a tendency to cover most activities in each set of instructional resources; and (3) subject knowledge was not included in the instructional resources in an elaborated manner as we expected teachers would integrate the NOS activities with the relevant subject matter knowledge that they have been competently teaching. The lack of elaboration of the subject knowledge gave an impression to teachers (except those attended our training workshops) that NOS teaching involves add-on activities separated from the teaching of the subject knowledge rather than appreciating that NOS understanding can enhance the learning of science concepts. These concerns posed much challenges to Hong Kong teachers who would prefer spending considerable time in examination drilling instead of on NOS teaching, especially when they are not yet certain about the mark allocation on questions assessing NOS understanding. This observation prompted the team to launch another 2-year teacher professional development project from September 2008 which might address the concerns.

Teach a Man to Fish and You Feed Him for a Lifetime

The new project aimed to cultivate a mutually supportive environment where teachers would collaborate and develop their pedagogical content knowledge of NOS. A key feature of the new project was the formation of study groups among teachers of the same subject discipline from different schools (subject-based approach) and science teachers of the same school (school-based approach). Siu Ling Wong and colleagues believe that putting like-minded teachers together in a study group is more likely to sustain and even enhance their commitment to teach NOS. While retaining the components that teachers participated in the previous project valued and treasured most (such as the SARS case study, the detailed review and discussion on their lesson videos with the university team members and their peers), they also encouraged teachers not to simply modify and adapt the available teaching resources but to proactively design their own instructional materials. This was indeed a goal that they wished teachers could ultimately achieve. Thus, at the outset of the current project, the intention was explained to the teachers by a Chinese proverb, 'Give a man a fish and you feed him for a day. Teach a man to fish and you feed him for a lifetime'. (In Chinese, it says 授人以魚，三餐之需；授人以漁，終生之用。)

Siu Ling Wong and colleagues reported earlier an example of two teachers who worked in the same school and collaboratively designed the teaching of NOS by situating the teaching and learning activities in a sports-related socio-scientific issues (SSI), the Ban of Shark Skin Swimsuits (Wong et al. 2011). As the more complex SSI like the SARS story in which they learned about NOS, the SSI chosen by them was also timely, relevant and familiar to the students which contributed to the effective learning of their students.

The team were encouraged by this favourable outcomes of and came to realise how we could better 'teach a man to fish' through provision of exemplars and modelling of the exemplars. The capability of identifying key features of exemplar materials and transferring such features to personalised teaching resources seems to be essential for future teachers' development. It involves careful identification and explicit communication of the key features to the learners and provision of opportunities for teachers to practise and develop such capability. Such a direction seems to be a way for our future teacher training programmes. Siu Ling Wong and colleagues envisage that such modelling is potentially applicable to other approaches of teaching NOS, e.g. doing inquiry and using history of science.

Bridging Policy, Research and Practice on Scientific Literacy

The comparison of the case studies of England & Wales and Hong Kong has revealed some common features favourable to bridging gaps in policy, research and classroom implementation on promoting scientific literacy. In both places, the governments have played an important role in supporting researchers and teacher trainers with expertise in the relevant new curriculum goal with considerable funding for conducting research which informs the design of professional development programme. In the training to implement inquiry-based science teaching methods in England and in the training to infuse the teaching of NOS in the teaching of subject knowledge in Hong Kong, exemplary classroom videos are used to illustrate how the target curriculum goals could be achieved.

The study by Wong et al. (2006) provides evidence that videos could reinforce and develop prospective teachers' conceptions of good science teaching in one or more of the following ways: (i) recognising exemplary practitioners in the videos as role models who can inspire them to formulate personal goals directed towards these practices, (ii) broadening their awareness of alternative teaching methods and approaches not experienced in their own learning, (iii) broadening their awareness of different classroom situations, (iv) providing proof of existence of good practices and (v) prompting them to reflect on their existing conceptions of good science teaching. The use of exemplary videos of in-service teachers pioneering the new teaching approaches towards the new intended curriculum goals should have similar effects. The involvement of 'pioneer' teachers could also help mitigate the perception of top-down policy on curriculum reform.

We have also identified from the comparison some potential pitfalls in achieving the intended curriculum aims. Although scientific literacy is advocated in both places, the foci of emphasis are different. A prominent distinctive difference is related to the ability in engaging in socio-scientific debate. In England, systematic and extensive training in promoting argument skills is well in place. In Hong Kong, the students' engagement in SSI is rarely on topics which are highly controversial which requires much debate. Such a lack of emphasis in developing argument skills is a reflection of the social, cultural and political environment of Hong Kong where people are contented with little engagement in debating on public policies. In the Chinese culture, obedience is regarded as a virtue. Politically, Hong Kong people have never been given the right to choose their governor, not in the colonial era and still not after Hong Kong has become part of China. Thus, the curricular aim of making informed decisions and judgments on physics (biology/chemistry)-related issues found in the Curriculum and Assessment Guides is more of a lip service than realty.

Lastly, any curriculum reform is not going to be successful if the assessment does not align with its intended learning outcomes. In this regard, England has been doing a good job in that the exam board syllabi are reflecting the components in the HSW framework. Hong Kong has much to catch up in this aspect as the sample examination paper of the new curriculum published earlier is very similar to the past and only assesses very minimally on the newly advocated learning outcomes (Wong et al. 2010). The team of science educators at the University of Hong Kong who have been working together with the Science Education Curriculum Officers in enhancing the pedagogical skills of science teachers for the new curriculum components have made a strong protest to the Hong Kong Examination Authority which has given a blow to a decade of effort on promoting teaching of nature of science.

Although the recommendations given above are based on the curriculum context related to scientific literacy in England and Hong Kong, we trust that they could be transferable to other curriculum initiatives, for example in relation to aligning assessment and curriculum goals.

Overall, our discussion in this chapter highlights the different senses of diversity in science education. First, we have articulated the definitions and the curricular contexts of the prominent 'Scientific Literacy for All' goal. In so doing, we have illustrated how the very conceptualisation of this international agenda has been situated in different national contexts. This sense of 'diversity' highlights how different educational systems are dealing with the incorporation and implementation of curricular goals to be more inclusive. Second, we have provided examples of instructional approaches such as 'argumentation' in the English curriculum and 'nature of science' in the Hong Kong curriculum which promote the inclusion of student voice in the learning environment. Here our emphasis has been on the classroom level implementation of goals that support the achievement of diversity of student engagement in science. Third, the broader descriptions of the curricula in England and Hong Kong illustrate how the very access to science education has been shaped in recent history. The Hong Kong case provides a stark example where the access to higher education has been dramatically improved in recent years. The contrast of the various iterations of the national curriculum in England illustrates how the very characterisation of 'science' can be instrumental in embracing diversity in the classroom. For instance, conceptualising science as a social activity involving different perspectives that are debated, evaluated and communicated is a step forward in engaging students from a range of dispositions, viewpoints and characteristics.

Acknowledgements Sibel Erduran's work on the Mind the Gap and S-TEAM projects was funded by the European Union FP7 Program. Erduran would like to thank Xiaomei Yan of the University of Bristol for assisting in data collection in the Mind the Gap project and Wan Ching Yee in the S-TEAM project.

Siu Ling Wong's work on the series of professional development and research projects was funded by two Quality Education Funds from the Hong Kong Education Bureau. The project outcomes could not be achievable without the great collaboration with my science colleagues, Dr Benny Yung, Dr Maurice Cheng, Dr Jeffrey Day, Dr. Zhihong Wan, Mr. Eric Yam, Miss Kwan Ling Chan of the University of Hong Kong and Prof. Se-yuen Mak of the Chinese University of Hong Kong. Wong is also grateful to the teacher members of the projects for their participation and the science colleagues in the Education Bureau for their advice.

References

Abd-El-Khalick, F., & Lederman, N. G. (2000). Improving science teachers' conceptions of the nature of science: A critical review of the literature. *International Journal of Science Education, 22*, 665–670.

Berliner, D. C. (1994). Expertise: The wonder of exemplary performance. In J. N. Mangieri & C. C. Block (Eds.), *Creating powerful thinking in teachers and students*. Fort Worth: Harcourt Brace.

Berliner, D. C. (2001). Learning about and learning from expert teachers. *Educational Researcher, 35*, 463–482.

Bray, M., & Kwok, P. (2003). Demand for private supplementary tutoring: conceptual considerations, and socio-economic patterns in Hong Kong. *Economics of Education Review, 22*, 611–620.

Brickhouse, N. W. (2007, May 28–29). Scientific literates: what do they do? Who are they? In *Proceedings of the Linnaeus Tercentenary 2008 symposium promoting scientific literacy: Science education research in transaction*, Uppsala, Sweden.

Brown, B., Reveles, J., & Kelly, G. (2005). Scientific literacy and discursive identity: A theoretical framework for understanding science education. *Science Education, 89*, 779–802.

Bybee, R. W. (1997). Towards an understanding of scientific literacy. In W. Gräber & C. Bolte (Eds.), *Scientific literacy. An international symposium* (pp. 37–68). Kiel: Institut für die Pädagogik der Naturwissenschaften (IPN).

CDC [Curriculum Development Council]. (1998). *Science syllabus for secondary 1–3*. Hong Kong: CDC.

CDC. (2002). *Physics/Chemistry/Biology curriculum guide (Secondary 4–5)*. Hong Kong: Curriculum Development Council. Retrieved August 15, 2011, from http://www.edb.gov.hk/index.aspx?nodeID=2824&langno=1

CDC-HKEAA. (2007). *Physics/Chemistry/Biology/Integrated science curriculum guide and assessment guide (Secondary 4–6)*. Hong Kong: Curriculum Development Council and Hong Kong Examinations and Assessment Authority.

Department for Education. (1995). *Science in the national curriculum*. London: HMSO. ISBN 0 11 270884.

Department for Education and Employment and Qualifications and Curriculum Authority. (1999). *Science. The National Curriculum for England*. London: HMSO.

Department for Education and Science. (1988). *Science for ages 5 to 16*. London: HMSO. x-10-171984-0.

Department for Education and Skills and Qualifications and Curriculum Authority. (1999). *Science. The National Curriculum for England*. London: HMSO.

Department for Education and Skills and Qualifications and Curriculum Authority. (2004). *Science. The National Curriculum for England*. London: HMSO.

Duschl, R., & Erduran, S. (1996). Modeling the growth of scientific knowledge. In G. Welford, J. Osborne, & P. Scott (Eds.), *Research in science education in Europe: Current issues and themes* (pp. 153–165). London: Falmer Press.

Education Commission. (2000). *Learning for life, learning through life: Reform proposals for the education system in Hong Kong*. Hong Kong: Education Commission.

Erduran, S. (Ed.) (2007). Editorial: Argument, discourse and interactivity. *Special Issue of School Science Review, 88*(324), 29–30.

Erduran, S. (2012). The role of dialogue and argumentation. In J. Oversby (Ed.), *Guide to research in science education* (pp. 106–116). Hatfield: Association for Science Education.

Erduran, S., & Jimenez-Aleixandre, M. P. (Eds.). (2008). *Argumentation in science education: Perspectives from classroom-based research*. Dordrecht: Springer.

Erduran, S., & Jimenez-Aleixandre, M. P. (2012). Research on argumentation in science education in Europe. In D. Jorde & J. Dillon (Eds.), *Science education research and practice in Europe: Retrospective and prospective*. Rotterdam: Sense Publishers.

Erduran, S., & Yan, X. (2009). *Minding gaps in argument: Continuous professional development in the teaching of inquiry*. Bristol: University of Bristol.

Erduran, S., & Yan, X. (2010). Salvar las brechas en la argumentacion: el desarrollo profesional en la ensenanza de la indagacion cientifica. *Alambique, 63*, 76–87.

Erduran, S., Simon, S., & Osborne, J. (2004). TAPping into argumentation: Developments in the application of Toulmin's argument pattern for studying science discourse. *Science Education, 88*(6), 915–933.

Erduran, S., Yee, W. C., & Ingram, N. (2012). *Assessment and practical inquiry in scientific argumentation*. CPD Resource. Bristol: University of Bristol. (www.apisa.co.uk).

Gitlin, A., & Margonis, F. (1995). The political aspect of reform: Teacher resistance as good sense. *American Journal of Education, 103*(4), 377–405.

Gott, R., & Roberts, R. (2004). A written test for procedural understanding: a way forward for assessment in the UK science curriculum? *Research in Science and Technological Education, 22*(1), 5–21.

Hodson, D. (2003). Time for action: Science education for an alternative future. *International Journal of Science Education, 25*, 645–670.

Holbrook, J., & Rannikmae, M. (2009). The meaning of scientific literacy. *International Journal of Environmental & Science Education, 4*(3), 275–288.

Jorde, D. (2009). *Mind the gap: Learning, teaching, research and policy in inquiry-based science education* (EU FP7, Science in Society, Project No. 217725). Oslo, Norway: University of Oslo.

La Velle, B. L., & Erduran, S. (2007). Argument and developments in the science curriculum. *School Science Review, 88*(324), 31–40.

Laugksch, R. C. (2000). Scientific literacy: A conceptual overview. *Science Education, 84*, 71–94.

Lemke, J. L. (2004). The literacies of science. In E. W. Saul (Ed.), *Crossing borders in literacy and science instruction* (pp. 33–47). Newark: International Reading Association.

Leung, F. K. S. (2008). In the books there are golden houses: Mathematics assessment in East Asia. *ZDM Mathematics Education, 40*, 983–992.

McComas, W. F. (2008). Seeking historical examples to illustrate key aspects of the nature of science. *Science Education, 17*, 249–263.

Morris, P., McClelland, J. A. G., & Yeung, Y. M. (1994). Higher education in Hong Kong: The context of and rationale for rapid expansion. *Higher Education, 27*, 125–140.

National Curriculum Council. (1991). *Science in the National Curriculum: A report to the Secretary of State for Education and Science on the statutory consultation for attainment targets and programmes of study in Science*. London: NCC.

National Research Council. (1996). *National science education standards.* Washington, DC: National Academy Press.

Norris, S., & Phillips, L. M. (2003). How literacy in its fundamental sense is central to scientific literacy. *Science Education, 87*(2), 224–240.

OECD. (1999). *Measuring student knowledge & skills: A new framework for assessment.* Paris: OECD Publications.

OECD. (2011). *Education at a glance 2011: OECD indicators.* Paris: OECD Publishing.

Roberts, D. A. (2007). Scientific literacy/science literacy. In S. K. Abell & N. G. Lederman (Eds.), *Handbook of research on science education* (pp. 729–780). Mahwah: Lawrence Erlbaum Associates.

Tang, L. F., Lam, C. C., & Ma, Y. P. (2010). Competition – A double-edged sword in educational change in Mainland China. *Educational Research Journal, 25*(2), 211–240.

Tao, P. K. (2002). A study of students' focal awareness when studying science stories designed for fostering understanding of the nature of science. *Research in Science Education, 32*, 97–120.

Tao, P. K. (2003). Eliciting and developing junior secondary students' understanding of the nature of science through a peer collaboration instruction in science stories. *International Journal of Science Education, 25*(2), 147–171.

Tao, P. K., Yung, H. W., Wong, C. K., & Wong, A. (2000). *Living science.* Hong Kong: Oxford University Press.

University Grant Committee. (2010). *Aspirations for the higher education system in Hong Kong.* Hong Kong: UGC.

Wong, S. L., Yung, B. H. W., Cheng, M. W., Lam, K. L., & Hodson, D. (2006). Setting the stage for developing pre-service teachers' conceptions of good science teaching: The role of classroom video. *International Journal of Science Education, 28*(1), 1–24.

Wong, S. L., Hodson, D., Kwan, J., & Yung, B. H. W. (2008). Turning crisis into opportunity: Enhancing student teachers' understanding of the nature of science and scientific inquiry through a case study of the scientific research in Severe Acute Respiratory Syndrome. *International Journal of Science Education, 30*, 1417–1439.

Wong, S. L., Kwan, J., Hodson, D., & Yung, B. H. W. (2009). Turning crisis into opportunity: Nature of science and scientific inquiry as illustrated in the scientific research on Severe Acute Respiratory Syndrome. *Science Education, 18*, 95–118.

Wong, S. L., Yung, B. H. W., & Cheng, M. W. (2010). A blow to a decade of effort on promoting teaching of nature of science. In Y.-J. Lee (Ed.), *The world of science education: Handbook of research in Asia* (pp. 259–276). Rotterdam: Sense Publishers.

Wong, S. L., Wan, Z., & Cheng, M. M. W. (2011). Learning nature of science through socioscientific issues. In T. D. Sadler (Ed.), *Socio-scientific issues in the classroom: teaching, learning and research* (pp. 245–269). Dordrecht, the Netherlands: Springer.

Part III
Science Teacher Education and Diversity

Part III presents studies where multicultural education is included as a key component of science teacher education programmes. Each of these studies have implications for the nature and goals of science education in general.

Science Teachers' Cultural Beliefs and Diversities: A Sociocultural Perspective to Science Education

Nasser Mansour

Introduction

A Sociocultural Perspective to Science Education

A key question for this study was asked by Lemke (2001), 'what does it mean to take a sociocultural perspective on science education?' Essentially, he answered, it means viewing science, science education and research into science education as human social activities conducted within institutional and cultural frameworks (p. 296). Another question from Lemke was, 'what does it mean to view the objects of our concern as "social activities"?' From a research perspective, he claimed, it means, first of all, formulating questions about the role of social interaction in teaching and learning science and in studying the world, whether in classrooms or research laboratories. Additionally, it means giving substantial theoretical weight to the role of social interaction: regarding it, as in the Vygotskyan tradition to be central and necessary to learning and not merely ancillary. Similarly, it means seeing the scientific study of the world as itself inseparable from the social organisation of scientists' activities (p. 296).

Essentially, science studies argue that we can know nature only through culturally constituted conceptual or epistemological frameworks, enabled and limited by local cultural features such as discursive practices, institutional structures, interests, values, cultural norms, and so on (Turnbull 2000). All cultures create their own stories or cosmologies that not only help explain but also provide a sense of wonder and awe about the universe and their place within it. This more inclusive view of science sees it as any systematic attempt to produce knowledge about the

N. Mansour (✉)
Graduate School of Education, University of Exeter, St Luke's Campus,
Heavitree Road, EX1 2LU Exeter, Devon, UK
e-mail: N.Mansour@exeter.ac.uk

N. Mansour and R. Wegerif (eds.), *Science Education for Diversity: Theory and Practice*,
Cultural Studies of Science Education 8, DOI 10.1007/978-94-007-4563-6_10,
© Springer Science+Business Media Dordrecht 2013

natural world, and so it makes room for other local/indigenous, indeed multiple, and previously excluded conceptualisations of scientific knowledge. Hence, scientific knowledge has arisen from local contexts and in response to local needs. Ash (2004: 857) assumes that the origins of everyday science lie in the lived cultural, historical, gestural and spoken practice of children and adults as they directly and indirectly interact with phenomena, including objects that are both living and dead (p. 857). In this respect, Kesamang and Taiwo (2002) identified that in Botswana, as in many African nations, specific local sociocultural factors shaped significantly the thinking of the average Motswana child. In the same vein, Ogunniyi (1988) (cited by Jegede and Okebkola 1988) states that since every human being:

> ... tends to resolve puzzles in terms of the meanings available in a particular socio-cultural environment, the baseline is that the meanings become firmly implanted in the cognitive structure and manifest themselves habitually and may act as templates, anchors, or inhibitors to new learning. (p. 276)

The role of context is important in learning and changing beliefs and practices; therefore, their impact should not be ignored (Mansour 2013). Moreover, Vygotsky focused on the roles that society plays in the thought development of an individual. Vygotsky (1978) believed that human thought developed from the social to the individual – humans beginning as social beings and culminated with inner individuality. In this respect, Wells and Claxton (2002: 2) argue that the way minds grow is not, fundamentally, through didactic instructions and intensive training, but through a more subtle kind of learning in which youngsters pick up useful (or unuseful) habits of mind from those around them and receive guidance in reconstructing these resources in order to meet their own and society's current and future concerns.

Science Teachers' Cultural Beliefs and Diversities

Teacher diversity deserves to be respected both on human grounds and for the sake of effective teaching. The diversity that is the concern of this chapter is not that of ethnicity, gender or age. It is the diversity of teachers' beliefs about teaching and learning, beliefs that guide the way we think about our teaching and the way we teach. In recent years, there has been an exponential rise in more socially and contextually oriented approaches to research, including the study of learners' beliefs in the contexts in which they emerge. In this view, Rust (1994) describes beliefs as 'socially-constructed representation systems' which are used to interpret and act upon the world. He acknowledges the role of context on mental processes. Here, beliefs are seen as fluid and dynamic, not stable entities within the individual. Socioculturally based studies on learner beliefs aim to bring students' emic perspectives into account including ethnographic classroom observations, diaries and narratives, metaphor analysis and discourse analysis (e.g. see Kalaja and Barcelos 2003).

Cultural beliefs filter into our understanding of the world. For example, a western view of isolating and controlling variables may contrast with an American Indian view of connectedness and living in harmony (Grotzer 1996). While paying attention to culture is important, students need to be treated as individuals who are influenced by the contributions of their culture first before treating them as part of a larger stereotyped cultural group (Grotzer 1996). Kathleen Roth's (1992) case study of a 'learning community' approach to teaching and learning is a good example of setting learning in a meaningful classroom community where learning is responsive to the students' voices and diversity. It is this type of learning that is gradually built up as part of the classroom culture based on supportive interpersonal relationships that engage and include all learners. In the Roth study, this included the use of personal writing by the students. Comparing it with the traditional 'work-oriented' classroom setting, where getting right answers, finishing tasks and listening to the teacher as expert, all in a depersonalised way, were the main goals, Roth argued for a 'conceptual change science learning community' where sense making and learning was the main goal and where 'everyone's ideas' voices and identities [were] valued and respected in the learning process' (p. 5). This kind of learning environment can enable diverse learners regardless their race, religion, background or even understanding of the nature of science to get engaged in science activities and to practise and exercise particular kinds of actions (inquiry, questioning, collaborating, etc.) surrounding knowledge that is connected and useful. Indeed, research indicates that students perceive the traditional approach to science education as largely irrelevant to the realities of their complex contemporary world and does not meet their diversity (Millar and Osborne 1998; Ogawa 2001; Fusco and Calabrese Barton 2001; Schreiner and Sjøberg 2004; Roth and Tobin 2007; Carter 2008). In traditional science classrooms that have diverse students, oppositions often exist between the uncomplicated way in which science is presented and the ways in which students' gendered, raced, views, identities and classed values are a part of their own construction of science (Fusco and Calabrese Barton 2001; Roth and Tobin 2007).

Hanrahan (2002) argues that journal writing on her experimental research changed both students' identity and relationships where it introduced a new way of students and teacher relating and hence changed the ethos in the classroom. These students were being treated as people with their own concerns and questions which needed to be addressed in the exchange of ideas involved in becoming enculturated into the new community. The affirmational dialogue journal writing was based on an appreciation of science learning as being part of a sociocultural process of change and, as such, as being a difficult process involving changes in attitudes, in beliefs and in relationships.

Research indicates that educational beliefs and practices are not context-free or separated from the wider sociocultural contexts that teachers are embedded in (Briscoe 1991; Rogoff and Chavajay 1995; Rogoff 2003; Ash 2004; Robbins 2005; Mansour 2010, 2011). These studies also argue that teachers' beliefs and practices cannot be examined outside of sociocultural context, but are always situated in a physical setting in which constraints, opportunities or external influences may derive from sources at various levels, such as the individual classroom, the school, the principal, the community or the curriculum. It is, therefore, necessary to take into account the contextual factors that have shaped and formed certain beliefs. The importance of studying this framework is supported by Olson (1988): 'what teachers tell us about their practice is, most fundamentally, a reflection of their culture and cannot be properly understood without reference to that culture' (p. 69). In this respect, a study by Briscoe (1991) points to the significance of the mental images as sources of knowledge that teachers use in constructing roles for themselves. In Briscoe's study, the teacher used images of typical schools which had constructed from his past experiences as a basis for developing his practices.

Most of the studies related to teachers' knowledge, beliefs and making sense have been carried out in western cultures, not in an Arab-Islamic culture such as Egypt. More importantly, many topics typically included in science education are acknowledged as controversial, e.g. evolution or cloning, and indeed, these issues pose problems for science teachers, especially in Islamic countries, owing to their perceived conflict with the Islamic religious view. In this respect, social constructivism emphasises the importance of culture and context in understanding what occurs in society (Derry 1999; McMahon 1997). Social constructivists view learning as a social process. It does not take place only within an individual, nor is it a passive development of behaviours that are shaped by external forces (McMahon 1997). Meaningful learning occurs when individuals are engaged in social activities. This approach assumes that theory and practice do not develop in a vacuum; they are shaped by dominant cultural assumptions (Martin 1994; O'Loughlin 1995). Social constructivists see as crucial both the context in which learning occurs and the social contexts that learners bring to their learning environment. In this respect, Engeström (1987) points out the relative danger of under-theorising context wherein experience is described and analysed as if consisting of relatively discrete and situated actions, while the system of objectively given context, of which those actions are a part, is either treated as immutable, given or barely described at all.

Teachers' Personal Religious Beliefs as a Case for Teachers' Cultural Beliefs

Teachers are 'agents of change' of educational reform, and their beliefs must not be ignored. Indeed, their pedagogical beliefs are at the 'core of educational change' (Mamlok-Naaman et al. 2007). Sociocultural research can enhance our understanding of science teachers' pedagogical beliefs and how science teachers learn from their experiences in different contexts, such as the university pre-service course, the practicum and the school in which they are employed (Goos 2008). This personal culture is constructed due to the interaction between the individual and the culture(s) around him. This personal culture reflects the individual beliefs system which is different from one to the other who is dealing with same culture(s).

The concept of culture has been identified as one of the two or three most difficult concepts in the English language. There are two fundamentally different senses in which the term has been used: (a) as a 'theoretically defined category or aspect of social life that must be abstracted out from the complex reality of human existence' and (b) as a 'concrete world of beliefs and practices' (Sewell 1999: 39).

Exploring the roles of sociocultural contexts in order to understand teachers' pedagogical beliefs and practice, in a manner grounded in empirical research, would enable evaluators to properly contextualise their findings and track cultural phenomena as both mediating and outcome variables and also would give programme planners and policymakers the tools to better understand their institutions and the ultimate effects of investments in reform (Hora 2008).

This section presents evidence for the sociocultural contexts in which ten Egyptian science teachers are embedded and how these teachers' experiences with these contexts can be used as a framework to understand their pedagogical beliefs. Teachers' sociocultural contexts are illustrated by examples of the verbatim quotations from the transcripts which are set out below in three sections. These show the sociocultural contexts which have formed the background of ten Egyptian science teachers and how these contexts can be used as a framework for understanding teachers' beliefs and practices.

The relationships between and evidence for these main categories are illustrated by examples of the verbatim quotations from the transcripts which are set out below. These show how the teachers' pedagogical beliefs were influenced by the formal and informal learning experiences, which shaped their personal religious beliefs, and how, in turn, these beliefs influenced their actual practices.

A Case Study: Personal Religious Beliefs

Data were collected by means of a semi-structured interview and qualitative observation.

In this study, a multi-grounded theory (MGT) was used to analyse the data. A multi-grounded theory (MGT) approach is a sophisticated model of grounded theory (GT) that deepens both inductive and deductive methods of theory generation (Ezzy 2002).

Table 1 illustrates how the theoretical coding emerged from the data. The initial process of data analysis was done inductively by using an incident-to-incident coding technique (Charmaz 2000). In 'conceptual refinement', a critical stance was adopted to examine the views that participants had expressed. At this point, a crucial one in the data analysis phase, every category that was developed was reflected upon with regard to its ontological status (Lind and Goldkuhl 2006).

The findings of the analysis suggested that it was not the religious context but it was mainly teachers' personal religious beliefs that shaped their pedagogical beliefs and practices. These are some comments of the participants:

Table 1 Emergence of the theoretical coding 'personal religious beliefs'

Inductive coding 'open coding'	Conceptual refinement	Building categorical structures 'axial coding'
Science as a creation of God	Teachers' religious understanding of the sources of science	Personal religious epistemology
Islamic view of science	Teachers' religious beliefs are reflected in how they understand science	
Gaining knowledge as an approach to God	The aim of any study is to approach God	
Stable religion	Teachers enact their Islamic belief	
Qur'an as starting point for scientific research	Understanding Qur'an precedes studying science	Personal religious scientific view of science
Discovering science by Qur'an	Starting searching science should start from studying Qur'an	
Science as wandering around nature	Studying science can be anywhere	
Moral principle in searching science	Teachers reflect their religious principles	
Qur'anic guide to science	Teachers' religious condition for searching science	
Moral principle in searching science	Teachers' religious moral as a framework of gaining science	
Muslim scientists	Teachers' emphasis on the characteristics of scientists	
Changeable science	Teachers' interpretation of the reality of science	
Abusing the nature	Teachers' reflection of Islamic-scientific responsibility towards the nature	
Science content as effective spirituality	Science content as effective spirituality	Religious view of curriculum
Cloning is destructive for society	Teachers' religious understanding of cloning	
Islamic science curriculum	Teachers' view of the development of the science curriculum	
Religious orientation of science content	Teachers' view of the integration between science and religion	
Explaining the supernatural of God by science	Spiritual aim of science	Religious view of the aim of science
Wondering about nature in Qur'an	Teachers' religious reflection of the physical laws of the nature	
Wondering about our body in Qur'an	Teachers' religious understanding of the concept of creation	
Science guide to a good Muslim	Religious attitude towards the role of science	

Item	Category	Theme
Science-religion war	Teachers' interpretation of the relationship between science and religion	Personal interpretation of religious view
Qur'anic motivation of science	Teachers' interests of science by influence of religion	
Islamic responsibility	Teachers' efforts to enact the religious principles	
Contradictions in science not in Qur'an	Teachers' knowledge about science and about the scientific processes	
Simplifying controversial issues religiously	Teachers' efforts to present the religious view	Religious view of teaching/learning science
Religious orientation of science content	Teachers' efforts to present the religious view	
Religion as a guide for teacher education	Teachers' saturation of the religious principles	
Biased religious views	Teachers' efforts to present the religious view	
Presenting science in Qur'an	Teachers' effort to present the religious view	
Islamic view of co-operation	Teachers' intervention of the morals and values based on religion	
Qur'anic verses as a guide for controversial issues	Teachers' religious perspective of teaching controversial issues	
Islamic approach of teaching	Teachers' religious reflection of teaching science	
Qur'anic approach for teaching controversial issues	Teachers' religious reflection of teaching science	
Helping in Islam	Teachers' religious motivation	
Expressing the right view	Evaluative role of teacher	Characteristics of Muslim science teachers
Changing the students' attitudes	Evaluative role of teacher	
Facilitator	Teachers' action	

Personal Religious Beliefs and Pedagogical Practices

Seven teachers articulated that if an opinion relates to religion, there is no controversy about it, and the religious belief is propounded directly. Aside from religious affairs, it was felt that the teacher should not impose his/her opinions on the students. Teacher D said:

> I myself don't force my opinion during the discussion with students, so students can express their opinions freely without being affected by my voiced opinion. However, if the opinions of scientists disagree with religious view, I present only the opinions that conform to our religion and society. (T/D)

Additionally, teachers' religious beliefs affected their pedagogical beliefs in general and their subject-specific pedagogical beliefs in particular. Teacher E had used co-operative learning within a 'drugs' lesson because she wanted her students to learn the Qur'anic concept of co-operation. In the classroom and at the end of discussion of 'the drugs issue', she again mentioned to them the importance of their co-operation in gaining and understanding new information when dealing with the problem from different aspects, and she referred to the verse:

> And help each other in righteousness and piety, and do not help each other in sin and aggression. And fear Allah. Surely, Allah is severe at punishment. (Qur'an, 5, part of verse 2)

Personal Religious Beliefs and Gaining Knowledge

Teacher H described how, upon finishing a lesson and leaving the classroom, he was challenged by one of his students, who asked why they were learning science. The teacher argued with him, putting forward the religious view that it is a responsibility and a duty to pursue knowledge. Teacher H mentioned that the Hadith literature (teachings of the Prophet Mohammed) is full of references to the importance of knowledge and to our duty to seek that knowledge. He pointed out that:

> The Prophet Mohamed said "Seek knowledge, even in China". It is worth mentioning that in the era of the Prophet Mohamed, people walked or used horses or donkeys to travel, so travel from Saudi Arabia to China might take many months. The Prophet Mohamed also said "Seek knowledge from the cradle to the grave", and "Verily the men of knowledge are the inheritors of the prophets". (T/H)

Personal Religious Beliefs and Teachers' Formal and Informal Experiences

Personal Religious Beliefs and Science Teacher Education

Some of the participants were critical of their pre-service teacher education experiences. For example, teacher C commented:

Teacher education was a waste of time, a joke. It didn't give us enough actual teaching experience; I couldn't see any relevance about what we were taught at university and what we teach now, especially regarding STS topics. (T/C)

He further clarified:

What I understand about these issues is that there is a relationship between science and religion but where is the role of teacher education here? (T/C)

Teacher F added:

When we went to do our school practice, we faced situations that were different from those, which we were trained for, especially when we started to teach something like cloning, which is sensitive and is related to our religious beliefs. We got nothing from university or in-service training. (T/F)

Personal Religious Beliefs and Past School Experiences

Other participants indicated that school teaching staff influenced not only their beliefs about science but also the way they later taught science to their students:

Here, teacher E commented:

The model I never forget is Mrs [name deleted by researcher] who taught us biology and who used to relate science to religion. On one occasion she taught us how important water is for everything. Here teacher E read a verse from the Qur'an[1]: "...God has sent down water down from the sky. With it We have produced diverse pairs of plants each separate from the others" [Qur'an 20: 53]. She also read the verse: "And Allah has created every animal from water:..." [Qur'an 24:45]. (T/E)

Then, she added:

I did like this way of teaching. I took her as a model for my own teaching, I do believe that everything is found in the Holy Qur'an and it is very easy to make the students understand or like science by using this Islamic approach to teaching science, especially with regard to controversial issues. (T/E)

Personal Religious Beliefs and Life-Out-of-School Experiences

The analysis revealed that early out-of-school experiences were potentially influential in shaping teachers' beliefs about learning and teaching. Teacher B saw his family as having a major effect on his teaching and his dealings with his students:

My parents are religious people. They brought me up according to the concepts of respect for the opinions of others, equality, responsibility, teamwork, trust, and patience. So, when I teach a lesson like 'pesticide use', which can be a controversial issue, I try to be objective, and stress to all my students that we should give everybody a chance to express his views freely and that all opinions are important. I try to teach my student what I learnt from my family – which is how to argue any controversial issue. (T/B)

---

[1] The English translations of the Qur'anic verses are based on Ali (2004).

The Teachers' Personal Religious Beliefs and Their Relationship to Pedagogy and Practice

Joint analysis of the interviews and the classroom observations revealed that teachers' beliefs regarding the epistemology of science, their roles, the students' roles, the aims of science and their teaching methods were strongly shaped by, and intertwined with, personal religious beliefs.

Personal Religious Beliefs and Epistemology of Science

Islamic religious experiences clearly influenced teachers' views concerning the nature and purpose of science. Science was not perceived as a divine revelation but as a means of promoting the wellbeing of humankind and providing a better understanding of the creation of Allah. Teacher F said:

> What I know about 'Ilm' [science] is that it means knowledge and we study it because our religion, 'Islam' encourages us incessantly to pursue knowledge. For example in the Qur'an, Allah ordained His servants to pray to Him thus: "High above all is Allah, the King, the truth! Be not in haste with the Qur'an before its revelation to you is completed, but say, O my Lord! Advance me in knowledge" [Qur'an 20: 114]. (T/F)

Teacher I views Islam as:

> ... a religion based upon knowledge, for it is eventually knowledge of the Oneness of God, combined with faith and total confidence in Him that saves man. (T/I)

I asked her why she chose inquiry to be her best way of teaching science. In reply, she said:

> The wording of the Qur'an is full of verses inviting man to use his mental powers, to wonder about things, to think, and to know, since the goal of human life is to discover the truth. (T/I)

Teacher H had a remarkable view about the relationship between science and religion:

> Science ... shouldn't be subordinate to culture but at the same time it shouldn't contradict religious concerns. If such a contradiction appears, it is merely an apparent contradiction that results from a misunderstanding of the scientific phenomenon of the religious text. The religious text is stable and untouchable. Thus, if science contradicts religion, the scientist should review the phenomenon and try to understand it correctly. Science can change a society's culture but not its religion. Rather, science can help people understand religion. (T/H)

Personal Religious Beliefs and Beliefs About Teaching/Learning

Teachers' personal religious beliefs or their interpretation of Qur'anic views clearly influenced their pedagogical beliefs, which in turn, powerfully affect their practices. For example, when asked about the teaching of cloning, the following comments were made:

The main consideration is our society is an Islamic one. For this reason, I should initiate the lesson on cloning with an Islamic introduction beginning with the Qur'anic verse that there should be husbands and wives or males and females. While I am explaining the lesson, I will confirm that we can take from cloning what is positive and leave what is negative. What is positive is that we can use cloning with plants and other living things rather than with humans. (T/C)

Teachers' personal religious beliefs also influenced their learning aims and how they achieved them. Teacher D, for example, said:

My main aim for an issue like cloning will be to analyse and show the students the scientific information on cloning, as well as evaluating the moral and ethical implications associated with it [cloning] from an Islamic point of view. (T/D)

Personal Religious Beliefs and Classroom Practices

Most of the teachers tried to start their science lessons with an appropriate verse (from the Qur'an) or a Hadith (sayings of the Prophet). For example, when he taught the unit about 'water' and in order to explain the idea that water was one of the most valuable natural elements on Earth, teacher B mentioned that God had said:

We made from water every living thing. [Qur'an 21: part of verse 30]

In another lesson on 'the Atom', teacher A wrote the following verse on the blackboard at the beginning of the lesson:

The unbelievers say, "never to us will come the hour": Say, "Nay! But most surely, by my Lord, it will come upon you; – by Him who knows the unseen, – from whom is not hidden the least little atom in the heavens or on earth: nor is there anything less than that, or greater, but is in the record perspicuous". [Qur'an 34:3]

Coloured chalk was used to highlight these words from the verse: 'atom', 'weight' and 'nor greater', and the teacher said that in light of modern scientific findings, 'the smallest possible part of matter' was called a molecule and began to explain the structure of an atom by pointing to the words 'less than' in the verse. He said that this meant that an atom included all particles, discovered or undiscovered:

I mean by 'discovered', nuclei, electrons and protons. And by 'discovering', I mean that by developing and advancing or through tools and methods of research, more parts or characteristics can be discovered in the future which we don't yet know. Then, he began to go through the verse in detail, relating it to the subject of the lesson, although:

... 'greater than that' – that includes chemical compounds, which I will discuss later.

He told his pupils that at the time of Dalton, an atom was the smallest, invisible particle of matter. This idea was no longer correct, so it was necessary to be very careful when trying to understand the Holy Qur'an in the light of scientific findings. He said:

The Holy Qur'an was a book that could not be doubted. (T/A)

Teacher H provided another example of how teachers' personal religious beliefs could affect the way they put their epistemological and pedagogical beliefs into practice. He began his lesson about 'how water is formed' with this verse from the Qur'an:

> See you not that Allah makes the clouds move gently, then joins them together, then makes them into a heap? Then will you see rain issue forth from their midst. And He sends down from the sky mountain masses (of clouds) in which is hail: He strikes with it whom He pleases and He turns it away from whom He pleases. The vivid flash of His lightning well-near blinds the sight. [Qur'an, 24:43]

At the end of the lesson, teacher H told students that the Holy Qur'an contained all knowledge about the universe. So when it was necessary to understand their environment, they also need to understand the Holy Qur'an very well. Also, water should be protected from pollution. He reminded them that the Prophet had lived in a harsh desert environment, where water was equal to life. As a gift from God, water is the source of all life on earth, as is confirmed in the Qur'an:

> Do not the unbeliever see that the heavens and the earth were joined together (as one unit of creation), before We split them apart? We made from water every living thing. Will they not then believe? [Qur'an 21:30]

Teacher H also pointed out that the Qur'an constantly reminded and encouraged individuals to keep water clean and not to abuse it and, in this connection, mentioned these verses:

> See you the water which you drink? [Qur'an 56: 68]
> Do you bring it down (in rain) from the cloud or do We? [Qur'an 56:69]
> Were it Our will, We could make it salt (and unpalatable): then why do you not give thanks? [Qur'an 56:70]

Discussion and Theoretical Grounding of the Personal Religious Beliefs (PRB) Model

This section presents the 'explicit grounding stage' of the multi-grounded theory approach. Using a theory-matching process, the empirically derived theory 'the personal religious beliefs (PRB) model' was compared with theories found in the literature. This process was used to seek internal and external validation of the PRB model (Goldkuhl and Cronholm 2003) and was an interactive comparison of the derived theory with existing theories.

The dimensions of the developed model include:

- Personal religious beliefs, teachers' experience and teachers' interpretation
- Teachers' interpretations of their experiences and the forming of their pedagogical beliefs
- Teachers' pedagogical beliefs, their framework for action and practice
- Knowledge and teachers' beliefs

- Teachers' identity as a product of the interaction between their personal religious beliefs, experiences, pedagogical beliefs and practices

The following paragraphs explain these dimensions and match the dimensions of the PRB model with the existing theories.

Personal Religious Beliefs, Teachers' Experiences and Teachers' Interpretation

Analysis of the interviews together with the classroom observations revealed that teachers' beliefs regarding their roles, students' roles, the aims of science and their teaching methods were strongly shaped by personal religious beliefs derived from the values and instructions inherent in the religion. The present study found that teachers' personal religious beliefs worked as a 'schema' which influenced what was perceived (McIntosh 1995). McIntosh defined a schema as 'a cognitive structure or mental representation containing organized, prior knowledge about a particular domain' (1995: 2). He also noted that schemas were built via encounters with the environment 'social context' and could be modified by experience.

The religious schemas of these teachers influence the way they perceive new experiences. Teachers arrange the elements of their social context to reflect the organisation of their own personal religious beliefs or religious schemas. A teacher with personal religious beliefs or religious schemas is more likely to force a religious interpretation on experience than a teacher without such personal religious beliefs or religious schemas. Moreover, teachers with particular personal religious beliefs may understand the situation or the experience very differently from those without these personal religious beliefs. These beliefs, in turn, work through the lens of past experiences, since they are translated into teacher's practices within the complex context of the classroom.

The study found, furthermore, that teachers' personal religious beliefs controlled the gaining of new knowledge and experiences. Ball-Rokeach et al. (1984) proposed that a person's value-related attitudes towards objects and situations and the organisation of values and beliefs about self formed a comprehensive belief system that provided an individual with a cognitive framework, map or theory. In this respect, the models explaining the influence of experiences on teachers' beliefs and practices (e.g. Knowles 1992) are largely supported by the findings of this study, which established that early and teacher-education 'formative experiences' were initially interpreted by individual teachers through their religious beliefs.

The influence of personal religious beliefs on other kinds of experience is represented in Fig. 1 by bold arrows that point from 'personal religious beliefs' to 'teachers' experiences' as well as to shaping teachers' beliefs and practices. The developed PRB model also shows that personal experiences can affect teachers' personal beliefs. However, the interactive influence between teachers' experiences and their personal religious beliefs is not equal. Personal religious beliefs are the stronger influence.

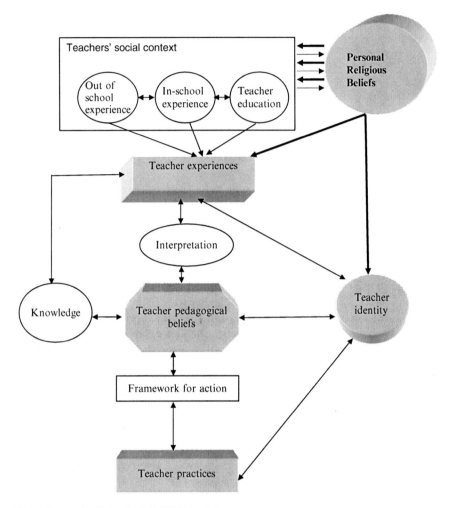

Fig. 1 Personal religious beliefs (*PRB*) model

Teachers' Interpretations as a Link Between the Experiences and Beliefs

The study supported the idea that teachers were not just simply formed or socialised by their lifetime experiences; they were, in fact, active participants in interpreting these experiences (Sexton 2004). According to Knowles (1992), the particular interpretation assigned to an experience was transformed to a 'schema', which he defined as 'a way of understanding or a cognitive filter and a basis for teacher-centred classroom practices' (1992: 138). In the present study, the term 'instructional schema' meant a settled system of pedagogical beliefs following the process of filtering by teachers' previous religious beliefs and experiences. In this respect, the results of the present study coincided with the arguments of Knowles and Holt-Reynolds (1991) that teachers' prior experiences had moulded their educational thinking and that through the interpretations of these experiences, teachers formed the beliefs that they used directly to evaluate their own teaching practices.

The findings of the study also agreed with Knowles (1992) that the interpretation and subsequent schema developed by an individual with regard to classroom practices and other relevant experiences were highly idiosyncratic; individuals experiencing a singular event would have multiple perspectives on that event. The schema or settled beliefs determine the manner in which teacher might take certain steps, so that the schema becomes an evaluative tool for examining teacher's practices and is transformed into a framework for action. As the study shows, teachers who view science as a body of knowledge rely on textbooks to assist them in transmitting science knowledge. Also, a teacher who believes that science is merely a body of knowledge to be acquired will have a very different approach to teaching science from one who believes science is a way of making sense of the world, of asking questions and seeking answers and of observing and exploring.

These findings concur with Richardson (1996) who found that teachers' beliefs were among the major constructs driving teachers' ways of thinking and classroom practices. Johnson (1992) reported research on literacy teaching that supports the notion of beliefs tending to shape teachers' instructional practices. That conclusion was also supported by Schoenfeld (1998), who claimed that teachers' beliefs shaped what they perceived in any set of circumstances, what they considered to be possible or appropriate in those circumstances, the goals they might establish in those circumstances and the knowledge they might bring to bear in them. So far, the developed PRB model (Fig. 1) has highlighted the idea that teachers' interpretation is the link or the transmitter between teachers' experiences and has formed teachers' beliefs. The PRB model also shows that interactive relationships either between 'teachers' experiences' and 'teachers' interpretations' or between 'teachers' interpretations' and 'teachers' beliefs' are in fact reciprocal relationships.

Teachers' Pedagogical Beliefs: Their Framework for Action and Practice

The analysis and interpretation of data on the process used by teachers to transfer teacher's beliefs into practice found that teachers tended to use the history of their own schooling and, in particular, specific teacher role models to guide their own practices. Maslovaty (2000) noted that a teacher's belief system, crystallised through a cultural context, resulted in the development of different educational ideologies. Maslovaty also found that teachers' social value orientation contributed to the choice of strategy to cope with socio-moral dilemmas (in the present study, the choice of strategy is called 'framework for action'). However, transforming this framework of action into real practice in the classroom depended on other contextual factors, e.g. constraints, school environment, teachers' personal religious beliefs and experiences and teacher's identity.

This conclusion was supported by Talbert and McLaughlin (1993) who defined the 'context effect' as a notion implying that conditions such as policies, resources, curricula, goals, values, norms, routines and social relations in the school influenced teaching and learning outcomes. The PRB model presents the idea of a 'framework for action' to indicate that teachers intend to enact their beliefs in the classroom. It also makes clear that other factors limit or facilitate the operation of teachers' plans or frameworks for action. Figure 1 shows a reciprocal interaction between teachers' practices and the future framework of action.

Knowledge and Teachers' Beliefs

The powerful influence of teachers' beliefs in general or teachers' personal religious beliefs in particular on gaining knowledge-related controversial issues was highlighted by the findings. However, the settled or developed teachers' beliefs 'schema' acted as an information organiser and priority categoriser and, in turn, controlled the way it could be used. In the interactions between knowledge and beliefs, beliefs controlled the gaining of knowledge and knowledge influenced beliefs. This suggested that teachers needed to create their own knowledge through a process of interaction between their existing beliefs and knowledge base, and the new ideas with which they came into contact (Richardson 1996). Dadds (1995) and Lichtenstein et al. (1992) suggested that increased content knowledge went hand in hand with increased confidence, while having knowledge about teaching carried its own kind of authority that had the potential to empower teachers.

Teachers' Identity as a Product of the Interaction Among Their Personal Religious Beliefs, Experiences, Pedagogical Beliefs and Practices

The study's findings concurred with those of Cole (1990) and Knowles (1992) that a teacher's role identity was determined by early family experiences, being young students, teacher role models, previous teaching experiences, and other significant prior experiences. However, the current study added teachers' personal religious beliefs as one of the main formative influences on teachers' identity. As long as a teacher's experience changes daily, his/her identity changes sequentially. This conclusion agrees with Yerkes (2004), who claimed that 'Identity is not set in stone'. Identity is always changing. A teacher's experiences play an essential role in his or her identity. Each teacher has different experiences, which is what makes all teachers unique. Thus, identity and identity construction are ongoing processes.

Concerning the dynamic relationship between teachers' identity, experience, beliefs and context, the study agreed with Wenger (1998) who pointed to five salient aspects of identity: (1) identity is related to one's personal history; (2) one's identity is also related to one's experience as negotiated within the context of existing cultural practices, complete with their categories and cultural histories; (3) identities are related to membership in communities; (4) people are members of multiple communities, and thus one's identity is at the nexus of these multiple memberships; and (5) one's identity at a given moment is an interaction between local and global contexts. This formulation provides bridge between the intensely personal nature of teaching and its very public and cultural aspects.

Not only do different experiences, and the belief systems that are subsequently formed, create the basis for teacher role identity; they also determine the (negative or positive) orientation of that identity. This study also suggests that the nature of teacher role identity (whether negative or positive) determined the extent of the influence on the teacher of social constraints or the school environment. The findings further indicated that pedagogical beliefs and practices were influenced by the results of the interaction between teacher's identity and social constraints or school environment. If a science teacher who has a positive teacher role identity works in a school environment which there are many constraints (e.g. pressure of examinations, large classes, lack of resources, students' family background, lack of time, school administration), his/her expected practice might be negative traditional practices or a mix of traditional and constructivist practices. The PRB model (Fig. 1) shows that teacher identity is a social product of the interaction among personal religious beliefs, teachers' experiences, teachers' beliefs and teachers' practice. However, teachers' personal religious beliefs produced the strongest influence on forming teachers' identity.

Conclusion and Implication

Teachers' Personal Religious Beliefs

Personal religious experience is one of the most influential social factors in gaining new experiences or interpreting these experiences, and this, in turn, influences teachers' pedagogical beliefs and practices (Roth and Alexander 1997; Shipman et al. 2002; Colburn and Henriques 2006; Stolberg 2008). Roth and Alexander (1997) explain that one's personal experience is mediated by the discursive practices of the community within which one lives, and they use this mediated experience as an example of the social construction of the personal dimensions of religion. Dagher and BouJaoude's (1997) study of college biology seniors revealed how students' worldviews, including their personal religious beliefs (PRB), aesthetic values and understanding of the nature of scientific theories, shaped their understanding and acceptance of the theory of biological evolution. Mansour (2008a) argues that PRB was one of the most powerful factors influencing science teachers' performance in the science classroom. PRB is a social construct, based broadly on the various experiences including in particular religious experiences that a person lives through (p. 1608). PRB is defined as 'views, opinions, attitudes, and knowledge constructed by a person through interaction with her/his socio-cultural context through her/his life history and interpreted as having their origins in religion. The PRB works as a framework for understanding events, experiences and objects on an individual level' (p. 1608) but also leads to diverse views about science and scientists in relation to religious discourse.

Teachers with particular personal religious beliefs may understand the situation or the experience in question very differently from those without such beliefs (McIntosh 1995; Knowles 1992). Reiss (2004) argues that within a particular society, there are certain characteristics among individuals (such as gender, religious beliefs, ethnicity, age, power, wealth and disability) that cause them to vary in their scientific understanding and conception of the world. Teachers' worldviews regarding science and religion also inform their own roles, practices and approaches to classroom teaching (Dagher and BouJaoude 1997).

Data analysis found that teachers' perceptions of the Islamic religious context that guided their criteria or basis for interpretation of their roles, the students' roles, the aims of science and their teaching methods were strongly shaped by personal religious beliefs derived from the values and instructions inherent in the religion. In addition, analysis of the interviews showed that teachers' personal Islamic religious beliefs imbued their beliefs concerning the role, nature, purpose and function of teaching science. In this respect, the study supported the idea that teachers were not just simply formed or socialised by their religious context; they were, in fact, active participants in interpreting these experiences (Sexton 2004). According to Knowles (1992), the particular interpretation assigned to an experience was transformed to a 'schema', which he defined as 'a way of understanding or a cognitive filter and a basis for teacher-centred classroom practices' (1992: 138). In this respect, the results of the present study coincide with the arguments of Knowles and Holt-Reynolds (1991) that teachers' prior experiences had moulded their educational thinking and that through the interpretation of these religious experiences, teachers formed the beliefs they used directly to evaluate their own teaching practices (Mansour 2008a). Loo (2001) argues that Islam, as one of the world's major religions, clearly has had, and will continue to play, a very important role in adjudicating the interaction between the philosophical and the social/cultural/religious environments of science.

From Teachers' Social Identity to Professional Identity

The current study emphasises that teachers' interactions with their sociocultural contexts formed their experiences, and the study supports the view that teachers were not just simply formed or socialised by the sociocultural contexts in which they operate, but they were, in fact, active participants in the interactions with these sociocultural contexts, which created the conditions for how they teach in schools. Teachers' interactions with, and internalisations of, their sociocultural experiences were transformed in many cases into teaching practices. This observation has validated many authors' arguments, e.g. Ringer (2001) who claimed that the educational system transmits, confirms, validates and perpetuates the knowledge, ideas and concepts that have emerged as dominant. By the same token, Shore (1996) argues that individuals internalise information and experiences from their physical and sociocultural environment; they become deeply embedded in the cognitive processes of the brain through repetition, reinforcement and attachment to key life events or emotions. In this respect, the results of the present study coincide with the arguments of Knowles and Holt-Reynolds (1991) in that teachers' prior experiences had moulded their educational thinking and that through the interpretations of these experiences, teachers formed the beliefs that they used directly to evaluate their own teaching practices. In this vein, the findings concur with Wertsch et al. (1993) that what individuals believe, and ultimately, how individuals think and act, is always shaped by cultural, historical and social structures that are reflected in mediational tools such as literature, art, media, language, technology and numeracy systems. In this sense, the study concludes that teachers are in a continuous process of constructing and reconstructing views about their pedagogical beliefs and practices in relation to others involved in a range of sociocultural contexts. This continuous process of interacting with the sociocultural contexts will lead to forming teachers' professional identities. Lave (1996) states it this way: 'Crafting identities is a social process, and becoming more knowledgeably skilled is an aspect of participation in social practice, who you are becoming shapes crucially and fundamentally what you know' (p. 57). However, transforming teachers' beliefs and ideas into real practice in the classroom depends on other contextual factors, e.g. constraints, school environment, teachers' personal religious beliefs and experiences and teacher's identity (Mansour 2008a). That conclusion is supported by Schoenfeld (1998) who claimed that teachers' beliefs shaped what they perceived in any set of circumstances, what they considered to be possible or appropriate in those circumstances, the goal they might establish in those circumstances and the knowledge they might bring to bear in them.

From a social perspective, teachers in this study are understood to enter their science classrooms with prior knowledge, beliefs and experiences that they can then employ to make sense of their students, instructional practices and school contexts (Saka et al. 2009). Abstracting from Vygotsky's (1978) notion of internalisation, it could be said that teachers themselves have internalised what a 'teacher' is and what a 'student' is in relation to how classes are conducted. Prior to starting on this career path, these individuals were students themselves and had assimilated, over time, similar assumptions about the roles of teacher and student in the institution of academia (Mansour 2008b). The outcomes and inferences of this study concur with Sexton (2007) in that entry-level teachers did remember their time spent in the classroom as students. It was those memories of actions, both taken and not taken, by teachers that influenced the type of teacher they did, and did not, want to become.

Sociocultural Contexts and RBR as a Framework of Teaching Science and Religion Issues

I endeavour to point out that the concept of science in a religion as shown on the PRB model (see Mansour 2008a) will depend on the interpretations of the religious principles as understood by its followers at a certain period. Religion influences science only to the extent that its interpreters could persuade other people to adapt their conceptions. In fact, it would be misleading for our purpose of teaching/learning science to consider the religious conceptions alone without taking into account the other sociocultural contexts in the situation that may collectively influence science.

By dealing and interacting with the sociocultural contexts, teachers create their own zone of understanding and interpretation of Islam related to science. This zone, as shown in Fig. 2, is the personal religious beliefs or 'PRB zone'. Teachers sometimes created a false contradiction between Islam and science due to their individual interpretations of the nature of Islam and science. That is why, as shown in the top left of Fig. 2, there is a big gap between teachers' understanding, interpretations, epistemology and ontology of the socio-scientific issue related to religion, on one side, and the religion's epistemology and religion of the same issue, on the other side. This gap might be created due to the lack of the awareness by the right understanding of religious beliefs (RB zone) of science or a controversial issue.

Most of the teachers' religious experiences related to teaching controversial issues were from informal sources (including family, previous teachers and the media). Educational decision-makers and science educators around the world should be made aware that teachers' personal religious beliefs within sociocultural context are a highly effective variable that can have a positive or negative influence on the entire educational process. It was also shown that teachers' personal religious beliefs could be considered a positive factor in developing positive attitudes among teachers towards science and teaching science. It is therefore suggested that decision-makers, curriculum developers and science educators should engage in thoughtful reflection and discussion about developing various study programmes. These would act as formal knowledge sources about the relationship between science and religion and would also train teachers how to debate issues related to science and religion.

To minimise the gap between the RB zone and the PRB zone, a formal experience about the relationship between science and religion should be based on the coordination among the scientific institutions, and the religious one is much needed with considering the other sociocultural contexts. I agree with the position that compatibility is needed between religious education and science education. In cultures where religion has a major influence on people's lives, the development of science curricula should be made in a partnership between science educators and religion scholars, especially with regard to socio-scientific issues associated with religion. This process would provide opportunities to challenge teachers' personal religious beliefs, to introduce appropriate perceptions of religious attitudes, and to leave the door open for different views and different understandings. By this

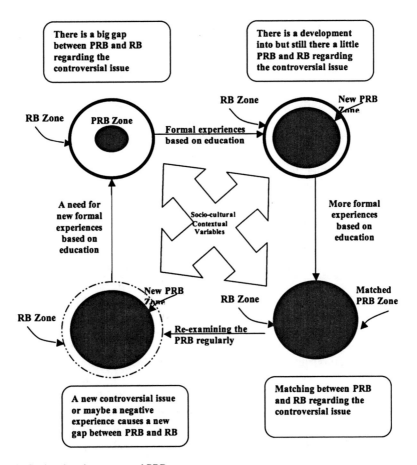

Fig. 2 Sociocultural contexts and PRB

educational process, PRB zone will get to the stage to match the RB zone. However, by the developing advanced technology and developing the scientific research, a new controversial issue may emerge which in turn will cause a new gap between PRB and RB. This will require a regular examination of the PRB and a regular training. Also, in cultures where religion has a major influence on people's lives, the development of science curricula should be made in a partnership between science educators and religion scholars, especially with regard to socio-scientific issues associated with religion.

Science Teachers as Socially Active Agents

In this chapter, I have argued that teachers as active agents in society might be restricted by sociocultural forces (e.g. the examination system, lack of time, work overload, high student density in the classroom, lack of resources or materials, the content) that set strict limits on what these teachers can achieve (Giddens 1984). Human actions are largely determined by the social structures that people inhabit (Hodkinson 2004). Similarly, Dirkx et al. (2000) argue that teachers are profoundly influenced by the social structures in which they operate and that these shape their future choices. Rogoff (2003) argues that 'people develop as participants in cultural communities. Their development can be understood only in light of the cultural practices and circumstances of their communities – which also change' (Rogoff 2003: 3–4). The study's findings showed that there were certain people with whom teachers dealt during the educational process, e.g. the school administration and science inspectors, educational decision-makers and their aims, the family and the learners themselves. Since all these people affected teachers' beliefs and practices in one way or another, there is therefore a need to investigate the role played by the beliefs of faculty/staff, administrators, principals and students' parents.

The study shows that science teachers are part of a complex dynamic; their beliefs, knowledge, values and actions shape and are shaped by the structural and cultural features of society and school cultures. While it is true that teachers are not simply pawns in the reform process – they are active agents, whether they act passively or actively – their actions are mediated by the structural elements of their sociocultural setting, such as the resources available to them, the norms of their school and the externally mandated policies (Lasky 2005; Datnow et al. 2002). Overall, this study provides science teacher educators with insights into how teachers view their professional roles, in the aspiration that this will help them determine the types of experiences that are important for these teachers as they enter the profession. The findings of the study showed that teachers alone cannot be responsible for the quality of their classroom practices; external contextual factors can be a barrier for teachers in putting their theories into practice. These constraints are socially constructed and can be modified, if not deconstructed and reconstituted. In order to achieve this change, many things should be changed at different levels, starting with the objectives of educational decision-makers, the examination system, teacher professional development, science curricula, etc. Also, decision-makers should consider teachers' views and perspectives of the educational policy and educational system when they implement changes or reforms related to science education.

References

Ash, D. (2004). Reflective scientific sense-making dialogue in two languages: The science in the dialogue and the dialogue in the science. *Science Education, 88*(6), 855–884.

Ball-Rokeach, S., Rokeach, M., & Grube, J. (1984). *The Great American values test.* New York: Free Press.

Briscoe, C. (1991). The dynamic interactions among beliefs, role metaphors, and teaching practices: A case study of teacher change. *Science Education, 75*(2), 185–199.

Carter, L. (2008). Sociocultural influences on science education: Innovation for contemporary times. *Science Education, 92,* 165–181.

Charmaz, K. (2000). Grounded theory: Objectivist and constructivist methods. In N. K. Denzin & Y. S. Lincoln (Eds.), *Handbook of qualitative research* (2nd ed., pp. 509–535). Thousand Oaks: Sage.

Colburn, A., & Henriques, L. (2006). Clergy views on evolution, creationism, science, and religion. *Journal of Research in Science Teaching, 43*(4), 419–442.

Cole, A. L. (1990). Personal theories of teaching: Development in the formative years. *The Alberta Journal of Educational Research, 36*(3), 203–222.

Dadds, M. (1995). *Passionate enquiry and school development: A story about teacher action research.* London: The Falmer Press.

Dagher, Z. R., & BouJaoude, S. (1997). Scientific views and religious beliefs of college students: The case of biological evolution. *Journal of Research in Science Teaching, 34*(5), 429–445.

Datnow, A., Hubbard, L., & Mehen, H. (2002). *Educational reform implementation: A co-constructed process.* London: Routledge.

Derry, S. J. (1999). A fish called peer learning: Searching for common themes. In A. M. O'Donnell & A. King (Eds.), *Cognitive perspectives on peer learning* (pp. 197–211). Mahwah: Lawrence Erlbaum Associates.

Dirkx, J., Kushner, J., & Slusarski, S. (2000, June 2–4). *Are we walking the talk? Questions of structure and agency in the research on teaching in adult education.* A paper presented on An International Conference "The Right Questions: Researching in a New Century" The 41st Annual, Adult Education Research Conference, University of British Columbia, Vancouver, BC, Canada.

Engeström, Y. (1987). *Learning by expanding: An activity – Theoretical approach to developmental research.* Helsinki: Orienta-Konsultit Oy. Retrieved from http://www.lchc.ucsd.edu/MCA/Paper/Engestrom/expanding/toc.htm

Ezzy, D. (2002). *Qualitative analysis: Practice and innovation.* London: Routledge.

Fusco, D., & Calabrese Barton, A. (2001). Representing student achievements in science. *Journal of Research in Science Teaching, 38*(3), 337–354.

Giddens, A. (1984). *The constitution of society: Outline of the theory of structuration.* Berkeley: University of California Press.

Goldkuhl, G., & Cronholm, S. (2003, March 20–21). *Multi-grounded theory – Adding theoretical grounding to grounded theory.* Paper presented at the 2nd European Conference on Research Methods in Business and Management (ECRM 2003), UK.

Goos, M. (2008). Towards a sociocultural framework for understanding the work of Mathematics teacher-educator-researchers (pp. 235–241). In M. Goos, R. Brown, & K. Makar (Eds.), *Navigating currents and charting directions* (Proceedings of the 31st Annual Conference of the Mathematics Education Research Group of Australasia).

Grotzer, T. A. (1996). *Math/Science matters: Issues that impact equitable opportunities for all math and science learners.* Cambridge, MA: Harvard Project on Schooling and Children, Exxon Education Foundation. Essay #1: Teaching to Diversity: Math and Science Learning for All Children.

Hanrahan, M. (2002, July 11–14). *Learning science: Sociocultural dimensions of intellectual engagement.* Paper prepared for the ASERA Conference, Townsville, QLD.

Hodkinson, P. (2004, November 12–14). *Theoretical constructions of workplace learning: Troubling dualisms and problems of scale.* A paper presented on European Society for Research on the Education of Adults (ESREA) Research Network Working Life and Learning, Northern College, Barnsley, UK.

Hora, M. T. (2008). *Using cultural models to understand faculty sense-making processes within the structural and socio-cultural context of a comprehensive university.* Paper presented at the 2008 American Educational Research Association Annual Meeting, New York, NY.

Jegede, O. J., & Okebukola, P. A. O. (1988). An educology of socio-cultural factors in science classrooms. *International Journal of Educology, 2,* 93–107.

Johnson, K. (1992). The relationship between teachers' beliefs and practices during literacy instruction for non-native speakers of English. *Journal of Reading Behavior, 24*(1), 83–108.

Kalaja, P., & Barcelos, A. M. F. (Eds.). (2003). *Beliefs about SLA: New research approaches.* Dordrecht: Kluwer Academic Press.

Kesamang, M. E. E., & Taiwo, A. A. (2002). The correlates of the socio-cultural background of Botswana junior secondary school students with their attitudes towards and achievements in science. *International Journal of Science Education, 24*(9), 919–940.

Knowles, J. G. (1992). Models for understanding pre-service and beginning teachers' biographies: Illustration from case studies. In I. F. Goodson (Ed.), *Studying teachers' lives* (pp. 99–152). London: Routledge.

Knowles, J. G., & Holt-Reynolds, D. (1991). Shaping pedagogies through personal histories in pre-service teacher education. *Teachers College Record, 93*, 87–113.

Lasky, S. (2005). A sociocultural approach to understanding teacher identity, agency and professional vulnerability in a context of secondary school reform. *Teaching and Teacher Education, 21*, 899–916.

Lave, J. (1996). Teaching, as learning, in practice. *Mind, Culture, and Activity, 3*, 149–164.

Lemke, J. L. (2001). Articulating communities: Sociocultural perspectives on science education. *Journal of Research in Science Teaching, 38*(3), 296–316.

Lichtenstein, G., McLaughlin, M. W., & Knudsen, J. (1992). Teacher empowerment and Professional knowledge. In A. Lieberman (Ed.), *The changing contexts of teaching* (pp. 37–58). Chicago: The University of Chicago Press.

Lind, M., & Goldkuhl, G. (2006). How to develop a multi-grounded theory: The evolution of a business process theory. *Australasian Journal of Information Systems, 13*(2), 1–10.

Loo, S. (2001). Islam, science and science education: Conflict or concord? *Studies in Science Education, 36*, 45–77.

Mamlok-Naaman, R., Hofstein, A., & Penick, J. (2007). Involving science teachers in the development and implementation of assessment tools for "Science for All" type curricula. *Journal of Science Teacher Education, 18*, 497–524.

Mansour, N. (2008a). The experiences and Personal Religious Beliefs of Egyptian science teachers as a framework for understanding the shaping and reshaping of their beliefs and practices about Science-Technology-Society (STS). *International Journal of Science Education, 30*(12), 1605–1634.

Mansour, N. (2008b). *Models of understanding science teachers' beliefs and practices: Challenges and potentials for science education.* Saarbrucken: VDM Verlag Dr. Mueller e.K.

Mansour, N. (2010). The impact of the knowledge and beliefs of Egyptian science teachers in integrating a STS based curriculum: A sociocultural perspective. *Journal of Science Teacher Education, 21*(4), 513–534.

Mansour, N. (2011). Egyptian science teachers' views of science and religion vs. Islamic perspective: Conflicting or complementing? *Science Education, 95*, 281–309.

Mansour, N. (2013). Modelling the sociocultural contexts of science education: The Teachers' perspective. *Research in Science Education, 43*, 347–369. doi:10.1007/s11165-011-9269-7.

Martin, R. J. (1994). Multicultural social reconstructionist education: Design for diversity in teacher education. *Teacher Education Quarterly, 21*(3), 77–89.

Maslovaty, N. (2000). Teachers' choice of teaching strategies for dealing with socio-moral dilemmas in the elementary school. *Journal of Moral Education, 29*(4), 429–444.

McIntosh, N. (1995). Religion-as-Schema, with implication for the relation between religion and coping. *The International Journal for the Psychology of Religion, 5*(1), 1–16.

McMahon, M. (1997, December). *Social constructivism and the world wide web– A paradigm for learning.* Paper presented at the ASCILITE Conference, Perth, Australia

Millar, R., & Osborne, J. (1998). *Beyond 2000: Science education for the future.* Retrieved August 28, 2006, from http://www.kcl.ac.uk/education

Ogawa, M. (2001). Reform Japanese style: Voyage into an unknown and chaotic future. *Science Education, 85*, 586–606.

Ogunniyi, M. B. (1988). Adopting Western science to traditional African culture. *International Journal of Science Education, 10*, 1–9.

Olson, J. (1988). Making sense of teaching: Cognition vs. culture. *Journal of Curriculum Studies, 20*, 167–169.

O'Loughlin, M. (1995). Daring the imagination: Unlocking voices of dissent and possibility in teaching. *Theory into Practice, 24*(2), 107–116.

Reiss, M. J. (2004). What is science? Teaching science in secondary schools. In E. Scanlon, P. Murphy, J. Thomas, & E. Whitelegg (Eds.), *Reconsidering science learning* (pp. 3–12). London: RoutledgeFalmer.

Richardson, V. (1996). The role of attitude and beliefs in learning to teach. In J. Sikula, T. J. Buttery, & E. Guyton (Eds.), *Handbook of research on teaching education* (pp. 102–119). New York: Macmillan.

Ringer, M. (2001). Education in the Middle East: Introduction. *Comparative Studies of South Asia, Africa and the Middle East, 21*(1&2), 3–4.

Robbins, J. (2005). Contexts, collaboration, and cultural tools: A sociocultural perspective on researching children's thinking. *Contemporary Issues in Early Childhood, 6*(2), 140–149.

Rogoff, B. (2003). *The cultural nature of human development*. Oxford: Oxford University Press.

Rogoff, B., & Chavajay, P. (1995). What's become of research on the cultural basis of cognitive development? *American Psychologist, 50*(10), 859–877.

Roth, K. J. (1992). *The role of writing in creating a science learning community* (Elementary subjects center series No. 56.). East Lansing: Michigan State University, The Centre for the Learning and Teaching of Elementary Subjects (ERIC Reproduction Service No. ED 352 259).

Roth, W.-M., & Alexander, T. (1997). The interaction of students' scientific and religious discourses: Two case studies. *International Journal of Science Education, 19*(2), 125–146.

Roth, W.-M., & Tobin, K. (2007). *Science, learning, identity: Sociocultural and cultural-historical perspectives*. Rotterdam: Sense Publishers.

Rust, F. (1994). The first year of teaching: It's not what they expected. *Teaching and Teacher Education, 10*, 205–217.

Saka, Y., Southerland, S., & Brooks, J. (2009). Becoming a member of a school community while working toward science education reform: Teacher induction from a Cultural Historical Activity Theory (CHAT) Perspective. *Science Education, 93*, 996–1025.

Schoenfeld, A. H. (1998). Toward a theory of teaching-in-context. *Issues in Education, 4*(1), 1–94.

Schreiner, C., & Sjøberg, S. (2004). *Sowing the seeds of ROSE. Background, rationale, questionnaire development and data collection for ROSE* (The relevance of science education) – A comparative study of students' views of science and science education. Oslo: Department of Teacher Education and School Development, University of Oslo.

Sewell, W. H. (1999). The concept(s) of culture. In V. E. Bonnell & L. Hunt (Eds.), *Beyond the cultural turn* (pp. 35–61). Berkeley: University of California Press.

Sexton, S. (2004). Prior teacher experiences informing how post-graduate teacher candidates see teaching and themselves in the role as the teacher. *International Education Journal, 5*(2), 46–57.

Sexton, S. (2007). Power of practitioners: How prior teachers informed the teacher role identity of thirty-five entry-level Pre-service teacher candidates. *Educate, 7*(2), 46–57.

Shipman, H. L., Brickhouse, N. W., Dagher, Z., & Letts, W. J. (2002). Changes in student views of religion and science in a college astronomy course. *Science Education, 86*(4), 526–547.

Shore, B. (1996). *Culture in mind: Cognition, culture, and the problem of meaning*. New York: Oxford Press.

Stolberg, T. L. (2008). Understanding the approaches to the teaching of religious education of preservice primary teachers: The influence of religio-scientific frameworks. *Teaching and Teacher Education: An International. Journal of Research and Studies, 24*(1), 190–203.

Talbert, J. E., & McLaughlin, M. W. (1993). Understanding teaching in context. In M. W. McLaughlin & J. E. Talbert (Eds.), *Teaching for understanding* (pp. 162–206). San Francisco: Jossey-Bass.

Turnbull, D. (2000). *Masons, tricksters and cartographers*. Amsterdam: Harwood Academic Publishers.

Vygotsky, L. (1978). *Mind in society*. London: Harvard University Press.

Wells, G., & Claxton, G. (2002). Introduction: Sociocultural perspectives on the future of education. In G. Wells & G. Claxton (Eds.), *Learning for life in the 21st century* (pp. 1–17). Oxford: Blackwell publishers.

Wenger, E. (1998). *Communities of practice: Learning, meaning and identity*. Cambridge: Cambridge University Press.

Wertsch, J., Tulviste, P., & Hagstrom, F. (1993). A sociocultural approach to agency. In A. Forman, N. Minick, & A. Stone (Eds.), *Contexts for learning sociocultural dynamics in children's development* (pp. 336–357). New York: Oxford University Press.

Yerkes, K. (2004). *Exploring teacher identity: A yearlong recount of growing from student to teacher*. Retrieved, May 14, 2005, from http://www.ed.psu.edu/englishpds/inquiry/projects/yerkes04.htm

Envisioning Science Teacher Preparation for Twenty-First-Century Classrooms for Diversity: Some Tensions

Norman Thomson and Deborah J. Tippins

The Arrival of the Anthropocene: Where Is Biodiversity Going?

Over the past several decades, our knowledge and concerns for climate change, especially in the context of twentieth- to twenty first-century global warming, is shaping the consciousness of scientists and global communities. Global warming is especially worrying for those peoples who inhabit low-lying oceanic coastal islands and for those who are immediately ocean resource dependent – but in a broader sense extending to all nations because (1) most nations have extensively populated coastal communities and (2) physical changes in the oceans can have a profound effect on weather and climate (National Research Council [NRC] 2010a). Presently, 40 % of the world's 6.5 billion people live within 100 km of a coastline and it is estimated that 2.75 billion people will be under threat from sea level rise by 2050; in other words, they will be forced to migrate from their islands or inland as coasts change their shorelines or all together disappear. Concurrently, the continental landmasses' climates will have greater variance in disruptive weather events including periods and areas with more extreme drought, rainfall patterns, temperatures, and other weather-related variables. And, as the nature of the oceans' water changes, a series of sequential and unanticipated events, biotic and abiotic, will most likely take place on land (NRC 2011a). Ehrlich and Holdren (1971) developed a model relating how population size (P), affluence or resource consumption per person (A), and the beneficial and harmful environmental

N. Thomson (✉) • D.J. Tippins
Department of Mathematics and Science Education, University of Georgia, Athens, GA, USA
e-mail: nthomson@uga.edu

N. Mansour and R. Wegerif (eds.), *Science Education for Diversity: Theory and Practice*, 231
Cultural Studies of Science Education 8, DOI 10.1007/978-94-007-4563-6_11,
© Springer Science+Business Media Dordrecht 2013

effects of technologies (T) [all inputs], help to determine the environmental impact (I) [output] of human activities. In their model equation:

$$\text{Impact (I)} = \text{Population (P)} \times \text{Affluence (A)} \times \text{Technology (T)}$$

What was missing from their discussion of the equation was consideration for a consequential feedback loop of how the Impact response will affect the inputs, for example, global warming. Of special concern then is that the current "business model" that demands unsustainable growth and development is leading toward a disastrous change in climate; scientists argue that on a finite planet, a model of sustainability is the only option that may save humans from extinction.

Most peoples are likely aware that climate and species have changed in the past through learning about dinosaurs and ice ages as a part of earth history, informally through pictures and movies and formally in early schooling. However, not all people are fully aware of the implications of contemporary climate change and, because it is the consequence of anthropogenic activities, that there are opportunities for its mitigation (International Panel on Climate Change [IPCC] 2011). Intensive research studies into contemporary climate change began in 1988 when the World Meteorological Organization and the United Nations Environmental Programme commissioned the formation of an International Panel on Climate Change (IPCC) with the purpose of evaluating the state of climate science, based on peer-reviewed published scientific literature, with the goal of formulating policies for action. One outcome of the panel's investigations has been a consensus of scientific opinion that "human activities ... are modifying the concentration of atmospheric constituents ... that absorb or scatter radiant energy ... most of the observed warming over the last 50 years is likely to have been due to the increase in greenhouse gas concentrations" (McCarthy and Canziani 2001, p. 21). These greenhouse gases are carbon dioxide (56 %), methane (16 %), tropospheric ozone (12 %), halocarbons (11 %), and nitrous oxide (5 %). Yet, despite the increasing and overwhelming knowledge that Earth's climate is changing, there remains a mixture of defining what actions should be taken as recently exemplified in the Global 2011 IPCC conference in South Africa. Unfortunately, the inaction concerning this global problem is being sidelined by a few countries whose own short-ranged economic and political interests disregard the majority of other nations (Kerr 2007). For science teacher educators, whose task is to prepare the next generation of science teachers, real-world issues that tend to be controversial, can be addressed at the intersection of science and cultures.

What Is Known About Current Climate Change?

The NRC (2011b) reports that (1) climate change is occurring, is caused largely by human activities, and poses significant risks for – and in many cases, is already affecting – a broad range of human and natural systems, and (2) the global community needs a comprehensive and integrative change in the science enterprise, one that not only contributes to our fundamental understanding of climate change but also informs and expands the world's climate choices. Some scientists consider that we may now be entering a new geological epoch, the Anthropocene (Crutzen and Stoermer 2000), mainly as a consequence of human activities resulting in an unprecedented and catastrophic environmental impact. Human activities such as "co-opting resources, fragmenting habitats, introducing non-native species, spreading pathogens, killing species directly, immersing indigenous peoples into developmental programs modeled on economic consumerism and infinite growth, changing global climate affecting food production are collectively seen as factors that are leading us towards a sixth mass extinction" (p. 51) both in rate and magnitude of species loss (Barnosky et al. 2011). Naomi Oreskes (2004), in her review of 928 refereed science articles, determined that there is a 100 % consensus among qualified scientists that current global warming is not caused by natural climate variation. And, documentation for the dynamics of the Earth's trophic downgrading (Estes et al. 2011) and rapid range shifts of species (Chen et al. 2011) in response to warming is clearly established.

In a World Bank (2010) survey of over 15,000 people in 16 countries, over 60 % thought climate change is already doing harm to people in their country; but in six countries, including Russia and the United States, only a minority thought climate change is having an effect now. Majorities in all countries thought that there would be widespread adverse effects if climate change were unchecked. All participants were asked whether they believe their country does or does not have a responsibility to take steps to deal with climate change. In all 16 countries, majorities said their country does have a responsibility. Most majorities were very large and ranged from 90 % or more in France, China, Indonesia, Vietnam, Senegal, Bangladesh, and Kenya to 80 % in the United States, Japan, Mexico, Turkey, Iran, Egypt, India, and Brazil. In Russia, a more modest but clear majority of 58 % said the country had a responsibility to deal with climate change.

On average across 16 countries, 87 % said their country has a responsibility and a majority thinks their national government is not doing enough. And, this perspective was reiterated in the 2011 United Nations' climate conference held in Durban, South Africa. Karl Hood, Grenada's foreign minister and chairman of the 43-nation Alliance of Small Island States (AOSIS) whose members are in the frontline of climate change, said the talks were going around in circles. "We are dealing with peripheral issues and not the real climate ones which is a big problem, like focusing on adaptation instead of mitigation," he said. "I feel Durban might end up being the undertaker of UN climate talks." The dragging talks frustrated delegates from small islands and African states, who joined a protest by green groups outside as they tried to enter the main negotiating room. Maldives' climate negotiator Mohamed Aslam lamented, "You need to save us, the islands can't sink. We have a right to live, you can't decide our destiny. We will have to be saved" he said (Chestney and Herskovitz 2011).

In contrast to the global consensus of climate scientists, the US public remains less convinced, and some are even vocally polarized and recalcitrant. And, although the United States is but one country with a minority of the world's people, many nations look toward the United States both as a major causal agent and for leadership in climate change mitigation (NRC 2011a). One argument as to why climate change remains a US publicly charged issue, in part, is because of a well-constructed "climate cover-up" perpetrated by major corporations, lobbyists, politicians, and a small group of influential "junk scientists" (Hoggan 2009) that continues the well-established tactics developed and learned through the cover-ups and litigation experienced in the tobacco industries while they invested 50 years denying any linkage of smoking to cancers and a multitude other health problems. In their attempts to thwart linkages between the natural and additive/addictive constituents with active carcinogens, the "merchants of doubt" (Oreskes and Conway 2010) introduced and flaunted the "uncertainty of science in an uncertain world" (Pollack 2003). This is contrary to the interpretation for the nature of science that scientists and science educators view as science literacy (Flick and Lederman 2004) in which "tentativeness with skepticism" is one of the hallmarks of science as a way of knowing. In addition, the cover-up consortium purposively ignores and manipulates scientific data, utilizes the media (including creating misleading web-based sites) and misrepresents and fabricates the qualifications of their spokespersons as another means of misleading the public among other deceitful malfeasant practices (Oreskes and Conway 2010).

Some characterizations of global warming finding their way into the media have their origins in quotes by individuals who seemingly do not have a grasp of basic science. For example, one recent aspirant candidate for the United Stares' Presidency (Rep. Michelle Bachmann, Minnesota) stated in the US House of Representatives on Earth Day: Rep. Michele Bachmann spent part of Earth Day arguing against a carbon "cap and tax" because carbon dioxide is a "natural by-product of nature." It is "portrayed as harmful, but there isn't even one study that can be produced that shows that carbon dioxide is a harmful gas… It is a harmless gas… And yet we're being told that we have to reduce this natural substance and reduce the American standard of living to create an arbitrary reduction in something that is naturally occurring in the earth" (Schmelzer, *Minnesota Independent*, April 2009). In part, the issues of global warming are basically economic. In the development of industrial-based societies, their foundations lie on a false assumption that on a planet of finite resources, an infinite requirement is exponential population growth and consumption (Morelo-Frosch et al. 2009).

The World Bank has been conducting numerous surveys to determine international awareness of climate change and global warming (World Bank 2010). In the United States, regardless of the several hypothesized causes associated with global warming, US adults remain divided on whether to take action or not. A recent national telephone survey of American adults reports that 69 % of the participants indicated that it is at least somewhat likely that some scientists have falsified research data in order to support their own theories and beliefs. Nevertheless, an overwhelming majority (72 %) believe that the United States is not doing enough to develop alternative sources of energy. While 40 % of the participants believe Americans should take immediate action to stop global warming, 42 % suggest waiting a few years. Out of three scenarios, 30 % of Americans say a period of dangerous global warming is likely to occur, while just four percent (4 %) say a dangerous ice age is more likely. Half of adults (50 %) say something in between is most likely to happen and 16 % are not sure what the consequences might be. Internationally, World Bank (2010) findings have changed little from surveys conducted over the past decade. For example, in the United States, 67 % of adults have been following news stories on global warming at least somewhat closely, while 32 % have not. In comparative public surveys, 44 % of the participants in the United States and Russia, and even fewer in China (30 %), consider global warming to be a very serious problem, whereas 68 % in France, 65 % in Japan, 61 % in Spain, and 60 % in Germany say that it is a serious problem (World Bank 2010). Accounting for the similarities and variance between countries seems to be multifaceted.

What Is Known About Past Climate Change?

Although all of today's extant organisms share an origin into antiquity in billions of years (Rogers 2012), humans are a relatively recent evolutionary arrival joining Earth's natural systems within, at most sensu latu 6–7 million years ago (*Sahelanthropus tchadensis*) and some could well argue only the last 1–2 million years (*Homo egaster/erectus*) as an assigned starting point (Cartmill and Simth 2009). And, although as members of our bipedal ancestors evolved in response to climate change for some reason, about 1.8 million years ago, some of our ancestral groups left Africa (Klein 2009). Why do people migrate? It seems most plausible that migration is a means of searching for and finding resources necessary for sustaining life when a local depletion has occurred (Kingston 2007). As our past ancestors meandered into new environments, nature acted as a selective filter among fortuitous adaptations that eventually emerged as cultures. Within these various peoples, distinctive cultures emerged with words, languages, and thoughts unique to the particular environments. And, it seems that it is only within the last 50,000 years that these characteristics of humans permitted them to successfully spread around the globe with a destructive and unsustainable impact on the entire global environment (Lieberman 2011).

Thus, as recent arrivals on Earth it seems that in many ways we humans have not ingratiated ourselves and are choosing to live "apart" from nature rather than recognizing how we exist as an "interconnected part" fully and integrally entwined within every ecosystem we inhabit. In being able to make choices, humans (*Homo sapiens*) are privileged in the sense that we are not only aware of the present, but through our collective cultures, languages, and experiences, we are able to share unique knowledge of our present experiences and contextually reflect upon our pasts, and perhaps more importantly dwell on the future. Furthermore, through technology, language, and culture, we are able to instantaneously share experiences with the other 6[+] billion people on our "island planet" that we inhabit. But, as previously described, our planet's biological diversities and cultural legacies are at risk of disappearing – extinction is really – forever.

Where Are the World's Cultures Going? Diversity at Risk!

It is estimated that if nothing is done, 90 % of the planet's 6,000+ languages spoken today will disappear by the end of this twenty-first century. With the disappearance of unwritten and undocumented languages, along with others less used, humanity will not only lose cultural wealth but also important ancestral ecological knowledge of localities embedded in the indigenous languages will disappear. While it is widely acknowledged that the degradation of the natural environment, in particular tradi-tional habitats, entails a loss of cultural and linguistic diversity, new studies suggest that language loss, in its turn, has a negative impact on biodiversity conservation. There is a fundamental link between language and traditional knowledge (TK) related to biodiversity. Local and indigenous communities have elaborated complex classification systems for the natural world, reflecting a deep understanding of their local environments. This environmental knowledge is embedded in indigenous names, practices, oral traditions and taxonomies, and can be lost when a community shifts to another language (Moseley 2011).

Ethnobotanists and ethnobiologists recognize the importance of localized sub-sistence, cultural attitudes and values, which have left us with some of the only remaining pristine areas on Earth. And, it is within these rich reservoirs that humans are provided some of the only hope that they might have to develop successful initiatives related to endangered species recovery and restoration activities. Every language reflects a unique worldview with its own value systems, philosophy, and particular cultural features. The extinction of a language results in the irrecoverable loss of unique cultural knowledge embodied in it for centuries, including historical, spiritual, and ecological knowledge that may be essential for the survival of not only its speakers, but also countless others. And, the impact of language colonization leads to a homogenization or extinction of not-knowing local environments.

One international project that demonstrates the need to maintain cultural heritage can be seen in food production in Africa. Africa has long been viewed as a continent unable to provide food for itself especially with constant instances of famine and starvation. Ironically, most external aid is used to promote the introduction and production of energy-demanding crops with exogenous origins, which create new forms of dependency (including corporate patented and protected genes) making them unsustainable food sources. This is just a new input into the cycles of children starving.

The National Research Council has conducted a series of studies into what has and is being lost throughout sub-Saharan Africa with respect to indigenous food sources: grains (NRC 1996), vegetables (NRC 2006), and fruits (NRC 2008). Why is this considered to be important? It is because we seemed to have evolved with grasses as a main food source. And currently, the world's six billion peoples' sus-tenance depends upon only three grasses: wheat, maize, and rice as sources of car-bohydrate. It is anticipated that genetic potentials of these crops will not be able to accommodate for change in climate. As with languages, there is a plethora of iden-tified lost African crops for which the elders lament. In international surveys of over

1,000 Africans the local peoples have identified and documented favorite grains, fruits, nuts, vegetables, legumes, and other food plants in significant numbers. Local people have identified crops that are not in commercial use and being ignored in development as including over 1,000 grains; over 3,000 roots, stems, leaves, bulbs, and fruits; and thousands of fruits they know and use, but are being displaced by introduced exportable fruits. These surveys have not even begun to explore the medicinal plants. Associated with climate change and seasonal rains seed germination, plant growth, flowering, and fruit production have become less predicable and dependable even for the indigenous foods resulting in marginal harvests (NRC 2005).

Why Do Languages Matter and Should We Care?

Indigenous communities make up one third of the world's 900 million extremely poor people whose existence is dependent upon a regional ecology. So where do we find our global heritage in the diversity of languages (Fig. 1)? Listed as a

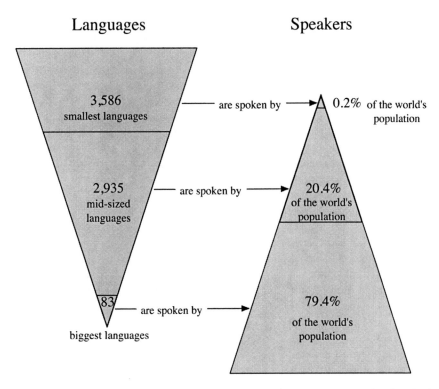

Fig. 1 The world's 6,000 languages and speakers represented as inverted *triangles*. Over half of the languages are spoken by a very few people placing many at risk (Harrison 2011)

Degree of Endangerment	Intergenerational Language Transmission
safe	A language is spoken by all generations; intergenerational transmission is uninterrupted
vulnerable	Most children speak the language, but it may be restricted to certain domains (e.g., home).
definitely endangered	Children no longer learn the language as mother tongue in the home
severely endangered	A language is spoken by grandparents and older generations; while the parent generation may understand it, they do not speak it to children or among themselves
critically endangered	The youngest speakers are grandparents and older, and they speak the language partially and infrequently
extinct	There are no speakers left.

Fig. 2 The world's languages are classified as to their status for intergenerational transmission (Moseley 2011)

percentage/estimated number of languages: Europe = 3 % /209; Americas = 15 % /949; Africa = 31 % /1,995; PACIFIC* = 21 % /1,341[*New Guinea has over 1,200 languages = 20 % of the world's languages]; and ASIA = 31 %/2,034. Traditional cultural and biodiversity losses share important threats, such as urbanization and exposure to globalized commercialization. Community participation in research has yielded data on the identification and distributions of new and previously described species. Indigenous cultural knowledge and know-how have informed and assisted in conservation research and practice. Indigenous peoples are contributing to and enforcing conservation policies (Cohen 2010, p. 30).

A culture and its language disappears when its speakers disappear or when they shift to speaking another language – most often, a larger language used by a more powerful group. Languages are threatened by external forces such as military, economic, religious, cultural, or educational subjugation, or by internal forces such as a community's negative attitude toward its own language. Today, increased migration and rapid urbanization often bring along the loss of traditional ways of life and a strong pressure to speak a dominant language that is – or is perceived to be – necessary for full civic participation and economic advancement. It is impossible to estimate the total number of languages that have disappeared over human history. Linguists have calculated the numbers of extinct languages for certain regions, such as, for instance, Europe and Asia Minor (75 languages) or the United States (115 languages) lost in the last five centuries, of some 280 spoken at the time of Columbus (Moseley 2011).

The most important thing that can be done to keep a language (i.e., local knowledge) from disappearing is to create favorable conditions for its speakers to speak the language and teach it to their children (Fig. 2). This often requires national policies that recognize and protect minority languages, education systems that promote mother-tongue instruction, and creative collaboration between community members and linguists to develop a writing system and introduce formal instruction in the language. Since the most crucial factor is the attitude of the speaker community toward its own language, it is essential to create a social and political environment that encourages multilingualism and respect for minority languages so that speaking such a language is an asset rather than a liability. Some languages now have so few speakers that they cannot be maintained, but linguists can, if the community so wishes, record as much of the language as possible so it does not disappear without a trace of its existence.

What Are the Language Options for Nations with Respect to Maintaining Diversity Versus Extinctions?

There are several choices that nations may make with respect to the languages they use, and the options are made in the context of the past, present circumstances, and the nation's future. A nation may:

- Remain uncommitted on the question of a language policy and allow things to change without interference.
- Use an ex-colonial language as the official (and national language) as it is perceived to be neutral at the expense of losing cultural heritages (British, French, Portuguese, etc.).
- Adopt the majority language, where such a language is predominant.
- Allocate to some of the major languages certain public roles at the regional or district level.
- Give nominal public roles or none to the smaller languages (which, e.g., most African countries have chosen).

As globalization, climate change, and languages of countries differentially affect peoples, internationally, science educators are facing daunting challenges. How do we make informed decisions about how to best prepare our science teachers for their own futures? As science educators, we are trying to find answers to this question in our own pre-service science teacher courses.

So, What Visions Do We Think Science Educators Need for the Twenty-First Century?

Science educators have an integral role in bringing "unity through diversity" in preparing the next generation of teachers who will be in classrooms preparing students who, in turn, will be their own decision-makers extending well into the twenty-second century. With respect to climate change we think that perhaps the most important role in preparing science teachers then, is not "what to think", but "how to think" as decision makers. The dimensions of "how to think" in a global context, in which our evolutionary past is integrally linked with changes in climate and, especially with reference to current global warming, brings together what we see as four important integrated phenomena: the geological record, climate change, human evolution, and culture and language. Today, as previously described, it is proposed that humans are in the process of forming a new geological epoch, the Anthropocene, that will bring transformative challenges and tensions in science teacher preparation that we suggest are essential for science educators to include in science teacher preparation for the twenty-first century. And, though our current data referents in our study are drawn mainly from the United States, from our own international experiences, and in working with many international science education colleagues throughout the world, we know they share our concerns.

Climate change is becoming a major topic at the forefront of secondary biology, earth and environmental science courses. Its study is of particular significance in light of the fact that most scientists' recognize that a basic knowledge of evolution is essential to understanding the processes that occur in the context of global climate change – speciation and extinction (NRC 2010b). Concurrently, science educators continue to give attention to polls, which suggest that more than 50 % of Americans reject evolution as a viable theory, supporting instead the teaching of creation/intelligent design science in public schools (Berkman and Plutzer 2010). And, non-evolutionary views seem to parallel the promulgation of religious fundamentalism globally. Researchers indicate that many students graduate from college without a basic understanding of evolution (NRC 2010b). This is of particular concern because most scientists (Cartmill and Simth 2009) contend that a basic knowledge of evolution is essential to understanding processes that occur in the context of what we have learned from paleoenvironmental data (NRC 2010b). In the past few years, through the efforts of the Intergovernmental Panel on Climate Change (IPCC), the issue of global climate change has shaped the consciousness of almost every citizen and is a topic of ever-increasing importance at the forefront of the curriculum of secondary biology (Wagler 2011) and earth and environmental science courses (Gautier et al. 2006).

The Standards for Science Teacher Preparation (NSTA 2010) emphasize the fundamental role of students' informed decisions about contemporary societal issues in developing scientific literacy and citizenship in a democratic society. In the context of understanding the relationship between evolution and global climate change, processes such as interpreting data, constructing hypotheses, evaluating alternatives, weighing evidence, interpreting texts, and evaluating the potential of scientific claims are all seen as essential components of meaningful learning and the construction of scientific arguments. An important part of the science teaching and learning process that has not been studied is the use of argumentation as a strategy for using evidence and observations of the real world to explain and understand climate's influence on evolution (NRC 2010b). There is an urgent need to prepare science teachers in ways that enable them to help their twenty-first-century students develop genuine understandings of global climate change as a factor in evolution.

What Do Our Current Twenty-First-Century Teachers Understand About Climate's Influence on Human Evolution?

Because of our concern for the introduction of climate's influence on human evolution to students in secondary science, we decided to conduct a study of our current pre-service teachers' (1) knowledge of, and, (2) what they might choose to teach as part of an assigned project in a school-based teaching practicum. Our study took place in a one-semester secondary science teacher preparation course that includes a field teaching practicum experience, and is required of all future secondary science teachers at a major university in the southeast United States. The students were introduced to human evolution and climate change through a 2-week curriculum unit (Thomson and Bealls 2008) that includes the use of replica cast skulls of extant vertebrates and fossil hominins (Bone Clones 2013), hands-on activities, power points, and background readings and the species are analyzed in the contextual interpretations of ecology and climate (Bobe et al. 2007). The students were asked to develop a two-lesson unit using the hominin casts (Fig. 3). The eight students chose to work as pairs with the earth science majors forming one group and the six biology majors the other three groups. The earth science majors developed and implemented a unit that focused on the oldest fossils (1). *Sahelanthropus, Ardipithecus,* and *Australopithecus spp.,* 3.0–6.5 Mya; and the biology groups focused on units that would cover (2) *Australopithecus spp., Paranthropus spp., Kenyanthropus,* 1.5–2.5 Mya; (3) *Australopithecus spp., Paranthropus spp., Homo spp., Kenyanthropus,* 1.2–3.2 Mya; and (4) *Homo spp.,* 0.0–2.0 Mya, respectively. They were encouraged to use an argumentation approach (observation/evidence to inference/claim) in their lesson design to promote active student participation (Erduran and Jimenez-Aleizandre 2008). They were allowed to modify and create their own lessons as they wished, but were asked to link human evolution to climate change in some way. The students used 1 week to develop their lessons and implemented in successive days, "Skull Groups I – IV" in two secondary classrooms.

The researchers included two science educators, a paleoanthropologist whose research focus is paleoclimate, and, three graduate students who participated in data-collection. Primary participants in the study were eight pre-service teachers (three males, five females) enrolled in the Science Teaching Curriculum course with majors in biology or geology. Secondary participants in the study were high-school students enrolled in two sections of tenth-grade biology classes in one school and an anatomy and physiology class in a second. A case study design utilizing interpretive research methodology (Patton 2002) was used in our study. Case study research (Hays 2004) involves the study of an issue or phenomenon explored through one or more cases within a bounded system. More specifically, the researchers explore the bounded system through detailed, in-depth data collection involving multiple sources of information (Creswell 2007). Patton (2002) stated that the purpose of a case study is "to gather comprehensive, systematic, and in-depth information about each case of interest" (p. 447) and that a case study illustrates

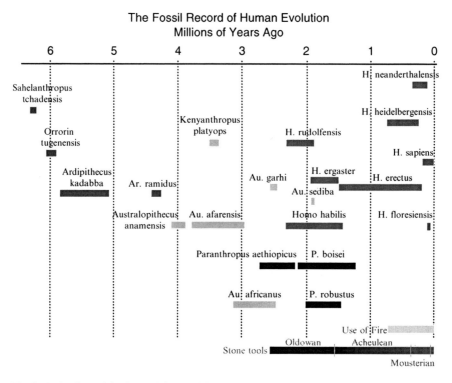

Fig. 3 A timeline of fossil hominins used in the development and implementation of the units designed by the pre-service teachers

"the value of detailed, descriptive data in deepening our understanding of individual variation" (p. 16). Data collection techniques consistent with interpretive research were used in this study. More specifically, the researchers used in-depth, semi-structured interviews, classroom observations, and the collection of artifacts and documents of the students.

We each used the following questions as our semi-structured interview framework:

1. What are the big ideas that science teachers need to be aware of in order to understand evolution?

   (a) What do you think science teachers need to understand about the relationship between _____ and evolution? [(i) Deep-time, (ii) Speciation and Extinction, (iii) Fossilization & Dating (absolute and relative), (iv) Phylogenetics, (v) Nature of science.

   (b) Describe the relative importance of these concepts for science teachers' understanding of evolution? Why did you make those choices?

   (c) What distinctions, if any, do you think science teachers should understand about the relationship between evolution and human evolution?

2. What are the big ideas that science teachers need to be aware of in order to understand climate change?

    (a) What do you think science teachers need to understand about the relationship between _____ and climate change? (i) Deep-time, (ii) Cyclical change / Orbital Forcing (cyclical change or Milankovic Cycles), (iii) Acyclical change (volcanism, plate tectonics, asteroids, etc.), (iv) Fossilization and dating, (v) Speciation and extinction, (vi) Nature of science.

    (b) Describe the relative importance of these concepts for science teachers' understanding of climate change? Why did you make those choices?

    (c) What distinctions do you make between historical climate change and modern climate change?

3. What do you think science teachers need to understand about the relationship between evolution and climate change?

The student pairs were interviewed on three occasions: (1) individually prior to initiating their lesson design, (2) as a pair during their lesson planning, (3) individually following their lesson implementation, and (4) on one final occasion, collectively as an entire group five weeks after their lesson implementations. Members of the research team met on a regular basis to plan and discuss what they were learning, as part of ongoing data collection and analysis. The research team used grounded theory and selective and axial coding to construct specific narratives and identify themes. Although we have copies of the students' curriculum materials, we feel that the students' comments made in the interviews best demonstrates what we have learned in this study.

What Did We Learn About Our Pre-service Teachers' (1) Knowledge and (2) Their Implementation of Climate Change and Human Evolution in the Design of Students' Lessons?

Four themes with specific relevance to the preparation of twenty-first-century science teachers who need to be prepared to teach evolution in the context of climate change are presented:

1. The pre-service teachers in their science content preparation courses are not experiencing interdisciplinary learning. Accordingly, the pre-service biology and geology majors are developing only partial and fragmented understandings of the evolutionary basis of climate change. Geology majors, for example, have a strong understanding of deep time but little knowledge of speciation, extinction and phylogenetics, and the consequences of climate change. Conversely, biology majors struggled to relate deep time, fossilization and dating, both absolute and relative, to climate change. This finding has strong implications, suggesting the need for the development of interdisciplinary science content courses for

our twenty-first-century teachers. This was particularly the case for the biology majors, who are not required to take a geology course. The geology majors do take one biology course, but the pre-service teachers indicated that because it was taught in isolation from their geology courses they were not able to see "the big picture:" A geology major feels that the biology students have not really developed a sense of deep time and fossilization even though they have all had a course in evolutionary biology:

> I think for understanding evolution deep time is the most important concept. If they don't understand deep time they're not going to be able to place when these evolutionary changes were happening. So I think deep time is a good stepping-stone to opening their mind to the fact that a billion years ago change was happening. Fossilization and dating I've worked with more because I'm in Earth science. But I would place it last only because I don't think other people perceive its importance. For example, in our group no one else talks about that kind of stuff. And they joke with us. When we start talking about rocks and geologic time scale they don't know what we're talking about. The biology people never had to take our geology classes and I don't think they understand how much science is involved in those classes. We Earth science majors have to take chemistry and physics, but they never have to take geology. It's interesting that there's not that interdisciplinary focus. I took astronomy and that was the first course when I really realized that I needed to grasp deep time. (Geology major, Interview #3)

However, one of the biology majors stated that it is not a problem of understanding, but the difficulty in representing the scale of deep time and strategies for teaching scalar concepts (time, matter, and space):

> Just to add on to what everyone has said, so teachers need to know how the students conceptualize deep time, evolution is such a broad topic, it seems like such a hard pitch out there, we don't see or experience it on a day-to-day basis, so we have to look back on it, so you have to look back and see what has occurred, and to get students to realize the time scales that we are referring to, so ... (Biology major, Interview #4)

2. In the process of designing and teaching lessons, the pre-service teachers struggled to create activities and experiences that reflect the most recent scientific understandings of the evolutionary consequences of climate change. They found most textbook resources useless, as these books were not able to keep pace with the exponential growth of science. For example, paleontology was not included in any of their textbook resources. At the same time, they had difficulty evaluating the credibility of Internet resources. As a consequence, the lessons they designed were more based on and limited by what they knew from their course work. In addition, this was the first time they had actually worked with 3-D hominin skull replicas, in contrast to seeing human evolution depicted only as pictures in their textbooks. Consequently, they spent much more time learning for their own professional growth and emphasized this topic in their lessons rather than moving onto understanding climate change and its link to human evolution. As a result, we feel that without more hands-on experiences addressing contemporary issues for the science-public interface not generally addressed at university level course work, the science pre-service teachers will be reluctant to address these issues in their future classrooms. On the other hand, they also viewed how rapidly science

generates new data, hypotheses, and methods as the reason why science teaching requires life-long learning:

Even when we taught the anthropology, the human skulls, I said "Wow"!, I had not even encountered that, and I was a biology major! I was really shocked that there was a whole other area that I didn't know about, it is fascinating to me about how much I did not know. (Biology major, Interview #4)

Well, I think that as a teacher you are constantly learning, I think that is what separates a good teacher from just any teacher ... you have to be really passionate about what you are trying to do, that is what you are going to have do. Well, that is what scientists are doing anyway. In fact, that is what we want to teach our students, that you are going to continually, continually learn ... you know, they shouldn't stop with what is in the classroom, they need to learn what is outside their classroom as well, that learning is a wondrous thing, I don't want to put the content aside, but it is that wonder -when we prepare to teach something we need to go out and discover what we don't know ... you do not what to stop that wondering, that curiosity, that the students don't want to learn just facts to regurgitate. (Biology major, Interview #4)

You do not want to get up in front of a group of students when you are supposed to be teaching and end up looking like a complete idiot like you don't know what you are talking about yourself! When you are teaching them and you don't know it – so, I spent a lot of time researching the different skulls that we have ... (Geology major, Interview #4)

3. With respect to the use of argumentation in the lesson design and teaching process, the results were mixed. The pre-service teachers were able to infuse some aspects of argumentation (weighing evidence, evaluating scientific claims) into their lessons with little difficulty. They experienced more difficulty in designing experiences that enabled students to interpret texts and construct viable explanations, important aspects of argumentation. Although the pre-service teachers planned their lessons working in pairs, they indicated that some large group planning sessions would have been helpful in making interdisciplinary connections:

Yeah and it is not just about knowing the content, that just took hours alone, it wasn't just spending hours learning what we needed to know, but then spending time thinking about how we were going to teach it. Delivery was very important for us and we did not what to do in a traditional kind of, you know just a bunch of details would be boring. We wanted to capture the students' attention, that was a key element for us, and to let them develop an understanding. (Biology major, Interview #4)

When we got together we were able to actually collaborate, for example my partner and I sort of approached it from an ecological perspective. And, from what I have learned at college, I know that when I came to college and took ecology it finally brought things together. It wasn't just organisms or organelles, so I think when you are teaching about evolution the important factor is to tie in all together. Well, we kind of constructed our own little ecology course. And made links to each other – even our lessons – we tried to connect them together. (Biology major, Interview #4)

4. The pre-service teachers entered the tenth grade classroom expecting to encounter some resistance to instructional lessons focused on evolution and climate change. Much to their surprise, they did not encounter the type of resistance that they expected. While there could be several explanations for this, including the

background of the host teachers, the tenth grade students were more amenable to learning about evolution and climate change than the media seems to portray to the "general public", especially in the context of the way in which the public is reported polarized on these topics. It is through teaching integrated subjects that our teachers seem to begin seeing the big picture for the nature of interdisciplinary science.

Well, we need to do it in our classrooms because they are going to be part of the general public. So, we need them to look at science and it is going to affect their lives. And, as a teacher the stuff we did in our other courses we need to teach them how to really look at issues and realize our impact on earth. And, then, how are we going to adapt to those changes. For example, global warming over time, we may not notice it everyday, but our skin may become more prone to cancer. You know any of your physiological features can change. And, we won't notice it until science brings it to our attention and then we will say "oh!" The atmosphere has changed, and now, so have we. I guess we need to keep the students aware on a daily basis and how things can affect their everyday lives. And, how they can make a difference and I think that will seep out into the public. Maybe not to the masses but at least even if only a few students, you never know where they may take it. (Biology major, Interview #4)

So, What Do We See as Some of the Tensions of Science Teacher Preparation for Diversity in Twenty-First-Century Classrooms?

In the preparation of science teachers for the twenty-first century, it seems to be essential that to be part of a global science education community that is concerned with the consequences of climate change, we include components of past and current climate change as a part of our curriculum. Paleoanthropology is reconstructing past environments, investigating the appearance and extinction of hominin species to provide some insight into biological responses to past changing environments in relation to other fauna and flora. Climate change is leading species' extinction and though we did not include it in this study, we wanted to draw awareness to language and cultural extinctions in our chapter. Languages and cultural diversity are integral to addressing issues of climate change and biological extinction. We have found out that our twentieth-century model for teacher preparation may no longer best meet the needs of our twenty-first-century science teachers. Not only is there diversity in learners, but also there is a need for a new diversity and combinations of integrated interdisciplinary science courses for effective science teacher preparation. We are constantly faced with issues of which courses are required to become an effective science teacher in one's discipline, but the number of courses for graduation and certification is generally fixed; what can be changed is the content of courses. Such a change might take place through multiple instructors in a modular course that includes a sequence of topics.

Science teachers need to be prepared to teach contemporary issues in which science knowledge is preceding opportunities for its inclusion into university science textbooks, current university science course structure and content, and the current science courses we think pre-service teachers need. We also need to change what and how we prepare our science teachers to think more holistically – beyond their individual science discipline – as our pre-service teachers brought to our attention: There needs to be more thoughtful integration of science learning for secondary science students through co-planning with teachers. Although our study is a glimpse into what our pre-service science teachers currently know about climate and evolution, we would like to suggest that twenty-first-century science educators have a critical role in ensuring that our future science teachers are prepared to teach important issues concerning climate change, human evolution, species and language/cultural extinctions, and possible consequences – but, more optimistically, offer solutions to our future generations.

Acknowledgement The authors wish to acknowledge the University of Georgia STEM Education Program, Improving Instruction and Enhancing Success in STEM disciplines, fir their support of this work.

References

Barnosky, A., Matzke, N., Tomiya, S., Wogan, G., et al. (2011). Has the Earth's sixth mass extinction already arrived? *Nature, 471*, 51–57.

Berkman, M., & Plutzer, E. (2010). *Evolution, creationism, and the battle to control America's classrooms*. New York: Cambridge University Press.

Bobe, R., Alemseged, Z., & Behrensmeyer, A. (2007). *Hominin environments in the East African Pliocene*. Dordrecht: Springer.

Bone Clones, Inc. Osteological Reproductions. (2013). Canoga Park, CA. Available: http://www.boneclones.com/

Cartmill, M., & Simth, F. (2009). *The human lineage*. Hoboken: Wiley.

Chen, I.-C., Hill, J., Ohlemuller, R., & Roy, D. (2011). Rapid range shifts of species associated with high level of climate warming. *Science, 333*, 1024–1026.

Chestney, N., & Herskovitz, J. (2011). Climate talks split on drafts, EU warns of collapse. (December 8), Reuters. Available On-line: http://www.reuters.com/article/2011/12/09/us-climate-idUSTRE7B41NH20111209

Cohen, J. (2010). In the shadow of Jane Goodall. *Science, 238*, 30–35.

Creswell, J. W. (2007). *Qualitative inquiry and research design: Choosing among five approaches*. Thousand Oaks: Sage Publications, Inc.

Crutzen, P., & Stoermer, E. (2000). *The Anthropocene* (IGPB Newsletter 41). Stockholm: Royal Swedish Academy of Sciences.

Ehrlich, P. R., & Holdren, J. P. (1971). Impact of population growth. *Science, 171*, 1212–1217.

Erduran, S., & Jimenez-Aleizandre, M. (2008). *Argumentation in science education: Perspectives from classroom-based research*. Dordrecht: Springer.

Estes, J., Terbough, J., Brashares, J., & Power, M. (2011). Trophic downgrading of planet Earth. *Science, 333*, 301–306.

Flick, L., & Lederman, N. G. (Eds.). (2004). *Scientific inquiry and nature of science: Implications for teaching, learning, and teacher education*. Dordrecht: Kluwer Academic.

Gautier, C., Deutsch, K., & Rebich, S. (2006). Misconceptions about the greenhouse effect. *Journal of Geoscience Education, 54*(3), 386–395.

Harrison, D. (2011). *Global language hotspots*. Available http://www.swarthmore.edu/SocSci/langhotspots/globaltrends.html

Hays, P. A. (2004). Case study research. In K. DeMarrais & S. Lappan (Eds.), *Foundations for research: Methods of inquiry in education and the social sciences* (pp. 217–234). Mahwah: Lawrence Erlbaum Publishers.

Hoggan, J. (2009). *Climate cover-up: The crusade to deny global warming*. Berkeley: Graystone Publishers.

IPCC (International Panel on Climate Change). (2011). *Special report on renewable energy sources and climate change mitigation*. Available http://srren.ipcc-wg3.de/

Kerr, R. A. (2007). Climate change: Global warming is changing the world. *Science, 316*(5822), 188–190.

Kingston, J. (2007). Shifting adaptive landscapes: Progress and challenges in reconstructing early hominid environments. *Yearbook of Physical Anthropology, 50*, 20–58.

Klein, R. (2009). *The human career: Human biological and cultural origins*. Chicago: The University of Chicago Press.

Lieberman, D. (2011). *The evolution of the human head*. Cambridge, MA: Belknap.

McCarthy, J., & Canziani, O. (Eds.). (2001). *Impacts, adaptation, and vulnerability: Contribution of Working Group II to the third assessment report of the Intergovernmental Panel on Climate Change*. Available http://www.ipcc.ch/publications_and_data/ar4/wg2/en/contents.html

Morelo-Frosch, R., Pastor, M., & Shonkoff, S. B. (2009). *The climate gap*. Los Angeles: USC Center for Sustainable Cities, University of Southern California.

Moseley, C. (Ed.). (2011). *Atlas of the World's Languages in Danger* (3rd ed.). Paris: UNESCO Publishing. Online version: http://www.unesco.org/culture/en/endangeredlanguages/atlas

National Research Council (NRC). (1996). *Lost crops of Africa: Vol. I, Grains*. Washington, DC: National Academies Press.

National Research Council (NRC). (2005). *The geological record of ecological dynamics: Understanding the biotic effects of future environmental change*. Washington, DC: National Academies Press.

National Research Council (NRC). (2006). *Lost crops of Africa: Vol. II, Vegetables*. Washington, DC: National Academies Press.

National Research Council (NRC). (2008). *Lost crops of Africa: Vol. III, Fruits*. Washington, DC: National Academies Press.

National Research Council (NRC). (2010a). *Advancing the science of climate change*. Washington, DC: National Academies Press.

National Research Council (NRC). (2010b). *Understanding climate's influence on human evolution*. Washington, DC: National Academies Press.

National Research Council (NRC). (2011a). *Warming world: Impacts by degree*. Washington, DC: National Academies Press.

National Research Council (NRC). (2011b). *Advancing the science of climate change*. Washington, DC: National Academies Press.

National Science Teachers Association (NSTA). (2010). *Standards for science teacher preparation*. Arlington, VA: NSTA.

Oreskes, N. (2004). Beyond the ivory tower: The scientific consensus on climate change. *Science, 306*(5702), 1686.

Oreskes, N., & Conway, E. (2010). *Merchants of doubt: How a handful of scientists obscured the truth on issues from tobacco smoke to global warming*. New York: Bloomsbury Press.

Patton, M. (2002). *Qualitative research & evaluation methods*. Thousand Oaks: Sage Publications.

Pollack, H. (2003). *Uncertain science … uncertain world*. Cambridge: Cambridge University Press.

Rogers, S. (2012). *Integrated molecular evolution*. Boca Raton: CRC Press.

Schmelzer, P. (2009, April 24). On climate science, Bachmann accused of 'making things up' on the House floor. *Minnesota Independent*. Retrieved July 5, 2010. Available On-line: http://minnesotaindependent.com/33294/on-climate-science-bachmann-accused-of-making-things-up-on-the-house-floor

Thomson, n., & Beall, S. (2008). An inquiry safari: What can we learn from skulls? *Evolution: Education and Outreach, 1*, 196–203.

Wagler, R. (2011). The anthropocene: An emerging curriculum theme for science educators. *The American Biology Teacher, 73*(2), 78–83.

World Bank. (2010). *Public attitudes toward climate change: Findings from a multi-country poll.* Available http://econ.worldbank.org/

Expanded Agency in Multilingual Science Teacher Training Classrooms

Silvia Lizette Ramos-De Robles and Mariona Espinet

Learning Science and a Foreign Language in the Same Classroom Activity

Since its foundation, the European Union has aimed to create an international space without borders where its citizens can enjoy greater opportunities for work, education, business, tourism, and cultural exchanges without compromising cultural diversity.

In this context, one of the most important demands is the need to accept that our society and schools are multilingual contexts. In consequence, the command of at least three languages is considered one of the most important basic competences that each European citizen should acquire and develop, basically through his or her compulsory education (Commission of the European Communities 2007). As a response, European educational institutions are developing new teaching methodologies. The acronym CLIL (Content and Language Integrated Learning) has become popular and constitutes a political platform that has been widely accepted and applied within the European Union to promote the learning of several languages. This platform allows learners to participate in socially contextualized activities and advocates the need to design learning environments in which both subject matter content and language content can be learned together.

Regarding the specific case of the Faculty of Educational Sciences at the Autonomous University of Barcelona, in 2002 the Teaching Committee of the

S.L. Ramos-De Robles
Department of Environmental Sciences, University of Guadalajara, Guadalajara, Jalisco, Mexico
e-mail: lramos@cucba.udg.mx

M. Espinet (✉)
Departament de Didàctica de la Matemàtica i de les Ciències Experimentals, Universitat Autònoma de Barcelona, Cerdanyola del Vallès (Barcelona), Spain
e-mail: Mariona.Espinet@uab.es; mariona.espinet@uab.cat

N. Mansour and R. Wegerif (eds.), *Science Education for Diversity: Theory and Practice*,   251
Cultural Studies of Science Education 8, DOI 10.1007/978-94-007-4563-6_12,
© Springer Science+Business Media Dordrecht 2013

Foreign Language Teaching in Primary Education degree proposed that nonlanguage subjects be taught in English. University teaching at this specific university is vehiculated in two languages, Catalan and Castilian (Spanish), since the autonomous region where this university is located, Catalonia, is officially bilingual. This initiative coincided with integrated teaching approaches for the promotion of multilingualism in at least three languages, Catalan, Castilian, and English.

Several years later, with an increase in both demands and university policies to adapt studies to the Bologna Plan, the promotion of this type of teaching methodology has become a priority issue (Masats et al. 2006). In the 2004/2005 academic year, the impetus for these initiatives gained strength through the creation of an interdisciplinary and interdepartmental group of lecturers, all of whom were determined to become part of the experience of teaching their courses in English. Indeed, it was in the 2004/2005 academic year that the Didactics of Experimental Sciences and Mathematics Department joined the initiative and took up the challenge of imparting the Science Teaching course in English, in accordance with the CLIL policy.

This represented the start of what is now our object of study. In 2007, we initiated a research project on the Science Teaching course, the general aim of which was to understand how meaning (understood as cultural production) is constructed through interaction in CLIL contexts for preservice science teacher training where three languages Catalan, Castilian (Spanish), and English are used (Ramos 2010).

The work presented in this chapter is part of a larger research project. In this chapter, we focus on the students' participation in terms of their agency to successfully use resources while interacting in small groups in the science laboratory. Our main question is *how do students expand their agency in the use of resources as a way of overcoming the difficulties derived from the need to construct a scientific explanation of natural phenomena using English as a foreign language?*

CLIL Contexts as Fields of Cultural Production

The students' interactions were analyzed using a sociocultural perspective as a theoretical and methodological framework. We explore science and science education as forms of culture, enacted in a variety of fields that are formally and informally constituted (Roth and Tobin 2006). In this sense, learning could be understood as a product of participation in collectively motivated activity systems (Van Eijck 2009). This participation could be interpreted as *agency*.

In cultural sociology and according to Sewell (1992), agency (or human action) is theorized in a dialectical relationship with structure, a construct relating to aspects lying both within and outside the acting human being.

We could define agency as the power to conduct social life (Tobin 2007) and the various ways in which individuals organize their participations to interact with material and human resources. According to Sewell (1992), agency is dialectically related to structures and involves access to and appropriation of structures, and it pertains to the fields in which culture is conducted. This dialectical relationship (agency/structure) implies that agents are capable of putting their structurally formed capacities to work in creative or innovative ways. And, if enough people or even a few people who are powerful enough act in innovative ways, their action may have the consequence of transforming the very structures that gave them the capacity to act (Sewell 1992, p. 4). This capacity to act (agency) requires access to the resources of a field and the cultural capital needed to appropriate them. According to Tobin (2010), the resources of a field, such as a science classroom, can be accessed and appropriated by participants as they exercise agency to reproduce and transform schema and practices (i.e., the culture of science). Using this perspective, agency could be documented analyzing the face to face interaction and specifically (in this case) the use of multimodal resources.

Given the multiple forms in which an individual's agency can manifest itself, it is essential to resort to strategies and concepts that enable them to be studied systematically. From a cultural sociology perspective, one of the concepts that has facilitated this task is the *social field*. The main aspect that enables social fields to be identified and determined is based on the *specificity of their cultural production*. According to Bourdieu and Wacquant (1992), the fields are:

> arenas of production, circulation, and appropriation of goods, services, knowledge, or status, and the competitive positions held by actors in their struggle to accumulate and monopolize these different kinds of capital. Fields may be thought of as structured spaces that are organized around specific types of capital or combinations of capital. (Bourdieu and Wacquant 1992, p. 97)

Thus, according to Bourdieu (1988), each field has its dominant and its dominated, its conservatives and its avant-garde members, its subversive struggles, and its mechanisms for reproduction, which are imposed on all agents entering the field. Likewise, every field presents conflicts, and these are the basis for legitimation and access to various types of cultural capital.

On the basis of field characteristics, we consider CLIL contexts for preservice science teacher training as social spaces for cultural production. The educational processes of which constitute forms of cultural action and representation that are enacted in three main fields: science, science teaching, and a foreign language. In these processes, each field provides and demands a specific use of resources. Therefore, agents are required to enact their agency to achieve the goal (*specific cultural production*).

On the basis of this vision, we consider that science teaching and learning always reflects the resources available for appropriation, and then, as culture is enacted, the enactment becomes part of a dynamic flux of resources supporting collective and individual agency as well as a resource for passive action (Tobin 2007). We assume that the enactment of agency in each field (science, science teaching, and foreign language) might be different for each student. Having a social space, such as the science education laboratory were to develop at the same time three different fields, might increase the complexity of the social space providing more opportunities for the enactment of agency.

Science teaching and learning requires the activation of a wide variety of resources that could be classified in three groups: the body and its activity with phenomena (natural and material resources), the signs (semiotic resources), and the pauses and silences (paralinguistic resources). The semiotic resources usually used in science education contexts are composed of different types such as graphs, signs, and gestures. We assume that the use of these resources might vary once a third language such as English is introduced in the science education learning environment. In the present study, we will focus on the role played by material, semiotic, and paralinguistic resources in increasing student teachers' agency in the science laboratory when using a third language such as English.

Methodology

The methodology used in this study adds conversational analysis tools (Drew and Heritage 1992) to the sociocultural approach in order to perform a microanalysis of the discursive interactions (Ramos and Espinet 2013). The data collection strategies included video recording of small-group interactions and a subsequent microanalysis of selective vignettes. For transcriptions, we used standard CA conventions (Atkinson and Heritage 1984). The initial transcripts were checked by at least two researchers in order to ensure reliability. Likewise, all videos were digitized to make them available for analysis using the professional version of QuickTime Player. This software allowed us to slow down and speed up the recording, which we interpreted image by image to capture phenomena at the microlevel, where we often observed patterned actions that the speed of everyday activity does not generally allow us to observe and become aware of in real time.

In this chapter, we shall first present an analysis of the interactions in which agency in the use of material resources such as plants and other relevant laboratory equipment brings about discursive changes, which in turn alter individual/collective relationships. We shall then analyze the various semiotic resources such as gestures that students use and the creativity with which they use them to overcome the difficulties that communicating observations and explanations on scientific phenomena on plant germination and growth in a foreign language entails.

Gestures are a very important aspect in the classroom's interactions; however, we recognize that although the number of studies on discourse in science classrooms has grown considerably in recent years, in comparison to those analyzing verbal discourse, very few have focused on investigating nonverbal communication resources such as gestures and body movements (Kelly 2007; Márquez et al. 2006). Nonetheless, the importance of these nonverbal resources for the construction of meaning is crucial. Since the postulates of Vygotsky (1986), which recognized that thought is not expressed in words alone, elements such as gestures have acquired particular relevance. It is only through the union between words and other available semiotic resources that richness in the construction of meaning can be achieved. Among the diversity of resources and modes of communication, gestures are the most widely used forms of nonverbal communication in the classroom (Goldin-Meadow 2004; Pozzer-Ardenghi and Roth 2007). It has been found that gestures play a crucial semiotic mediation function because they act as connectors for other resources intervening in teaching and learning (Kress et al. 2001).

For the analysis, we use McNeill (1992, pp. 12–18) proposal, in which gestures can be classified into four major types: *iconic, metaphoric, deictic,* and *beats. Iconic* and *metaphoric* gestures are pictorial illustrators that provide a visual representation of the elements to which they refer. To be more precise, iconic gestures present images or visual representations of specific objects, whereas metaphoric gestures refer to abstract aspects. *Deictic* gestures indicate or point to referents or contextualized objects in a similar way to verbal deictics (e.g., demonstratives), which, incidentally, are often accompanied by gestures. *Beats* are gestures that seem to mark time or units. Regarding their form, they are relatively simple, they comprise two sequential movements of execution and retraction to the initial posture, and they mark the pragmatic relevance of the discourse that they accompany. The analysis will identify the different types of gestures used by the students in order to make meaning in relation to plant germination and growth and the possible changes in their use which could be associated to the language and more specifically to English.

Finally, we analyze how the paralinguistic resources identified as pauses and silence emerge in the interaction. These resources are the nonlinguistic elements that accompany linguistic utterances. Paralinguistic resources constitute signs that facilitate the contextualization and interpretation of linguistic information (volume and pitch of voice, speed of utterances, laughter, rhythm, etc.). In our case the analysis focuses on the pauses and silences that modify both the speed and fluidity of utterances. Paralinguistic resources can be associated to participants' specific emotional states, as well as to the level of confidence in relation to the different fields. In our case paralinguistic resources appear to be resources mostly associated to the field of English. It became necessary for the analysis to identify in what moments pauses and silences emerge, whether they are associated to the use of English as a foreign language, and finally what implications this use has in relation to the capacity to act and the accomplishment of activity goals. In addition we are interested in looking at how silences and pauses within the discourse originate moments of synchrony between students, in which collaborative completion is evidence of a collective commitment to achieving the science teacher education goals.

The Classroom Context and the Activity

The data presented here is from a first-year Primary Education Science Teaching course offered as part of the Foreign Language Teaching for Primary Education degree program. The course was offered once a week for two and a half hours during the second semester of the first year. The first hour was oriented to the whole group where students were offered lectures, were asked to engage into small-group reflection on the readings or lectures, and were encouraged to design or analyze primary science education activities, projects, or classroom situations. The whole group was split in two during the second and a half hour to undertake practical work in the laboratory or outdoors. In addition students were offered the possibility to visit an environmental education center, a science education resources center, and a primary science classroom to undertake small-group tutoring in CLIL primary science classrooms.

The activities included in the course were designed using a model-based approach to science education (Espinet et al. 2012) and were guided by the following aims: (a) to become acquainted with basic ideas and English-speaking literature in the field of Didactics of Science in primary education which use a model-based approach to science education; (b) to know, value, and analyze some classroom resources, activities, textbooks, and science projects that use a model-based approach to science teaching in primary education; (c) to develop a positive attitude towards teaching and learning science through direct experience with natural phenomena; and (d) to develop English reading, speaking, and writing competences.

In relation to the language management strategies, the teacher trainers in charge of the course were always speaking in English, whereas students were using the three languages, Catalan, Castilian, and English, during small-group work and English when participating in whole-group activities. The teaching materials and readings used in the course were written in English, and students were asked to use English when filling up worksheets, writing assignments, and doing assessment activities.

Fifty-four trainees participated in the course activities in which they work in small groups in the classroom, in the science laboratory, and outdoors. For the purpose of the study, one activity was selected based on the openness of the proposal and the richness of interactions. This was an exploratory activity where students started to build the model of living beings. All the interactions were videotaped and transcribed. We selected one group (Group 1) to perform a microanalysis of interactions and of specific moments to illustrate ways in which students use resources as a way of overcoming activity-related difficulties in the science laboratory. The spatial disposition of students in Group 1 is shown in Fig. 1. The selected activity was framed within two moments: experimental work at home and subsequently in the university science laboratory. The task at home was very open since the students were given three weeks to grow five beans at their own. During this time, they were asked to write a diary collecting observations and thoughts in English. After that, they brought the plants into the classroom and shared their results in a small-group discussion and later in the whole group for a period of 90 min.

The episodes presented here come from this small-group discussion in the laboratory. The main questions that were guiding the discussion were: (a) How can I compare the different beans? (b) What are the essential factors that help bean seed germination and plant growth? Student teachers were encouraged to write down in English the consensus reached in each of the two questions. The first question was requesting the sharing of the experimental results in the form of bean plants that each individual student brought to the science laboratory. Through the lenses of our

Fig. 1 Spatial disposition of students (Group 1)

framework, students could be seen as producers of a scientific phenomenon, the individual bean plant, grown within an everyday context such as home. Observation and comparison among all plants were the main scientific processes activated through the first question. The second request, on the other hand, put on student teachers the demand for collective abstraction so that they could engage into a scientific modeling process. It was expected that students would initiate a language transition between an everyday language register supporting observational processes towards a more abstract language register which would facilitate the building of scientific explanations through the development of the basic ideas around the model of living beings. The generalization was encouraged through a process of meaning negotiation where students had to agree on the essential factors for seed germination and plant growth based on the comparison of their individual plants.

CLIL Contexts: An Opportunity to Expand Agency in the Science Laboratory

What follows is the analysis of several fragments from Group 1 (Ana, Emma, Laura, Josep, and Sonia) that illustrate relevant interactions in which the use of resources is a salient feature. The chosen fragments illustrate the use of material resources such as plants, semiotic resources such as gestures, and paralinguistic resources such as silences and pauses. Along the development of interaction within this group of students, we could identify associations in the use of languages while building an initial model of living beings. Whereas Catalan and Spanish were mostly used during the first part of the interaction focusing on the descriptions of the plants' physical properties, plants' changes, and experimental designs (Figs. 2 and 3), English appears to be mostly used during the construction of hypothesis, the use of evidences, and the establishment of generalities (Figs. 8 and 10).

Original text	Translation
125. Emma: [però: vols di:r/] si estan plantades a dalt\	125. But.. are you sure.. they are planted from the top
126. Sonia: cla:r (.) és que hi ha terra (.) les mongetes a dalt \ ((va señalando con las manos))	126. Of course.. there is some soil, the beans are on the top
127. Emma: és com aquí: (.) ella \	127. *It is like in here,… she*
128. Emma: >[mira jo] crec que<-	128. *Look.. what I think is*
129. Josep: =ho has mirat/	129. *Have you looked at it…*
	130. *No…I want to see*
130. Laura:no (..) [vull mirar?]\\	131. *I think the problem is*
131. Emma: [jo crec que] el problema é:s-	132. ((Asks Sonia)) but did you put the bean seeds inside or on the top
132.Ana: ((pregunta a SONIA)) **però tu les teves mongetes les vas ficar per sobre o dintre/**	133. in the middle
133. Sonia: al mig\	134. in the middle
134. Ana: al mig/	135. *But I think my problem is… there is cotton, bean seed, and a lot of cotton… and it has less cotton*
135.Emma: *pero yo creo que mi problema es (..) hay algodón moncheta y mucho algodón (.) y esa tiene menos algodón\ [xxxx]*	136. *I did cotton, cotton, cotton, seeds on the top and a total disaster*
136. Ana: *[yo hice] algodón, algodón, algodón* **mongetes** *arriba (.)[y desastre total eh(.)]!*	

Fig. 2 Transcriptions of communicative exchanges developed in the first moments of interaction

Original text	Translation
187. Ana*: total ahí* **estan totes fetes** [¿](.)	187. Here **they are all done.**
188. Sonia: *tenemos las totalmente cerradas*	188. *We have those totally closed*
189. Emma: *vamos a poner las que han crecido*	189. *Let's group all that have grown*
190. Laura: *la de:l serrín_*	190. *The ones having sawdust*
191. Sonia: *luego están estas que no:_*	191. *then we have these, they have not…*
192. Emma: *hombre éstas?* ((toman todos los recipientes y observan cada una)) **están crecidita** *eh :(..) que tiene* **arrel** *y todo.*	192. *These? (they take all flasks and start observing each plant) .. this is quite grown.. it has* **roots**
193. Sonia: *pero esas-/*	193. *but these…*
194. Josep: **aquestes també**	194. **these too**
195. Sonia: *éstas están abiertas*	195. *these are open*
196. Josep: ((toma dos frascos y los pone al centro)) **aquestes també ho intenten** ((clasifican: las que crecieron y las que no))	196. ((take the flasks and put them in the center)) **these are also trying** ((they classify between those that grew and those that did not))
197. Emma: **com aquesta: _**	197. **Like this**
198. Ana: *he perdido uno.*	198. *I have lost one..*
199. Ana: *=ésta también.*	199. *This too*
200. Emma: **aquesta (.) com aquesta\ i com aquesta** ((se dirije a ANA))	200. This like this and like this ((looking at Anna))
201. Ana: y luego *está:n-*	201. *and then we have got*
202. Emma: *nuestros fracasos*	202. *ours failures*

Fig. 3 Transcription used to illustrate the dialectical relationship between individual/collective voice

When the Use of Plants as Material Resources Facilitates Crossing the "I" Border to Build the "We"

After the teacher trainer had indicated to the students that it was time to share their results, the group members initiated their dialogue. The direct exploration and manipulation of different phenomena (grown beans) brought by each individual student constituted the first moment to enact their agency. The communicative exchanges developed in their first languages (**Catalan** and *Spanish*) and were characterized by descriptions of the experimental conditions that each individual student teacher arranged to make bean seeds grow (Fig. 2).

Each student presented and explained the strategies used to make seeds grow, as well as the results obtained. The dialogues included in the previous fragment (Fig. 2) are characterized by first-person narratives where each student describes his or her own experience and compares it to the other students' experiences. The individual/collective distance is strongly marked as can be seen in line 135. The difference between "mine" (the self) and "yours" (the other) constituted the first fundamental step towards the collective challenge. Once students had made several observations and comparisons, they decided to classify experimental designs based on the structural changes observed in grown beans. They organized phenomena into three groups: (a) seeds that are just germinating, (b) seeds that grow into plants, and (c) failures (Fig. 3). This was the first collective construction which used the plants as material resources. Student teachers had just built new collective plant phenomena as a result of their interaction.

The previous fragment (Fig. 3) shows the students' interactions while moving the seed flasks from their initial positions in order to form three groups (Figs. 4 and 5). This change in object position is accompanied by a significant change in the students' discourse. For the first time, students shift from descriptions in the first-person singular ("I") to the first-person plural ("we"). In row 188, Sonia introduces the verb *"tenemos"* (*we have*) and Emma follows in row 189 with the expression *"vamos a poner las que han crecido"* (*let's group all that have grown*). The group continues using the first-person plural. At the end, in line 202, Emma assumes the failures as being *"nuestros"* (*ours*).

The action of classifying experimental phenomena facilitated a voice change within the group, thus illustrating individual/collective dialectics. Initially, each student brought an individual scientific production which facilitated his individual agency within the group. Only when the phenomena became collective, that is, classified, new scientific entities became evident as a result of new cultural productions, the groupings of plants into meaningful categories. At this phase English was irrelevant as a resource, whereas individual phenomena became crucial. From this moment on, students interactions were oriented towards the search for explanations as a group led by one specific student, Emma: *Now, let's see ((pointing at each group)) what has been done here, what has been done here, and what has been done here?*

Fig. 4 Initial position of phenomena on the table

Fig. 5 New position and classification of phenomena

When the Use of Semiotic Resources Such as Gestures Becomes the Other's Voice

While students were undertaking the activity, the need both to find a better explanation of the phenomena and to make use of English (oral and written) altered the use of resources. At this point of the activity, semiotic resources were more

Original text 393. Josep: que llavo- fem esqueixos (.) els esqueixos els poses en aigua (.) crea una arre:ls_ ((gesticula con las manos para representar las raíces)) Translation: 393. Josep: then-we make slips (.) you put slips in water (.) roots are created_ ((uses gestures with the hands to illustrate the roots))	
	Josep illustrates the shape of roots

Fig. 6 Iconic gesture used by Josep to illustrate the shape of roots

necessary than the material ones. The change from using first languages (Catalan and Spanish) towards using a foreign language (English) made possible situations such as (a) slower rhythm of speech, (b) fewer overlapping turns, (c) greater attention to the student holding the turn, and (d) greater use of gestures and pauses. Without a microanalysis, these signs might be seen as indications of communication shortcomings. However, the detailed interactional analysis undertaken from a sociocultural perspective allowed several interesting phenomena to be identified, which are described below.

A detailed observation of the students' interaction allowed a substantial change in the use of gestures to be identified in relation to the languages being used. Gestures were one of the most frequent semiotic resources used to keep communicative sequences in English going at this point. The students' use of gestures differed depending on whether they were speaking in their first languages or in English. The functions of gestures while speaking in Catalan or Spanish dealt with establishing connections between verbal and nonverbal aspects of communication or between words and material artifacts present in the activity. When students spoke in English, as we will see below, the use of gestures increased and their functions diversified. During the students' interactions, we identified interactions in which gestures (a) represent a property of phenomena (iconic), (b) represent non-observable phenomena and explanations (metaphorical), (b) request a specific word (beats), and (c) point directly to phenomena (deictic).

Presented below is a series of vignettes to exemplify the diversity of ways in which the students used gestures while interacting.

Iconic-type gestures providing visual representations of specific objects were used the most. Generally speaking, their presence is related to descriptions of both the experimental design and the physical changes that took place in the seeds from germination to plant growth. For example, when one of the students (Josep) describes the roots of the plant, he indicates their shape by making a downward movement with his hands (Fig. 6).

Another example is when Emma tells her fellow students about the way she dosed the amount of water to make the seeds germinate. In this case, the gesture describes an experimental action (Fig. 7).

Original text

335. Emma: *pero ésta lo puso el último con tanta agua eh/ los primero- los primeros días me mojaba los dedos y hacía así\ ((mueve los dedos para mostrar cómo dosificaba la cantidad de agua))*

336. Sonia: *ah sí/*

Translation

335. Emma: *but this was put the last with so much water / the first- the first days I watered my fingers and I was doing this\ ((she moves the fingers to show how she was dosing the amount of water))*

336. Sonia: *Oh! Yes/*

Emma shows how to put water.

Fig. 7 Iconic gesture used by Emma to show how to put water

The presence of gestures of this type does not appear to be associated with the use of languages, that is to say, the forms they use are practically the same in the first languages (**Catalan** and *Spanish*) and in the foreign language (English).

Metaphoric gestures occurred mainly when the students referred to non-observable processes. In this case and in accordance with the phenomenon that they were studying, these processes were related to germination, plant growth, transpiration, and so on. The presence of gestures varies in relation to the use of the different languages. We found that when the students mentioned and explained processes of this type in their first languages, they did not usually accompany their descriptions with gestures yet; when they were mentioning or describing them in English, they resorted to gesture as a semiotic support resource.

In the following vignettes accompanied by images (Fig. 8), we can see how, when Emma and Josep are communicating in English and referring to processes of germination and plant growth, they move their hands to make the message they are trying to communicate more obvious.

Regarding *deictic*-type gestures, we were able to find a significant difference in the ways they were used and, consequently, in their meanings. While the aim of all of them is to point to or highlight the presence of something, the construction of meaning achieved by them is diversified. For example, when a student pointed to something while using a first language, the act of pointing was accompanied by the spoken word (word + gesture). However, when using the foreign language, a deictic gesture was used in most cases to point to the object or the nature of the object for which the English word was unknown. In other words, the gesture had the meaning of the element to which a student was pointing (the gesture as a substitute for the word).

An example of this, presented in Fig. 9 in the first instance, is the use of deictic gestures, while experimental designs were being described in first languages.

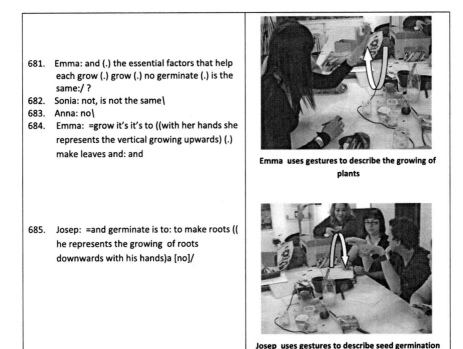

681. Emma: and (.) the essential factors that help each grow (.) grow (.) no germinate (.) is the same:/ ?
682. Sonia: not, is not the same\
683. Anna: no\
684. Emma: =grow it's it's to ((with her hands she represents the vertical growing upwards) (.) make leaves and: and

Emma uses gestures to describe the growing of plants

685. Josep: =and germinate is to: to make roots ((he represents the growing of roots downwards with his hands)a [no]/

Josep uses gestures to describe seed germination

Fig. 8 Metaphoric gestures used by the students to describe the growth and germination processes

Regarding *deictic* gestures while using the foreign language, presented below are the interactions in which it was possible to observe how the students pointed to objects for which *the word was missing*. In the vignette below (Fig. 10), we can see that Josep and Emma are discussing the essential factors for plant germination and growth and are concentrating specially on the role of light. In order to identify if light is essential or not, they start comparing the various tones of green that can be observed on the leaves of the plants. To do that, they use the expression "very green" (Josep) and "strong green" (Emma) to refer to dark green leaves or the greenest leaves. But, when Emma needs to describe a leaf that is not as green, she begins the phrase and pauses when she cannot find the adjective she needs at that time and then points insistently at the leaf. At the end of the phrase, she resorts to the insertion of a word in her first language. At the same time, however, Josep provides the adjective in English: pale. Consequently, her gesture served to obtain the word.

Finally, the use of *beat*-type gestures occurred mainly when the students were communicating in the foreign language and did not know the word they needed to keep the communication sequence going. On top of this, the element to which they were referring was not physically present (as we saw in the previous example). These gestures are, therefore, a sign of an emotion that we could associate with desperation. They are *gestures that substitute or solicit the unknown word*. In these cases, gestures are usually accompanied by fillers, the repetition of words and/or pauses. Let us take a look at the examples below in Fig. 11.

Original text

119. Emma: >Sí pero mira en algodón< ((señala una de las plantas que más han crecido y que está en algodón))

Translation

119. Emma: >Yes but look at the cotton< ((she points at one of the plants that has grown the most within cotton))

Emma points at an entity from the experimental design

Original text:

62. Sonia: les meves ((las señala con el dedo)) han estat
63. Laura:sí:/ ((con interés))
64. Sonia: >tanca- bueno [?] les vaig a tapar i les vaig a ficar dintre un armari (.) i en dos dies (.) les-[les vaig destapar i estaven ja sortint per aquí] ((señala con el dedo el recipiente donde tienen la planta y el punto hasta donde había crecido en dos días))

Translation

62. Sonia: Mines ((points with her fingers)) have been
63. Laura:yes:/ ((with interest))
64. Sonia: >close- well [?] I covered them and I put them in the closet(.) and in two days (.) [I uncovered them and they were sprouting here] ((points with her finger the plant container and the point where they had grown in two days))

Sonia points at her experimental design while she is describing it

Fig. 9 Deictic gestures used by students to point the experimental designs

Finally, it should be underscored that the presence of gestures increased significantly when the students were communicating in the foreign language. In addition, gestures fostered moments of greater cooperation and synchrony among team members because, when someone did not know a word and resorted to using them, fellow students picked up on it. Consequently, the student who did not know the word received support from fellow students to continue the communication sequence without abandoning the use of the foreign language.

When the Use of Paralinguistic Resources Such as Pauses and Silences Facilitates a Collective Search for New English Words

Like gestures, the use of pauses constituted one of the most frequent resources that the students used to keep their communication exchanges in English going.

527. Josep: ok but I- (.) I think that ehh- (.) in order to germinate (.) the seed the s- (.) light is is not a: an important factor (..) but when (.) when the plant is growing up then we need the sun\

528. Emma: yes because of the color and the size

529. Josep: yeah but more: [?]

530. Josep: also for the phot- photosynthesis (.) **pot ser** [?]/ (.)

531. Ana: xxx *clorofila*\

532. Emma: the chlorophyll.

533. Laura: there is more: green:

534. Ana: yeah\

535. Josep: and the chlorophyll has-

536. Ana: have a: (..) a good color\

537. Josep: yeah yeah (.) that's very green. ((points at one of her plants))

538. Emma: that green is strong. ((points at one of her plants))

539. Josep: strong green.

540. Emma: but this this green is (..) hum ((points at one of the plants)) *dèbil/*

541. Josep: pale (.) pale

Emma points at the plants while she is comparing their color

Fig. 10 Deictic gesture used by Emma in order to research an unknown word in English

735. Josep: ok (..) emm (..) the growing process (.) of the bean can help you/ in order to: become: or to improve your: your science teaching: ((he moves his hands as if he was searching for a word)

Josep uses gesture jointly with a pet word in search of a word

94. Emma: I think that because of the: (.) hummm ((she moves her hands))

Emma uses gestures in search of a Word

Fig. 11 Beat gestures used by students when they had communication problems

However, far from representing a shortcoming in the participants' communicative competence, it became a resource for collaborative completion and construction of explanations. In other words, a pause made by one of the participants was an opportunity for another to participate (take turns) to construct the utterance.

This opportunity also involved or demanded attention to and a grasp of the speaker's idea, since that was the only way that the students could provide the speaker with the word (or potential word) that he or she needed. Consequently, the use of pauses fostered moments of synchrony between the participants, which originated collaborative completion, as shown in the three examples that follow.

The dialogue below took place when the students were discussing where seeds should be placed in the ground to enable them to germinate. They tried to explain the best place for them and, as a result, explained both their successes and failures. In this case, Josep picks up one of the pots in which the seeds had not germinated and begins to give his explanation. Throughout his explanation, however, he resorts to pauses when he does not know a word: *it's it's too (..)*; this pause encourages his fellows Laura and Ana to give him the help he needs to complete the phrase. Finally, A4 gratefully exclaims: yeah! (.) too deep in the ground\.

573. Josep: it's it's too (2.45)
574. Laura: =more time\
575. Josep: it's too (2.53)
576. Ana: too deep.
577. Josep: yeah! (.54) too deep in the ground\
578. Ana: yeah\

In this second example, we can see an exchange between Josep and Emma, in which Emma provides support for the construction of the idea when she hears Josep's pauses.

642. Josep: no: (1.23) they are emm (3.01)
643. Emma: big places/
644. Josep: big places with eh sunlight and with wa- a warm place

Finally, presented below is a longer sequence in which there is an occurrence of collaborative completion and construction of ideas about the various or alternative methods that can be used should an experimental design fail. We can see how Ana initiates the idea, which Emma and Josep enrich (turns 803 and 804). Subsequently, Emma and Josep establish another sequence of collaborative completion, which is interspersed with pauses. In these sequences, each pause marks an opening for a new turn, a new voice that eventually reaches a consensus on multiple voices.

802. Ana: try it again \ *(..)*
803. Emma: you can try [another thi:ng]_
804. Josep: [yeah you can try] another thing and other methods_
805. Emma: =xxxx methods_

806. Josep: and maybe- (..)
807. Emma: maybe will- will go:-
808. Josep: better \ *(..)*
809. Emma: better (.) and and will grow up or will I don't know (.)
       something\
810. Sonia: because xxxx you follow another method_
811. Emma: =exactly \

This type of involvement in pauses within an interaction was also identified by Tobin and Roth (2006) and Tobin (2008), who, from a sociocultural perspective, assert that: *"the silent pause was a resource for changing speaker, affording the agency of speakers and presumably the listeners as well"* (Tobin 2008, p. 90). In other words, long pauses are signs of a speaker's completion of a turn of speaking or of an opportunity for the speaker to be interrupted by another turn (Tobin and Roth 2006).

Rethinking Preservice Science Teacher Training in CLIL Contexts

In our context, the incorporation of a foreign language into disciplinary content learning was an added value rather than a disadvantage. The examples show how students enact their agency and make use of a wide diversity of resources to overcome the difficulties that Content and Language Integrated Learning (CLLIL) in a science teacher education classroom taught in English entails. Consequently, we would like to conclude this analysis by stating that the experience documented in this chapter provides some evidence to support the idea that the introduction of a foreign language within science teacher education courses is not an obstacle for students' learning but an opportunity to improving the quality of students' interactions.

Our initial question was *how do students expand their agency in the use of resources as a way of overcoming the difficulties derived from the need to construct a scientific explanation of natural phenomena using English as a foreign language?* In framing the question this way, we assumed that agency could be observed at the level of social interactions and that it would affect the use of resources that were available such as material, semiotic, and paralinguistic resources. As in any other science learning environment, the design of the task was very important since it provided the structure in which students would enact their agency. The task was constituted in two important moments so that each individual student could build his or her own cultural capital in science during the first part of the activity and could participate in a collaborative exchange so that an explanation could be jointly built during the second part. The resulting cultural production of the first part was a joint phenomenon (the classification of individual plants as a result of a group negotiation), and the resulting cultural production of the second part was a joint explanation (the identification of some essential factors of plant germination and growth). The difficulties each student experienced in these two parts were different both in terms of science and English.

In this particular task, the introduction of English increased the use of semiotic resources when building an explanation of plant germination and growth to enrich the layers of meaning. Thus, a wide variety of gestures were used to indicate the characteristics and actions undertaken during the experimental designs as well as to represent plants' properties and changes such as germination and growth processes, among other things. We consider CLIL-type educational spaces to be complex conversational spaces because they foster the use and enactment of every possible semiotic mode. One of the postulates of Lemke (1998), on the use of the various semiotic modes that intervene as mediators in the construction of meaning, leads us to the following reflection:

> Speech co-evolved as part of interactional synchrony: the bodily and material integration of individual organisms into their ecosocial environments. The intonational patterns of speech and the musical patterns of song descend from common ancestral modes of behavior. The synchrony, not just in individuals but across dyads and groups, of verbal action with other body movements and rhythms signals the participation of gesture and movement in the unitary communication system from which we abstract the semiotic patterns we call language or gesture. Our perception as well as our production of semiotic interaction makes use of visual and kinesthetic information and responsiveness as much as it does of the auditory channel. (Lemke 1998)

Likewise, we found that using the foreign language led students to alter some paralinguistic aspects of discourse, such as a slower rhythm of speech, fewer overlapping turns, more gestures and pauses, and so on. All of these aspects fostered moments of greater synchrony among team members because they listened to each other more carefully and paid greater attention to the speaker. These situations highlight the collaborative effort that was made; support was given to a fellow student with difficulties communicating in English to enable the construction of utterances (collaborative completion), and as a result, the final constructions were the product of a veritable collaborative effort.

Analyzing the complexity inherent to CLIL science classrooms from a sociocultural perspective has helped us to identify the students' diversity in enacting their agency. The uneven distribution of their own social and cultural capital, as well as the use of resources in a creative manner, transformed communicative difficulties into opportunities (Ramos and Espinet 2011). These were characterized as leading to communicative synchrony and solidarity where partially distributed competences enriched the individual/collective dialectic.

If European science teachers need to be able to simultaneously teach science and a foreign language such as English, we need to change our vision of what a competent communicative student in science is. This study provides evidence to support that multilingualism in science classrooms as a consequence of today's globalization can be a richer context for science learning than we might think.

Some implications for the improvement of science teacher education programs can be drawn from this study. These implications deal primarily with the role given to language in science teacher education and open a space for reflection. We see important challenges for science teacher education that can be framed in the following goals. The first challenge deals with the introduction of language and foreign language learning goals in science teacher education programs. In fact, teaching for the improvement of students' communicative competence in most European science teacher education programs is at present a prerequisite. In order to learn, student teachers need to master the languages to be used in classrooms; otherwise, learning cannot take place satisfactorily. In addition there is a common belief held by newcomers entering into CLIL approaches that teaching a foreign language detracts science teachers and science teacher educators from teaching science or science teacher education, their main mission.

The second challenge deals with the design of learning environments that promotes both science education and foreign language learning in science teacher education programs. These learning environments should be complex enough so that foreign language and science teaching learning progressions could be interwoven. The third challenge is related to the strengthening of co-teaching strategies that would involve a science teacher educator and the foreign language teacher educator in the same classroom. The normalization of co-teaching strategies in higher education more generally, and in science teacher education more specifically, would facilitate the true development of interdisciplinary programs which are at the heart of CLIL approaches. Finally, the fourth challenge addresses the need to start working with schools to develop CLIL approaches in infant, primary, and secondary science education so that this experience is used in science teacher education programs. It is sometimes disturbing to see how many schools embrace CLIL approaches to teach science and a foreign language and how few science teacher education institutions show interest in its development. It would be important to open a true forum on this issue so that we can really help schools in this challenging endeavor.

Acknowledgments Supported by Spanish MCYT grant (EDU-2012-38022-C02-02; Catalan PRI 2009SGR1543 and Spanish MICINN grant EDU2010-15783- subprogram EDUC).

References

Atkinson, J. M., & Heritage, J. (1984). *Structures of social action: Studies in conversation analysis*. Cambridge: Cambridge University Press.

Bourdieu, P. (1988). *La Distinción. Criterio y bases sociales del gusto*. Madrid: Taurus.

Bourdieu, P., & Wacquant, L. (1992). *An invitation to reflexive sociology*. Chicago: University of Chicago Press.

Commission of the European Communities. (2007). *High level group on multilingualism. Final report*. Belgium: Official Publications of the European Communities.

Drew, P., & Heritage, J. (1992). Analyzing talk at work: An introduction. In P. Drew & J. Heritage (Eds.), *Talk at work: Interaction in institutional settings* (pp. 3–65). Cambridge: Cambridge University Press.

Espinet, M., Izquierdo, M., Bonil, J., & Ramos, L. (2012). The role of language in modeling the natural world: Perspectives in science education. In B. Fraser, K. Tobin, & C. McRobbie (Eds.), *Second international handbook of research in science education.* New York: Springer.

Goldin-Meadow, S. (2004). Gesture's role in the learning process. *Theory into Practice, 43,* 314–321.

Kelly, G. (2007). Discourse in science classrooms. In S. Abell & N. Lederman (Eds.), *International handbook of research on science education.* Mahwah: Lawrence Erlbaum.

Kress, G., Jewitt, C., Ogborn, J., & Tsatsarelis, C. (2001). *Multimodal teaching and learning: The rhetorics of the science classroom.* London: Continuum.

Lemke, J. (1998). Multiplying meaning: Visual and verbal semiotics in scientific text. In J. R. Martin & R. Veel (Eds.), *Reading science* (pp. 87–113). London: Routledge.

Márquez, C., Izquierdo, M., & Espinet, M. (2006). Multimodal science teachers' discourse in modeling the water cycle. *Science Education, 90*(2), 202–226.

Masats, D., Feixas, M., Couso, D., & Espinet, M. (2006, July). *La docència en anglès en assignatures no-lingüístiques a la titulació de Mestre Especialitat Llengua Estrangera.* Presented at the 3rd International Congress of University Teaching and Innovation, Barcelona.

McNeill, D. (1992). *Hand and mind: What gesture reveals about thought.* Chicago: Chicago University Press.

Pozzer-Ardenghi, L., & Roth, W.-M. (2007). On performing concepts during science lectures. *Science Education, 91,* 96–114.

Ramos, S. L. (2010). *Contextos CLIL para la formación inicial del profesorado de ciencias: análisis de la interacción desde una perspectiva sociocultural.* Doctoral thesis, Universidad Autónoma de Barcelona. www.tesisenxarxa.net/TESIS_UAB/AVAILABLE/TDX//slrr1de1.pdf

Ramos, L., & Espinet, M. (2011, September 5–9). *Using English to build scientific explanations: A sociocultural analysis of interactions.* In European Science Education Research Association (ESERA) Conference, Lyon, France.

Ramos, S.L., & Espinet, M. (2013). Una propuesta fundamentada para analizar la interacción de contextos AICLE en la formación inicial de profesores de ciencias. *Enseñanza de las Ciencias, 31*(2).

Roth, W.-M., & Tobin, K. (2006). Announcing cultural studies of science education. *Cultural Studies of Science Education, 1,* 1–5.

Sewell, W. (1992). A theory of structure: Duality, agency and transformation. *The American Journal of Sociology, 98,* 1–29.

Tobin, K. (2007). Collaborating with students to produce success in science. *Journal of Science and Mathematics Education in Southeast Asia, 30*(2), 1–43.

Tobin, K. (2008). Structuring success in science labs. In A. J. Rodríguez (Ed.), *The multiple faces of agency: Innovative strategies for effecting change in urban contexts* (pp. 83–102). Rotterdam: Sense Publishers.

Tobin, K. (2010). Reproducir y transformar la didáctica de las ciencias en un ambiente colaborativo. *Enseñanza de las Ciencias, 28*(3), 301–314.

Tobin, K., & Roth, W.-M. (2006). *Teaching to learn: A view from the field.* Rotterdam: Sense Publishers.

Van Eijck, M. (2009). Scientific literacy: Past research, present conceptions and future developments. In W.-M. Roth, & K. Tobin (Eds.), *The world of science education: Handbook of research in North America* (p. 245). Rotterdam/Boston: Sense Publishers.

Vygotsky, L. S. (1986). *Thought and language.* Cambridge, MA: MIT Press.

Transcription conventions

---

1. Intonation types:
      (a) Falling: \
      (b) Rising /
      (c) Maintained ‗
2. Pauses, timed (in seconds):
      (a) Short: (.)
      (b) Long: (..)
3. Overlapping: [text]
4. Latching: =
5. Interruption: text-
6. Lengthening of a sound: text:
7. LOUD
8. °soft°
9. <slow>
10. >fast<
11. emphasized
12. Transcriber's comments: ((text))
13. Incomprehensible fragment: xxxx
14. Continuation of a previous turn: speaker>
15. *Spanish Catalan* English
16. [?] transcriber's best guess
17. Turns:
      1:
      2:
      3:

Part IV
Cultural Issues in Science Education

Part IV provides examples of how we can respond to cultural issues (e.g. religion, gender and language) in science classrooms.

Reconceptualizing a Lifelong Science Education System that Supports Diversity: The Role of Free-Choice Learning

Lynn D. Dierking

Mansour and Wegerif's introduction to this edited volume recognizes the high value being placed on science education by governments around the world with the understanding that the skills embodied in science, technology, engineering, and mathematics (STEM) disciplines do not merely allow nations to compete in the world of work but also enable their citizens to be active participants in a global society. The value of STEM literacy is only heightened by the twenty-first-century challenges nations face, challenges which are increasingly complex and global in nature, often requiring an understanding of STEM to solve. This issue becomes only more sobering in light of decreasing engagement in science in school and declining enrollments in university science courses, indicators that suggest declining societal vitality and less informed democratic participation in science.

It is important to appreciate though that less participation in science in school and university is just one set of indicators of the STEM literacy problem. The societal changes and global challenges we face warrant fundamentally changing the approaches educators and educational researchers take to reforming science education globally. I agree with Mansour and Wegerif's notion outlined in the introduction to this book that diversity is key to any reform, not only the traditional notion of diversity as gender and ethnicity but also "a new understanding of diversity ..., diversity as a way of thinking about science and about education into science, which does not define specific diversities in advance of practice but embraces openness, responsiveness and responsibility in the nature of the practice of science education itself" (Mansour and Wegerif 2013, p. X). I also agree with their perspective that to take such a diverse approach requires viewing learning and education as not only the transmission of knowledge to prepare learners for further education and a career, but as processes by and through which learners construct their identity and, in so doing, develop their *own* lifelong relationship to science.

L.D. Dierking (✉)
College of Science and College of Education, Oregon State University, Corvallis, OR, USA
e-mail: dierkinl@science.oregonstate.edu

N. Mansour and R. Wegerif (eds.), *Science Education for Diversity: Theory and Practice*,
Cultural Studies of Science Education 8, DOI 10.1007/978-94-007-4563-6_13,
© Springer Science+Business Media Dordrecht 2013

However, focusing only on school and university indicators and, as a consequence, framing the issue within this activity space diminish one's ability to approach this challenge in a more innovative and comprehensive manner, particularly given current realities regarding school and university participation. Increasing numbers of youth, particularly from low-income, disenfranchised groups underrepresented in STEM, do not graduate from high school, and the vast majority of adults either are not privileged enough or do not choose to further their schooling beyond high school (Falk and Dierking 2010). Solutions that focus only on school- or university-aged children and youth as is the case for the majority of chapters in this volume also do not recognize a vast group of middle-aged and older adult learners who could benefit from continued STEM learning. After all, since science changes so rapidly, science literacy cannot be achieved in any society by merely focusing reform efforts on the young.

Despite these realities though, current approaches to science education reform at least in the USA rarely address disenfranchised youth and adults, and most solutions are neither complex nor innovative, centering solely on improving science teaching in K-12 schools or, if enlightened, extending reform efforts down into the pre-K years or up into the university (Carnegie Corporation 2009; National Academies of Science 2006; National Research Council 2011). These approaches neglect the fact that a "quiet revolution" is underway in societies worldwide, resulting in increasing opportunities for diverse forms of education and learning. The centers of this revolution are not the traditional educational establishment of schools and universities but rather a community network of educational entities: libraries; print and broadcast media; the Internet, personal games, podcasts, and social networking media; and museums, zoos, aquariums, and science centers (Horrigan 2006; Falk et al. 2007).

Taking a school-first approach also neglects the contributions of the workplace as another venue for science learning. Although a relatively small percentage of the public (3.8 %) are employed in jobs requiring a science or engineering *degree* (National Science Board 2004), the percentage rises dramatically to 40 % if one considers the number of people who work in "middle-skill" science- and engineering-related occupations that require technical training such as an associate's degree or occupational certificate. In addition to the free-choice learning arena, the workplace is a neglected yet important third educational sector in our society (Falk and Dierking 2002), a sector that supports the learning of adults and older youth not in school or university.

Given these realities, I think we need to step much further back. It is not enough to simply frame the task as a need to redesign schools and curriculum for school-age children and youth. We need to seriously take on the challenge posed by Mansour and Wegerif by fundamentally changing the *practice of science education*, envisioning a lifelong science learning system that supports diversity broadly. This system would support the lifelong STEM learning of citizens of all ages, backgrounds, and stations of life, recognizing the myriad places and ways in which they engage and participate in STEM, as well as the many reasons they might choose to engage and participate. Most importantly, this system would be designed in a way that acknowledges what is most important for science educators, whether teachers in schools, educators in free-choice science learning settings and environments, or university professors in science and education are to create opportunities for the diverse learners with whom they interact to construct a relationship with and to science that meets their needs and makes sense to them in their everyday lives. This would be the case whether the person is or plans to be a scientist, a technician, someone who engages in science-related hobbies and pursuits, or a citizen with other types of expertise and interests, but whom we hope will have some basic understanding of science in order to lead a healthy life and make sound decisions based on evidence.

This is not meant as a condemnation of school-based learning. The point is merely to emphasize that improving schools and curriculum though important is only one piece of a comprehensive approach to educational reform. Certainly the authoritarian one-size fits all model of the traditional science curriculum is no longer appropriate. We need to be seeking new ways to engage science learners of all ages in co-constructing their own learning of science, not only in school, but throughout their lifetimes. To do this well, we must understand how to more effectively connect science learning opportunities across settings and the life span by working with educational colleagues in the myriad science settings in which science learning occurs. If we understand the connections and interrelationships that learners make within this lifelong science learning web and work collaboratively with learners and colleagues engaged in science education across settings and the life span, we are far more likely to be able to build a system that better leverages and contributes to lifelong science engagement and learning, to a citizenry that identifies at some level with science.

This is a huge, long-term undertaking, requiring careful thought, collaboration across settings, deliberation, and much formative testing, certainly not the purview of a book chapter. What I can do in a chapter though is build a case for such an approach, in particular describing the component of this lifelong STEM learning system that I know best: the free-choice science learning sector. First I will document this growing free-choice science learning movement and the often hidden infrastructure supporting it. Then I will share findings from a US National Science Foundation (NSF)-funded retrospective study of the long-term impacts of gender-focused, free-choice STEM learning experiences on 213 young women from diverse backgrounds underrepresented in STEM who had participated in free-choice programs 10–20 years before. I close the chapter by discussing the need for a research and education infrastructure that embraces this comprehensive view and attempts to understand learning across settings and time.

The Free-Choice Learning Revolution

Much evidence supports the contention that the public learns science in settings and situations outside of school. A 2009 report by the National Research Council, *Learning science in informal environments: Places, people and pursuits* (NRC 2009), describes a range of evidence demonstrating that even everyday experiences such as a walk in the park contribute to people's knowledge and interest in STEM. For example, in any given week, a person might watch a television program on evolution, research a diagnosis of high cholesterol by her physician, and build a model rocket with a child. Each of these is an example of free-choice science learning activity—the learning that individuals engage in throughout their lives when they have the opportunity to choose what, where, when, how, why, and with whom to learn (Dierking and Falk 1998). Children *and* adults are spending more of their time learning, not just in classrooms or on the job, but through free-choice learning at home, after work, and on weekends.

The question arises though do communities have the resources and expertise to support lifelong science learning among their citizens not engaged in school or university or school-aged children and youth outside of school (Dierking in press-a)? Without hesitation I say it is not only possible, but rich examples already exist in many communities. Science programs take place in parks, shopping malls, during scouts, in senior communities, YMCA/YWCAs, libraries, even cars, and restaurants (CDs featuring current research conducted on site can be borrowed while visiting national parks and French fry wrappers and recyclable paper cups at the Pacific Northwest fast-food chain, Burgerville, feature ecological information about rough-skinned newts, and sockeye salmon).

There are also science-related museums and other free-choice science education settings such as zoos, national parks, aquariums, and science-technology centers. Museums, particularly science-technology-oriented ones, currently rank as one of the most popular out-of-home leisure experiences in the world; the Association of Science-Technology Centers (ASTC) estimates that there were 89.6 million visits to their member science centers and museums worldwide in 2009, with 62.9 million of those visits made in the USA (ASTC 2010).

One can even learn about science in a pub! Science Cafés and Science Pubs, first developed in Europe in the early 1990s, have flourished in the USA for nearly a decade. Although they began in major urban areas, due to their success, they are now being replicated in "less usual" communities: rural areas in Montana and South Dakota in the USA and Cockermouth in the Lake District, UK; on islands, Corfu, Greece, and Orkney, Scotland; within immigrant and gypsy communities in Europe; and even in Palestine (Dierking in press-a).

The above examples are community-based, but expanding beyond one's local community, there are books. Despite the hype about declining literacy, the number of books sold in the USA in 2006 was up from 2005, and with the increasing adoption of e-books (the share of adults in the USA who own an e-book reader doubled from November 2010 to May 2011), the number of books sold is at an all time high; many of these are science and/or technology-related (Purcell 2011; U.S. Bureau of the Census 2010).

There is also television and radio. Not only is television viewing up (U.S. Bureau of the Census 2010) but, so too, are the number and diversity of information-oriented programs, many of them are science and/or technology-related (Miller et al. 2006). And then there is radio, a medium in which there are many science-related programs.

For instance, the 20-year-old *Talk of the Nation: Science Friday* radio program has a weekly audience of 1.3 million listeners. *Science Friday* presents STEM news and policy analysis, as well as the interplay of science and society. Scientists and science policymakers debate and disagree, argue, and analyze, live and unedited—just as they do in their working lives. Listeners hear about science as a work in progress and are offered a unique opportunity to speak directly with the show's guests, so that the conversations on the program become distinctively relevant to the lives of the general public.

There is also the staggering growth of the Internet—and science and technology topics are being communicated there also; data shows that once people turn to the Internet for science news and information, they learn to rely on it as a source, especially young people (Horrigan 2006). And these media are not silos either. For instance, *Science Friday* described above is active on the Internet and social media sites. It was the first national radio program to broadcast on the Internet and to introduce the then unfamiliar concept of the "World Wide Web" to its listeners. It was also the first National Public Radio (NPR) program to produce podcasts (they register 24 million podcast downloads per year) and remains the second most popular NPR program available in podcast. Listeners can submit questions via Twitter, Second Life, and www.sciencefriday.com, and also can engage in science discussions via a Facebook community and website blogs. In 2006, as YouTube's popularity with younger audiences became apparent, *Science Friday* began including STEM videos on their website. SciFri Videos, available in both English and Spanish, are designed to appeal to a younger demographic and were viewed more than one million times in 2009. Due to this success, *Science Friday* has recently received US NSF informal science education funding to expand in new and innovative ways via broadcast, portable media (e.g., smartphones, tablets), and the Internet (e.g., websites, Facebook, and other social communities), all with a focus on reaching youth, in particular Latino/Latin, between 12 and 24 years of age.

These are but a few examples of a vast and vibrant free-choice science education infrastructure which is unseen, undervalued, and underfunded (certainly by public dollars), because the window through which most science educators and policy-makers gaze is focused on K-12 (Dierking in press-a). From the growth of the Internet to the proliferation of gaming and educational programming offered by IMAX, educational television, and museums, there are more opportunities for self-directed, free-choice learning than ever before, much of it science and health related. Most of the examples I have shared are indoor activities but people also engage in such learning outdoors every day—hiking, visiting national parks, and engaging in other nature activities—tapping into a vast science learning infrastructure available 7 days a week, 24/7, across a life span. These opportunities are important, in fact, essential ways that people learn. Even more critical, these modes of learning allow individuals to *contextualize* and *personalize* their science knowledge, interest, and understanding throughout their lifetimes. It is hoped that these science experiences contribute, along with schools and the workplace, to building science identities that meet the needs of lifelong learners and enable them to become science-informed citizens, perhaps even *engaged science participants*, broadly defined, to include scientists, engineers, mathematicians, and technicians, as well as youth and adults who choose to engage in science-related hobbies, pursuits, and habits of mind.

The Lifelong Science Education Infrastructure

Well over a decade ago, St. John and Perry (1993) proposed that science educators rethink the entire learning enterprise, suggesting that school and free-choice learning sectors be considered components of a single, larger educational infrastructure which supports and facilitates science learning in a society. John Falk and I expanded upon this idea, positing that there were actually three educational sectors in society: schools and universities, the free-choice learning arena, and the workplace (Falk and Dierking 2002). We argued that in the twenty-first century, society needs a broad-based and richly integrated educational infrastructure capable of supporting millions of unique individuals attempting to meet widely varying learning needs at any point in their lives, any time of day. The educational entities that compose this basic infrastructure form the fundamental backbone of a learning society and provide citizens with current and accurate information about health, politics, economics, the arts, and sciences. As suggested, this infrastructure already exists in communities and ideally all the entities work together to support and sustain learning across the life span (Dierking and Falk 2009). From this perspective, the educational/learning infrastructure is vital to a nation's economic well-being—as well as its intellectual and spiritual well-being.

Each educational sector—schooling, workplace, and free-choice learning— contributes to the science learning of the public. However, of the three, the free-choice sector is far and away responsible for providing more people educational opportunities, more of the time, than either of the others. The free-choice learning sector also is the most diverse, fastest growing, and arguably the most innovative. The explosion of the Internet and World Wide Web provides significant evidence for the perceived value of having a readily accessible tool that can provide virtually anyone, anywhere, with any information, any time. The Web, though, is just one aspect of an ever expanding, and hopefully improving, network of learning resources available to the general public.

One consequence of taking a broad-based approach to science education is that one begins to notice science teaching and learning in novel places (like cafes and pubs!). For example, over the last 20 years, the Astronomical Society of the Pacific, based in San Francisco, California, has explored and experimented with ways to tap into the vast teaching potential of amateur astronomers. With initial funding from the informal science education program of the NSF, they have involved amateur astronomers in elementary and middle school teaching in classrooms through *Project ASTRO* (Dierking and Richter 1995) and through *Family ASTRO* have provided engaging astronomy experiences for families through a network of museums, science-technology organizations, and community-based organizations such as amateur astronomy clubs. They are now providing more focused astronomy training to free-choice learning educators in small science centers, museums, and planetariums. Although NSF funding ended several years ago, programs remain in communities around the country supported by existing networks of educational partners.

As the US population ages, there are also significant and increasing numbers of young elders. All of these adults have the potential to participate in science-related special interest groups and leisure pursuits, watch nature or science specials on television, and/or participate in noncredit university courses or Road Scholar programs; many focused on STEM (formerly called Elderhostel; "Road" connotes a journey and real-world experience, and "Scholar" reflects a deep appreciation for learning). Research shows that many adults also visit settings such as national parks, science centers, and botanical gardens to satisfy their intellectual curiosity and stimulation, as well as fulfill a need for relaxation, enjoyment, and even spiritual fulfillment (Ballantyne and Packer 2005; Brody et al. 2002; Falk 2006, 2007; Azevedo 2004). Some of these elders also have STEM expertise that they can share as free-choice educators and mentors.

School-age children also spend a significant amount of time outside of school (current estimates are 80–90 % of waking hours are *outside* of school), and some of this nonschool time is devoted to free-choice science learning, most often with family: they visit parks, zoos, and libraries and participate in various after-school and extracurricular experiences, including scouting and summer camps (e.g., Korpan et al. 1997; Dierking and Falk 2003; Dierking in press-b; Rounds 2004). A small but growing movement of home educators also values science and mathematics learning for their children and engages in it regularly (Bachman 2011). As noted earlier in the chapter though, free-choice learners are not always school-aged children and their families. They include post-high-school adults (some of whom did not further their schooling), as well as those who did not graduate from high school at all (Falk and Dierking 2010).

Free-choice learners sometimes "choose" to engage in science learning for very different and sometimes profound reasons that may not even occur to us. Recently I heard about Safecast, a small not-for-profit citizen science organization whose mission is to empower people with data, primarily by building a sensor network which empowers nonspecialist citizens to collect and interpret data and freely use it through an Internet portal (Penuel 2011; *Scientific American* 2011). The organization was featured in a recent issue of *Scientific American* as a citizen science project exemplar.

The impetus for Safecast's creation was a real-world catastrophe. After the earthquake and resulting radiation leak at Fukushima Diachi in March 2011, it became clear that people in the area wanted more data about the earthquake, resulting tsunami and damage to nuclear power facilities than was available from the Tokyo Electric Power Company (TEPCO). With the support of Safecast, citizens build Geiger counters, measure radiation levels, making existing data more robust since multiple sources are better and more accurate when aggregated, and make the data available to the public through maps, a website, and data feeds to citizens and scientists around the world. As of July over 300,000 data points had been collected; while Japan and radiation are the primary focus of Safecast at present, this work has made the organization aware of a need for more global environmental data, and their long-term goal is to engage in collaborative research with citizens in additional areas.

Most people find the radiation data collected difficult to read and interpret, but the disaster has created a need for ordinary citizens to engage in deep science learning about safe and unsafe levels of radiation in water and food, using complex scientific instrumentation. It is a visible and striking example of the point of this chapter, namely, that as a field we include organizations like Safecast and other citizen science movements that have been in operation for many years (Audubon bird counts, Cornell's Pigeon Watch), as critical entities supporting lifelong STEM learning even though they support the learning of nonspecialist citizens, 18 years or older. Some citizen science projects even involve participating citizens in professional science practices such as conceptualizing research studies and analyzing data, even collaborating on peer-reviewed publications.

In projects like Safecast, science and learning about science are central, but new cultural practices around science are being constituted. The goals of these practices are not focused on learning science in an abstract manner. They are about living and surviving in a particular place and learning that science is a tool which allows one to do so, a tool for empowerment and identity building. To my mind Safecast is an excellent example of authentic science practice. Although that term has become a popular buzzword in science education reform, often used as a synonym for science activities that resemble scientists' everyday practice (Martin et al. 1990; McGinn and Roth 1999; Roth 1995, 1997), I agree with Rahm et al. (2003) that this focus often neglects acknowledging what authenticity means, to whom, and according to whom. I prefer to define authentic science practice as what emerges as facilitators, citizens, and scientists interact, make meaning of, and come to own the activities they engage in collaboratively. Thus, authenticity is not viewed as only the scientist's science but instead as an emergent property of those engaged, the task, and the environment, as they interact in complex ways (Barton 1998a, b, 2001a, b; Fusco 2001; Hodson 1998; Wellington 1998). Not only does Safecast embody authentic science practice but it also is a rich example of Roth's notion of the *Fullness of Life* (or *Total Life*) unit of analysis, an approach that suggests one must understand STEM learning from the perspective of life as a whole, rather than focusing on STEM learning as abstract, decontextualized activity (Roth and Van Eijck 2010).

Such efforts test the very roots of authoritarian science. Safecast is releasing data openly and pushing the Japanese government, as well as universities and researchers, to share their data. They argue that open data is an important trend, which adds a new layer of robustness and democratic participation in scientific research that the Internet and data science affords. However, pushing scientists to release their data as well as their results and findings, particularly to the public, is likely to be controversial and contested even though it represents a willingness to share the authority and power of science with citizens of all ages, walks of life, and backgrounds.

Evidence of the Potential for Free-Choice Learning Opportunities

For more than a decade, the US National Science Foundation (NSF) has funded more than 300 free-choice/informal education projects focused on enhancing girls' interest, engagement, and understanding in STEM, a total investment of more than $100 million. Despite this significant investment, little is known about the strategic impact of these efforts, in part because their impacts have been conceptualized and measured differently, often in ways not generalizable across projects (Darke et al. 2002).

Modest existing research suggests that free-choice learning settings can be beneficial, providing influential experiences and building capacity and confidence in science among girls and women (Dierking and Martin 1997; Fehrer and Rennie 2003; Fadigan and Hammrich 2004). The influence that significant adults/facilitators bring to these experiences is also an important factor. Female scientists often cite family members, youth program leaders, or contexts outside of school as significant influences in their career choices (Baker 1992; Campbell 1991; Fort 1993) and reflect on the importance of informal networks and supporters. While every girl may not have support from her family or school, the presence of a significant adult or mentor in an out-of-school setting can make a vital difference. Studies have begun to investigate the impact and nature of experiences in which influential adults are involved (Crowley et al. 2001; McCreedy 2003; Jones 2006). For instance, McCreedy (2003) investigated the informal context of community-based organizations such as museums and scouting, identifying social/cultural factors that led to adult engagement in science learning and to their role in transforming young women's science-related identity (and their own). Findings suggest that informal contexts can offer unique opportunities for youth to engage in science practices, supported by caring and influential adults, and when designed and facilitated well, these experiences come close to the Safecast model of authentic science practice.

Studies within schools and family contexts and outside of schools have also begun to examine how personal frameworks and identities in science can be influenced by free-choice science experiences (Davis 1999, 2001; Ellenbogen 2002; Katz 1998; McCreedy 2003); however, there is much more to learn. In an effort to fill this research gap and better understand the processes and strategies that enhance opportunities for girls and women to meaningfully participate and identify with science, I am completing an NSF-funded retrospective research study with a colleague at Franklin Institute Science Museum, Dr. Dale McCreedy, to investigate the long-term impact of gender-focused free-choice STEM programs. Based on Clewell and Burger's (2002, p. 249) perception that "Quantitative data can only take us so far; it will be the words of the young women themselves that will inform our future programs and projects to make science and technology careers more welcoming for women," the study was designed to explore girls' long-term science involvement in rigorous but more qualitative ways.

The overarching research question for the study was: What role do free-choice STEM experiences play in girls' interest, engagement, and participation in science communities, hobbies, and careers? Sub-questions included: (1) How do girls describe their relationship to science and their sense of themselves (identities) as science-interested learners and advocates? (2) How does participation facilitate and lead to additional opportunities for engagement? (3) What role, if any, do significant adults play in facilitating these impacts? Research goals included:

- Document the long-term impacts of girls' participation in free-choice science programs and their perception of the ways if any of these experiences influenced their future choices.
- Determine the ways in which free-choice contexts contribute to girls' science learning and achievement broadly defined to include careers and education but also hobbies and habits of mind.
- Share the results of this research within and across the free-choice science learning community in order to influence program policy.

The study was framed within a sociocultural perspective that posits that human learning and development are best understood within their cultural, historical, and institutional contexts. Thinking and doing are intertwined within these contexts (Vygotsky 1981; Wertsch 1985, 1998; Rogoff 2003). The specific sociocultural framework of the study was Community of Practice (CoP). CoP identifies three key elements of participation in a learning community (Wenger et al. 2002). In the case of these programs, they included: (1) a *domain of knowledge,* the content or focus of the free-choice STEM activity; (2) *shared practices,* the science practices and processes in which girls and women in these programs engaged; and (3) *community,* who was involved and how was individual and community learning supported. This study explored whether and, if so, how participation within a free-choice science Community of Practice (CoP) led to learning, broadly defined to include interest, engagement, and participation in science communities, hobbies, and careers. The study also probed how this learning related to an individual's perspective about herself, her relationship to science, and issues related to gender and culture. Since CoP theory also posits that identity and community are interconnected with the individual evolving as a result of her participation in the community and the community evolving through her participation and influence, we were also interested in observing these impacts as well.

Methodology

Sample Research participants were recruited from five successful initiatives whose focus was to engage girls in informal science education practices. All projects from which we recruited women met the following five criteria:

1. Were informal programs, targeting girls, particularly from communities under-represented in STEM
2. Represented long-standing efforts, initiated more than 10 years ago
3. Had access to participants who are now 18 years or older and had participated at least 5 years ago
4. Had staff and/or evaluators who were willing to facilitate contact with these girls and to share existing evaluation and research efforts
5. Were diverse programs with regard to the three elements of engagement described by the CoP literature

The projects were: (1) *Women in Natural Science* (WINS), developed and implemented at the Academy of Natural Sciences in Philadelphia, is a year-long natural science enrichment program with opportunities to work in science labs and conduct research, offered to academically talented females who are entering grade 9 or 10, enrolled in public school, maintain a C or better average in all major subjects, live in households where one or both parents are absent, and demonstrate financial need (free or reduced school lunch); (2) *National Science Partnership* (NSP), initiated and piloted at Franklin Institute and then disseminated nationwide through Girl Scouts of the USA and informal networks, prepares and supports Girl Scout leaders to do science badge work with girls ages 6–11, as well as involves older girls as mentors in these efforts; (3) Girls, Inc.'s *Eureka* and *Project SMART* engage women in under-resourced communities nationally in STEM experiences and mentoring projects; (4) *Techbridge*, an after-school and summer program in the San Francisco-Oakland area of CA, encourages girls underrepresented in STEM in technology, science, and engineering activity, as well as field trips to STEM-related businesses; and (5) the *Rural Girls in Science* project developed and implemented at University of Washington, Seattle, engaged rural young women, primarily Latina, in community-based STEM projects to improve their communities. Based on previous evaluation studies, each of the programs from which we recruited research participants was a highly successful free-choice science learning program, characterized by social, open-ended, voluntary, and noncompetitive structures.

Design The study had two individual investigations:

Investigation 1 (I#1)—Personal Meaning Mapping/in-depth interviews with active/core girls
Investigation 2 (I#2)—Web-based survey

Data Collection: Investigation #1

In keeping with the sociocultural perspective of the study and to ensure that young women's own ideas and terminology were centermost, Investigation #1 explored the ways in which young women discussed their early free-choice science learning experiences and identity. Two data collection approaches were utilized, both face to face: Personal Meaning Mapping (PMM) and an in-depth qualitative interview with each young woman to discuss the maps she had created. Each program identified

2–3 young women who were past active/core participants and had kept in touch with the program. They may, or may not, have continued to engage in science-related activities in their lives (college science classes, science-related hobbies, clubs, careers, etc.). The goal was to interview a range of women, although in this phase of the project, McCreedy and I were not concerned about whether the young women interviewed constituted a representative sample. All participants (or their parents if they were minors) received information about the study and were asked to complete a consent/assent form before participating in the study.

PMM is an approach designed specifically for use in free-choice learning settings (Falk et al. 1998) and is grounded in a relativist-constructivist paradigm, which recognizes that individuals participating in programs in informal, free-choice settings bring varied backgrounds and knowledge to the experience. This varied background and knowledge, as well as the social/cultural and physical context of the experience itself, shapes how a person perceives and processes the experience. The approach measures the unique conceptual, attitudinal, and emotional impacts of a specified learning experience on an individual and the community in which she participated, focusing both on *the degree of the change* but equally on *the nature of the change*. Importantly, PMM provides a valid way to understand the personal meaning people construct from learning experiences.

The protocol for PMM involves asking individuals to write down on a piece of paper as many words, ideas, images, phrases, or thoughts that come to mind related to a specific concept, picture, or word. Similar to the concept mapping approach from which PMM was adapted, the word, picture, or phrase prompt is placed in a circle at the center of the page. The words, ideas, images, phrases, or thoughts written down by the individual in response to the initial cue then form the basis for an open-ended interview in which research participants are encouraged to explain why they wrote down what they did and to expand on their thoughts or ideas relative to the circled concept. The discussion allows participants to elaborate in their own words and from their own frame of reference on their perceptions and understandings of the prompt. The researcher records participants' responses on the same piece of paper using the participant's own words and conceptualizations. To permit discrimination between unprompted and prompted responses, the follow-up interview data is recorded in a different color ink.

In this study we asked young women to complete two personal meaning maps, the first with the prompt "me" and the second initially with the prompt "science," later with a prompt of the program in which they participated, for example, *Women in Natural Science*. All sessions were face to face and tape recorded. After the first "me" map was completed, the young woman was interviewed about that map. Then they made their second map and were interviewed. Finally with the two PMM's side by side, each young woman was interviewed about how the ideas expressed on the two maps overlapped in their lives—if at all. The purpose was to get participants to articulate what makes them "me" in their own words and whether science (or ultimately the program) had played any role in their life decisions and personal identity.

Data Collection: Investigation #2

Findings from phase 1 were the foundation for creating a valid and reliable questionnaire, as well as helping us develop a baseline sense of the range of possible outcomes that might result from these experiences. In keeping with our theoretical framework, young women's own ideas were used to help focus item development within the three dimensions of CoP. Given that young women in our study were scattered around the country (and in a few cases around the world), the instrument was web-based.

Item Development Since the questionnaire was the primary tool for understanding the long-term impacts of gender-based programs, the questionnaire's organization needed to reflect a general understanding of the programs and the CoP framework as well as utilize the language and concepts learned in phase #1. In close collaboration with a national (US) Research Advisory Committee, McCreedy and I created a matrix of data categories and subcategories which should be included in the questionnaire. Questions were then developed to explore each of these concepts at least once. In the interest of reducing response time, instrument items related to descriptions of the program or organization were asked only once. Questions related to performance, planning, and outcomes were asked in a minimum of two different locations with at least two question types (i.e., open ended vs. forced choice). Where possible, drop-down menus were created to reduce the time required for response. WebSurveyor was used to develop a logic-based questionnaire constructed around a core framework which allowed young women to answer items tailor-made for the program in which they participated.

Usability and Reliability Usability, reliability, and validity of the instrument were thoroughly tested. First drafts were circulated to project team members for comments and suggestions. A close-to-final draft of the web-based questionnaire was completed by our advisors to identify questions that were unclear, missing choices from drop-down menus, and other problems.

Survey Implementation Administering the questionnaire involved five major tasks: (1) receiving Institutional Review Board (IRB) approval of the final version of the questionnaire; (2) development of an initial database of young women (18 years or older) with which to recruit based on information collected from program leaders; (3) making initial contact with young women and program leaders announcing the study via e-mail and providing IRB information; (4) creation of a survey invitation and a link, with necessary follow-up by e-mail and phone; and (5) additional follow-up phone calls and use of Facebook to increase the sample size. The web-based questionnaire was launched in January 2009 and closed in March 2011.

Data Analysis: Investigation #1

Personal Meaning Mapping, along with the accompanying in-depth interview, is generally used to measure four dimensions of impact—extent, breadth, depth, and mastery. For the purposes of this study, the first three parameters were used, and responses were analyzed through content and cross-case analysis. *Extent* in this study was defined as whether a young woman developed a STEM-rich life and identified strongly with STEM, be that through education, career, or hobby choices. By coding the vocabulary an individual used to discuss a concept or experience, the extent of a person's awareness and understanding of STEM was documented, and because young women made two PMMs, the overlap between the vocabulary/ideas presented in the two maps was also assessed.

The second parameter assesses the *breadth* of a person's sense of impact, for instance, how many different ways did participants discuss or explore relationships between their lives and the informal science program in which they participated. The interviews of the young women were carefully reviewed, to quantify and qualify the kinds of connections that were made in order to identify the full range of relevant ideas/connections possible.

The third parameter investigates the *depth* of a person's knowledge to document how deeply and richly someone understands a particular concept. The connections that young women saw between their lives and the program were more intensively probed, as well as a specific either negative or positive experience in STEM that they recalled to determine whether they felt the experience shaped their understanding/perception of STEM.

Data Analysis: Investigation #2

Data Compilation, Coding, and Analysis To ensure the safety and integrity of the data during the period that the survey was being implemented, data from completed questionnaires were downloaded from WebSurveyor nightly to an Excel file stored on a hard drive and to a removable memory stick. Upon the closing of the survey, data were backed up in a similar manner. Questionnaire data included demographic (age, current educational status, career) and psychographic data (interest in science, science-related hobbies, etc.), as well as open-ended questions. It also included young women's course selection and other evidence of academic achievement and leadership in science.

Data were analyzed with SPSS; quantitative, non-numeric responses were assigned numeric codes and responses converted numerically. Coding rubrics were also created for qualitative items and responses coded numerically, utilizing a CoP lens (participants' level of perceived engagement in the program, and if relevant, other informal science learning experiences, as well as their current participation in science communities, be that education, career, or avocation). The numeric coding system distinguished between questions that were skipped due to programmed logic in the questionnaire and those that were asked but not answered. An extensive code sheet with codes, field names, and question text for each survey question was developed. Findings were reviewed and validated by core/active youth who had not participated in these programs and by science-engaged adults (either in education, careers, or avocations).

Findings: Investigation #1

Investigation #1was completed in 2008. Twelve young women were interviewed: two who had participated in the *Eureka* project in the San Francisco Bay area; three in the *NSP* project, primarily in Philadelphia and New Jersey; three in *WINS*, primarily in Philadelphia; two in *GAC*; and two in *Rural Girls in Science*. Young women ranged in age from 20 to 29 and most had participated in the program 10–15 years before.

Extent Findings revealed the range and power of these programs as memorable and lasting, even for young women who had not pursued science careers. Most of the 12 young women (all active participants) were able to describe specific activities they had engaged in with great detail. Most impacts were positive though not all. Young women did not discuss much traditional "science content" per se but were able to talk about doing science and engaging in science processes. Science-related hobbies suggested a broad perception about "what science is," and impacts were not only related to science but included leadership skills and positive changes in self-esteem.

Breadth demonstrated how many different ways participants discussed or explored relationships between their lives and the science within the free-choice science program in which they participated. For example, girls/women were asked about saved items from the programs in which they participated and were invited to share ways in which the program made a difference in their lives that included a wide range of options—confidence, relationships, community, a safe haven, etc. Although the sample size was small, there were some differences in impact observed between women who had pursued science and those who had not. For example, there was evidence for the role of community, access to STEM networks, and the building of social capital, particularly for women who had pursued further science education and careers. These women indicated that the network established by the program had been beneficial in their later pursuit of STEM, informing/reinforcing their choices and ensuring that they had "no stereotypes about a future in science." For young women who were not pursuing STEM careers or education, they indicated that the program had expanded their world, sense of self, and their awareness of science, helped them be successful in future nonmajor science courses and influenced their interest in STEM generally and in STEM-related hobbies specifically.

Depth Findings again demonstrated that these programs were memorable. Years after, women were able to describe experiences they had in great detail, as well as their connection to other STEM experiences in which they had engaged, both in school and out of school. Women engaged in STEM felt the program added to their science portfolio. Women now engaged in STEM, as well as those not engaged, also indicated that the program built self-esteem/self-efficacy, developed their leadership skills, and helped them be empowered and proud to be smart. One young woman who had participated in *Eureka* both as a participant and later as a mentor commented that "[The program] made me a better person. I was a very angry young woman and the program helped me channel those feelings in positive ways."

It was also possible to use data to develop trajectories of impact for women engaged in STEM now (either through a vocation, education, or avocation) and those who were not engaged, which reinforced that the approach being taken was a valid one which could effectively frame Investigation #2.

Findings: Investigation #2

Research participants in the web-based survey ($n = 213$) were from multiple, diverse sites, including urban, suburban, and rural communities, representing different cultures, ethnicities, socioeconomic status, and participation levels (Girls, Inc.— *Eureka* and *Operation SMART*; $n = 102$; *WINS*; $n = 44$; *NSP*; $n = 34$; *Techbridge*; $n = 30$, and *Rural Girls in Science*; $n = 3$). All were 18 years of age or older and had participated in one of the programs at least 5 years earlier. Not surprisingly, girls who participated more recently were easier to locate, possibly because they were more accessible via the web-based avenues we used to recruit: 107 were in the 18–23 age range, 79 were between 24 and 30 years old, 22 were between 31 and 35, 3 were between 36 and 40, and 2 were over 40. One hundred forty (140) were from urban settings, 61 suburban, and 12 rural. Eighty women identified themselves as Black/African American; 74 were White/Caucasian, 29 Asian/Asian American, 3 American Indian/Alaskan Native, 1 Hawaiian/Pacific Islands, and 9 identified themselves as other (some chose more than one category with which they identified).

Findings from the web-based survey reinforced the findings from Personal Meaning Mapping and in-depth interviews. Programs were memorable and lasting, even for women not pursuing STEM careers or education currently. Most impacts were positive though not all. While the majority were early in their careers, some not only reflected on how program participation influenced their career and education choices but also specifically how it had affected hobbies/interests and even parenting. Long-term impacts of participation with a representative quote demonstrating how women expressed these outcomes included:

1. Increased understanding, appreciation, and enjoyment of science
   "*Eureka* inspired me to actively participate in science and math because I found it could be fun when it pertained to me." (Existing program that engages women in under-resourced communities nationally in STEM experiences and mentoring projects)

2. Increased interest/choices around STEM
   "It gave me a chance that no other program or my school did. I was a poor white girl in a good school who no one paid attention to and was dying for a different type of science than what school offered (only lab sciences). I craved environmental and animal science programs. *WINS* opened that door for me and allowed me to take part in free programs." (Existing year-long natural science enrichment program with opportunities to work in science labs and conduct research)

3. Enhanced skills/performance more broadly (organizational/leadership skills; opportunities to be a mentor)

"It was through this program that I was able to get my first exposure to the work field. Through these experiences, I was able to shape my leadership and interpersonal skills for future jobs and interviews. It was because of the staff members' support and help that I made it to college today."

(*National Science Partnership*, existing program that prepares and supports Girl Scout leaders to do science badge work with girls ages 6–11, as well as involves older girls as mentors in these efforts)

4. Enhanced social networks—long-term friends, mentors, and program leaders, who offer support, advice, access to communities of interest (STEM and other), etc.

"It gave me mentors, especially female mentors. It also gave me a network of professionals that helped me grasp how to be professional and the opportunities that science has for women. No one in my family or immediate circle had gone to college or worked in science so these introductions were invaluable."

(*Rural Girls in Science* engaged rural young women, primarily Latina and low-income White girls, in community-based STEM projects)

5. Increased sense of agency—increased confidence, self-esteem and aspirations, and self-initiating new behaviors or considerations

"It influenced me to have the confidence to be smart, and to own my intelligence. It also allowed me to find out that I deserve to be smart." (*Techbridge*, existing San Francisco-Oakland, CA, after-school/summer program that encourages girls underrepresented in STEM in technology, science, and engineering activity)

6. Evidence of changes in identity (changes in trajectory, interests, and sense of self—both in STEM and more general)

"I can't express enough how much the program helped me. I wouldn't be who I am today. I'm more aware and involved with my kids in every way, both nurturing their education and their physical activities because I know how important that is. Now that I'm a mother of three, looking back at my years in the program, I wish my parents were more involved in my education and in my growing up as a teenager because it is so important."

(*Eureka*, existing program that engages women in under-resourced communities nationally in STEM experiences and mentoring projects)

7. Increased awareness, recognition, and pride around gender and race-ethnicity-specific issues

"I received support and motivation, which I did not receive from others. The program gives young girls an opportunity to participate in activities schools do not offer. It helps girls set aside any stereotypes set for women in the field of science and engineering."

(*Operation SMART*, existing program that engages women in under-resourced communities nationally in free-choice STEM activity)

A series of individual quantitative items that captured a variety of possible program impacts were also collapsed into eight outcome scales, and Cronbach's alpha was used to calculate their reliability. Scales were (1) academic/career interest in science (.95), (2) social skills development (.94), (3) self-Awareness/self-confidence (.92), (4) awareness/understanding of science careers (.91), (5) leisure interest in science (.91), (6) science identity (.85), (7) critical thinking (.84), and leadership (.73).

Comparative analyses were conducted for these scaled outcomes and significant differences found for 7 of the 8 scales as a function of the program young women participated in and whether they lived in an urban setting, for 5 scales if young women had a current science hobby (including significance for the academic/career interest in science outcome scale), 4 scales if they currently use science websites, 3 scales if they currently watch science-related television, and 2 scales based on their current job status (those currently working outside their field rated the science identity scale higher than those working in the field and those working in a science-related job rated the academic/career interest in science outcome higher than those who were not so employed).

Findings strongly support the notion that participation in gender-focused free-choice STEM programs contributes to lasting effects on young women's interest, engagement, and participation in science communities, hobbies, and careers, influencing their identities and relationship to and with science. Program participation also supported participants' interest in STEM and their appreciation for the diversity of disciplines and practices embodied within it and built social capital, such as long-term mentors and friends who could further interest and persistence in STEM, both while participating in the program, but also long after, and increased agency, influencing future careers, education, and hobbies/pursuits. Noteworthy, these program effects were particularly significant and impactful for girls living in urban areas when compared to those in suburban areas (unfortunately the sample size for rural girls was too small for statistical comparisons).

The focus of this study was to determine how participation in these programs *contributed* to women's long-term understanding of science and most importantly to their relationship to and with science, so it is important to reinforce that these programs alone were not the reason for these impacts; participation in them *contributed* to these impacts. There was evidence throughout the data that young women's experiences in these programs were not isolated but connected to their activities at school, home, and in other free-choice learning settings and programs. There was also evidence that these science experiences contributed, along with schools and the workplace, to building science identities that met the needs of these young women, encouraging them to become science-informed citizens, perhaps even *engaged science participants*. As a result of participating in these programs, many of the women have an idea for and appreciation of what science is, not an abstract, decontextualized activity, but as a useful tool for life, reinforcing Roth's notion of the *Fullness of Life* (or *Total Life*) unit of analysis (Roth and Van Eijck 2010).

Results reinforce the thesis of this chapter that free-choice science learning should be a critical player in a comprehensive, whole life approach to science education reform for diversity, one component of a lifelong science learning system in which learning of a variety of kinds are respected and supported. By participating in these diverse communities of free-choice STEM learning, these young women were able to *contextualize* and *personalize* their science knowledge, interest, and understanding over the long term. Although they learned about science, it was not merely as a body of knowledge but as processes by and through which many of these young women were able to construct their identity and, in so doing, develop their *own* lifelong relationship to science.

In a later analysis, my colleague and I hope to be able to show that when meaningful connections were made across settings, even greater impact resulted. To that end, further analyses will determine at a more fine-grained level if there are patterns in outcomes as a function of a host of variables that emerged as important or that the literature suggests are important:

1. Girls' motivation for participating
2. Their childhood interests
3. The age at which they participated
4. The length of time they were in the program
5. The intensity of their participation
6. The length of time since they participated
7. The type of program and "community" afforded by the experience
8. Whether they had opportunities to mentor other girls
9. Whether there were opportunities for outdoor activities, including camping, hiking, and/or physical leadership experiences
10. Whether the program included academic and college preparation activities such as monitoring grades, helping with schoolwork, facilitating trips to visit colleges, assistance in preparing for college testing and the completion of applications
11. The influence of significant adults in science thinking

Although this study focused on young women, I am certain that similar outcomes would result for young men and hope to have the opportunity to extend this research and justify that claim. It is hoped that these findings inform science education practice and research and provide useful information to educators designing and implementing free-choice science experiences and teachers in schools striving to achieve more diverse, in-depth outcomes. I also hope they provide evidence for how more purposeful articulation and collaborations between and among educators in schools, universities, free-choice learning settings, and the workplace could create strategic impact for learners.

An Infrastructure to Support Free-Choice Science Education and Research

It is not only the learners who are different in this new system. The traditional boundaries and roles that have distinguished various groups of science educators are changing also. In the twenty-first century, free-choice learning institutions such as museums, the Internet, and broadcast media are assuming ever more prominent roles in the science education of the public—but the facilitators of free-choice science learning are not classroom teachers. They include nontraditional teachers and mentors, such as after-school youth leaders, professional and amateur scientists, museum educators, educational web developers, and even parents. This point is not trivial. To make a comprehensive lifelong science education system work, science educators who have traditionally only considered schooling must embrace free-choice science learning institutions and organizations, as well as the educators who work within them, as equal partners.

Unfortunately, the value of these educators (or in some cases even their presence) is not recognized nor is there a broad-based realization that they require expertise in teaching science in different ways and configurations than classroom teachers, with learners of all ages (Tran 2006, 2008). Typical teacher education programs only are effective for these educators if they plan to work within schools or at free-choice learning institutions that primarily serve schools. The vast majority of such educators work with other learners or children and youth in out-of-school time. Value also should extend to compensation. Although the public discusses how underpaid public school teachers are, a little-known fact is that most free-choice science educators work year-round, yet earn less annually, receive more modest benefit packages if at all, and have less job security than their counterparts in classrooms (Biggs and Richwine 2011a, b; Bureau of Labor Statistics 2011a, b; Grayson 2011).

There have been some efforts to recognize this sector. For instance, two leading science education organizations, one focused on research and another on practice, have provided some leadership. A free-choice/informal science learning strand was formed in 1995 by the National Association for Research in Science Teaching (NARST), after years in which research in this area was in an "other" strand. In 1999, the NARST board established an ad hoc committee focused on informal science education with the goal of exploring interest among NARST members for additional leadership in this arena. A major product was a policy statement in the area of out-of-school (free-choice) science education research published in the *Journal of Research in Science Teaching* (Dierking et al. 2003).

In 1998, the National Science Teachers Association (NSTA) published a policy statement on informal (free-choice) learning, and in 2000 NSTA leadership established a board seat representing this community of science educators, allowing them to play a larger role in developing policy. And in 1998, the American Educational Research Association also created a Special Interest Group focused on Learning in Informal Environments, and though multidisciplinary in focus, this group has provided an outlet for scholarship in free-choice science education also.

Significant funding also was given to a few consortia in the early 2000s through the US NSF's Centers for Teaching and Learning effort enabling two small research communities to be fostered: the Center for Informal Learning and Schools (CILS), a collaboration among The Exploratorium, San Francisco; King's College, London; and the University of California, Santa Cruz, focused on the intersection of informal and formal education institutions and provided graduate education for a handful of free-choice learning researchers, and the Center for Inquiry in Science Teaching and Learning (CISTL), at Washington University in St. Louis, devoted some of its resources to studying inquiry in informal learning environments. Both of these centers were part of teacher education programs and thus focused primarily on free-choice learning research designed to improve schooling. Neither center was refunded by the NSF, although CILS, through The Exploratorium, has successfully procured funding focused on after-school science. Unfortunately, neither of the academic programs continued; thus, no full-time faculty members solely committed to free-choice learning remain at the three universities.

A handful of graduate programs support free-choice learning educators and researchers. In particular, programs at the University of Pittsburgh, the University of Washington, and Oregon State University (OSU) now exist. The University of Pittsburgh's Center for Learning in Out of School Environments (UPCLOSE) supports students through doctoral programs in the Learning Sciences and Policy. One of the PIs at the University of Washington's Learning in Formal and Informal Environments (LIFE) Center, one of the first of four Science of Learning Centers funded by the US NSF, is focused on free-choice learning, and although he primarily engages in teacher education, the program has been able to support a few graduate students solely interested in free-choice science learning.

Probably the most extensive effort is being undertaken at my own university, OSU, in Corvallis, Oregon (Dierking 2010). With leadership from Oregon Sea Grant and initial funding from the National Oceanic and Atmospheric Administration (NOAA), a graduate program in lifelong STEM learning has been established by the College of Science, in partnership with the College of Education. The program is the first comprehensive, lifelong learning research program in the USA. The free-choice learning area of concentration offers an online master's program which is supporting the education of practitioners working in this arena. Students have included national park interpreters, museum and science center educators, and public health program managers among others. Even a veterinarian technician, who appreciated that much of her job was about communicating science and health information to the owners of her animal patients, was a student. She did not need to know how to develop a lesson plan or manage a class; she needed to understand how to motivate and educate adult learners to take better care of their pets. In the doctoral program, core courses are taken by all students together (there are K-12, college teaching, and free-choice learning options), building a community of researchers that crosses settings, ages, and backgrounds, fostering cross-disciplinary and cross-institutional learning. Each area of concentration also builds specific knowledge base and expertise.

The OSU group is trying to make a difference in terms of the type of research conducted and the practices scaffolded and supported. Current or recently completed research by free-choice learning doctoral candidates includes studies of (1) handheld use in a marine science center, (2) STEM learning activity across a range of home-educating families, (3) STEM learning among Koi fish hobbyists, (4) whale-watching tours in the context of ecotourism, (5) a citizen science project, and (6) informal staff-family interactions in a science center.

Our group is also trying to understand STEM interest development and learning ecologically, across multiple settings and time. For example, with funding from the Noyce Foundation, I am collaborating with OSU colleague John H. Falk and University of Colorado, Boulder colleague, William R. Penuel, on a four-year longitudinal study of the STEM learning of 10-year-olds, in school and out of school, in a diverse Portland community. The premise of the project is that if one more fully understood how and why people, in particular early adolescent children within poor, under-resourced communities, develop STEM-related interests (or not) in their everyday lives, it would be possible to create a more synergistic and effective STEM education system, a system that more successfully supported STEM learning for all. A key feature of this effort, called *Synergies*, is defining the "STEM education system" as the STEM learning resources/assets of the entire community, including schools, but not exclusively. Researchers are also actively involving members of the community in a collective effort to first understand and then try to enhance children's STEM interest and engagement. Currently we are preparing 12 high-school-age youth to be community ethnographers and ambassadors.

Also underlying this effort is the development of a comprehensive agent-based model (ABM) of STEM interest and engagement that will allow key STEM educators in Portland (in school and out), as well as community members themselves, to better visualize and understand the STEM resources/assets available and the complex, multidimensional dynamics of a child/youth's lifelong and life-wide STEM learning. In years 3 and 4 of the project, this model will serve as a tool to formulate alternative strategies for enhancing the effectiveness of educational interventions to improve this community's STEM learning system.

These efforts though ambitious are still small and nascent. Currently lacking is a critical mass of established programs, each with sufficient resources to attract clusters of faculty and graduate students, each cluster pursuing long-term and sustained research aimed at answering basic and applied questions fundamental to the field. This landscape is changing as evidenced by this growing research community, but there is still much that needs to happen.

Ultimately also, taking such a comprehensive approach to whole life science teaching and learning has implications for funding. Currently at the national, state, and local levels, more than 95 % of all public resources for education are spent on schooling, and research monies studying whole life learning are equally scanty. Building a more comprehensive educational system suggests rethinking what constitutes public education. If a comprehensive educational system encompasses all community resources that citizens access for learning across their life span, including those in the workplace and free-choice learning sectors, we should also consider how federal funding for education (and educational research) is allocated. Data certainly supports the claim that free-choice learning is vitally important, in particular for youth and families living in poverty (Bouffard et al. 2006). Yet most of the institutions/organizations supporting such learning are either small underfunded not for profits or institutions that have to charge fees for their use. Equal access to free-choice learning resources, particularly for communities that could benefit most from them, is a tremendous issue that societies worldwide face.

In summary, in order to actualize a comprehensive whole life science education system, I argue for increased efforts that document (and fund) the cumulative and complementary influences of both in- and out-of-school science learning. Given that school-based science education efforts and research currently receive an order of magnitude more resources than free-choice or workplace learning, even a modest change in this ratio could make a huge difference for practice and research. The data suggests it would be a wise investment.

Author Bio

Lynn D. Dierking is Sea Grant Professor in Free-Choice Science, Technology, Engineering and Mathematics (STEM) Learning, College of Science, and Interim Associate Dean for Research, College of Education, Oregon State University. Her research expertise involves lifelong STEM learning, particularly free-choice, out-of-school time learning (in after-school, home- and community-based organizations), with a focus on youth, families, and communities, particularly those underrepresented in STEM. She is currently working on two research projects: an NSF-funded investigation of the long-term impact of gender-focused free-choice learning experiences on girls' interest, engagement, and involvement in science and a Noyce Foundation-funded four-year longitudinal study of the STEM learning ecologies of 10-year-olds, in school and out of school. Lynn has published extensively and serves on the Editorial Boards of *Journal of Research in Science Teaching* and the *Journal of Museum Management and Curatorship*. She received a 2010 John Cotton Dana Award for Leadership from the American Association of Museums.

References

Association of Science-Technology Centers. (2010). *Science center and museum statistics.* Washington, DC: Association of Science-Technology Centers.

Azevedo, F. S. (2004). *Serious play: A comparative study of learning and engagement in hobby practices.* Unpublished doctoral dissertation, University of California, Berkeley.

Bachman, J. (2011). *STEM learning activity among home-educating families.* Unpublished doctoral dissertation, Oregon State University, Corvallis, OR.

Baker, D. (1992). I am what you tell me to be: Girls in science and mathematics. *ASTC Newsletter, 20*(4), 5, 6 & 9.

Ballantyne, R., & Packer, J. (2005). Promoting environmentally sustainable attitudes and behavior through free-choice learning experiences: What is the state of the game? *Environmental Education Research, 11*(3), 281–296.

Barton, A. C. (1998a). Teaching science with homeless children: Pedagogy, representation, and identity. *Journal of Research in Science Teaching, 35*, 379–394.

Barton, A. C. (1998b). Reframing "science for all" through the politics of poverty. *Educational Policy, 12,* 525–541.

Barton, A. (2001a). Capitalism, critical pedagogy and urban science education: An interview with Peter McLaren. *Journal of Research in Science Teaching, 38,* 847–859.

Barton, A. (2001b). Science education in urban settings: Seeking new ways of praxis through critical ethnography. *Journal of Research in Science Teaching, 38,* 899–917.

Biggs, A. G., & Richwine, J. (2011a). Public school teachers aren't underpaid (Column, 11/16). *USA Today,* McLean, VA.

Biggs, A. G., & Richwine, J. (2011b). *Assessing the compensation of public-school teachers: A report of the Heritage Center for Data Analysis.* Washington, DC: The Heritage Foundation

Bouffard, S., Little, P., & Weiss, H. (2006). Demographic differences in patterns of youth out-of-school time activity participation. *Journal of Youth Development, 12*(1/2), 2–6

Brody, M., Tomkiewicz, W., & Graves, C. (2002). Park visitors' understanding, values and beliefs related to their experience at Midway Geyser Basin, Yellowstone National Park, USA. *International Journal of Science Education, 24*(11), 1119–1141.

Bureau of Labor Statistics, U.S. Department of Labor. (2011a). *Occupational outlook handbook, 2010–11 edition: Archivists, curators, and museum technicians.*

Bureau of Labor Statistics, U.S. Department of Labor. (2011b). *Occupational outlook handbook, 2010–11 edition: Teachers—kindergarten, elementary, middle, and secondary.*

Campbell, P. (1991). *Math, science, and your daughter: What can parents do? Encouraging girls in math and science series* (ERIC Document Reproduction Service No. ED 350 172). Newton: Women's Educational Equity Act Publishing Center.

Carnegie Corporation of New York. (2009). *The opportunity equation: Transforming mathematics and science education for citizenship and the global economy.* www.Opportunityequation.org. Accessed on 14 May 2013.

Clewell, B. C., & Burger, C. J. (2002). At the crossroads: Women, science, and engineering. *Journal of Women and Minorities in Science and Engineering, 8*(3&4), 249–254.

Crowley, K., Callanan, M. A., Jipson, J. L., Galco, J., Topping, K., & Shrager, J. (2001). Shared scientific thinking in everyday parent-child activity. *Science Education, 85,* 712–732.

Darke, K., Clewell, B. C., & Sevo, R. (2002). A study of the National Science Foundation's program for women and girls. *Journal of Women and Minorities in Science and Engineering, 8*(3&4), 285–303.

Davis, K. S. (1999). Why science? Women scientists and their pathways along the road less traveled. *Journal of Women and Minorities in Science and Engineering, 5*(2), 129–153.

Davis, K. S. (2001). "Peripheral and subversive": Women making connections and challenging the boundaries of the science community. *Science Education, 85,* 368–409.

Dierking, L. D. (2010). A comprehensive approach to fostering the next generation of science, technology, engineering & mathematics (STEM) education leaders. *The New Educator, 6,* 297–309.

Dierking, L. D. (in press-a). A view through another window: Free-choice science learning and generation R. In D. Zeidler (Series Ed.) & M. P. Muelle, D. J. Tippins, & A. J. Stewart (Eds.), *Assessing schools for generation R (responsibility): A guide to legislation and school policy in science education* (Contemporary trends and issues in science education).

Dierking, L. D. (in press-b). *Museums, families and communities: Being of value.* Walnut Creek: Left Coast Press.

Dierking, L. D., & Falk, J. H. (1998). Free-choice learning: An alternative term to informal learning? *Informal Learning Environments Research Newsletter.* Washington, DC: American Educational Research Association.

Dierking, L. D., & Falk, J. H. (2003). Optimizing out-of-school time: The role of free-choice learning. *New Directions for Youth Development, 97,* 75–89.

Dierking, L. D., & Falk, J. H. (2009). Learning for life: The role of free-choice learning in science education. In K. Tobin, & W.-M. Roth (Series Ed.) & W. M. Roth, & K. Tobin (Eds.), *World of science education: Handbook of research in North America* (pp. 179–205). Rotterdam: Sense Publishers.

Dierking, L. D., & Martin, L. M. W. (Eds.). (1997). Special issue of *Science Education, 81*(6).

Dierking, L. D., & Richter, J. (1995). Project ASTRO: Astronomers and teachers as partners. *Science Scope, 18*(6), 5–9.

Dierking, L. D., Falk, J. H., Rennie, L., Anderson, D., & Ellenbogen, K. (2003). Policy statement of the "Informal Science Education" Ad Hoc Committee. *Journal of Research in Science Teaching, 40*(2), 108–111. Netherlands: Sense Publishers.

Ellenbogen, K. M. (2002). Museums in family life: An ethnographic case study. In G. Leinhardt, K. Crowley, & K. Knutson (Eds.), *Learning conversations in museums* (pp. 81–102). Mahwah: Erlbaum.

Fadigan, K., & Hammrich, P. L. (2004). A longitudinal study of the educational and career trajectories of female participants of an urban informal science education program. *Journal of Research in Science Teaching, 41*(8), 835–860.

Falk, J. H. (2006). The impact of visit motivation on learning: Using identity as a construct to understand the visitor experience. *Curator, 49*(2), 151–166.

Falk, J. H., & Dierking, L. D. (2002). *Lessons without limit: How free-choice learning is transforming education.* Walnut Creek: AltaMira Press.

Falk, J. H., & Dierking, L. D. (2010). The 95% solution: School is not where most Americans learn most of their science. *American Scientist, 98*, 486–493.

Falk, J. H., Moussouri, T., & Coulson, D. (1998). The effect of visitors' agendas on museum learning. *Curattor, 41*, 107–120.

Falk, J. H., Dierking, L. D., & Storksdieck, M. (2007). Investigating public science interest and understanding: Evidence for the importance of free-choice learning. *Public Understanding of Science, 16*(4), 455–469.

Fehrer, E., & Rennie, L. (2003). Special Informal Learning issue of *Journal of Research in Science Teaching*, Guest Editorial, *40*(2), 105–107.

Fort, D. C. (1993). The consensus. In D. C. Fort (Ed.), *A hand up: Women mentoring women in science* (pp. 121–144). Washington, DC: Association for Women in Science.

Fusco, D. (2001). Creating relevant science through urban planning and gardening. *Journal of Research in Science Teaching, 38*, 860–877.

Grayson, L. (2011). How much does a historian get paid? *eHow.* Accessed on 21 May 2013.

Hodson, D. (1998). Is this really what scientists do? In J. Wellington (Ed.), *Practical work in school science: Which way now?* (pp. 93–108). New York: Routledge.

Horrigan, J. (2006). *The Internet as a resource for news and information about science.* Washington, DC: Pew Internet & American Life Project.

Jones, K. R. (2006). Relationships matter: A mixed methods evaluation of youth and adults working together as partners. *Journal of Youth Development, 1*(2), 31–47.

Katz, P. (1998, April). *Mothers as informal science class teachers: Voluntary participation, motivation, and outcomes.* Paper presented at the meeting of the National Association for Research in Science Teaching, San Diego, CA.

Korpan, C. A., Bisanz, G. L., Bisanz, J., Boehme, C., & Lynch, M. A. (1997). What did you learn outside of school today? Using structured interviews to document home and community activities related to science and technology. *Science Education, 81*(6), 651–662.

Mansour, N., & Wegerif, R. (2013). Why science education for diversity? In N. Mansour & R. Wegerif (Eds.), *Science education for diversity: Theory and practice.* New York/Heidelberg: Springer.

Martin, B., Kass, H., & Brouwer, W. (1990). Authentic science: A diversity of meanings. *Science Education, 74*, 541–554.

McCreedy, D. (2003). *Educating adult females for leadership roles in an informal science program for girls.* Unpublished Dissertation, University of Pennsylvania.

McGinn, M. K., & Roth, W.-M. (1999). Preparing students for competent scientific practice: Implications of recent research in science and technology studies. *Educational Researcher, 28*, 14–24.

Miller, J. D., Augenbraun, E., Schulhof, J., & Kimmel, L. (2006). Adult science learning from local television newscasts. *Science Communication, 28*(2), 216–242.

National Academies of Science. (2006). *Rising above the gathering storm: Energizing and employing America for a brighter economic future*. Washington, DC: The National Academies Press.

National Research Council. (2009). *Learning science in informal environments: Places, people and pursuits*. Washington, DC: The National Academies Press.

National Research Council. (2011). *A framework for K-12 science education: Practices, cross-cutting concepts, and core ideas*. Washington, DC: The National Academies Press.

National Science Board. (2004). *Science and engineering indicators: 2004*. Washington, DC: U.S. Government Printing Office.

Penuel, W. R. (2011, September 28). *Analyzing the mutual constitution of persons and cultural practices in STEM learning research*. Center for Lifelong STEM Learning 2020 Vision Presentation, Oregon State University, Corvallis, OR, USA.

Purcell, K. (2011). *E-reader ownership doubles in six months: Tablet adoption grows more slowly*. Washington, DC: Pew Internet & American Life Project.

Rahm, J., Miller, H. C., Hartley, L., & Moor, J. C. (2003). The value of an emergent notion of authenticity: Examples from two student/teacher–scientist partnership programs. *Journal of Research in Science Teaching, 40*(8), 737–756.

Rogoff, B. (2003). *The cultural nature of human development*. Oxford: Oxford University Press.

Roth, M.-W. (1995). *Authentic school science*. Boston: Kluwer Academic.

Roth, M.-W. (1997). From everyday science to science education: How science and technology studies inspired curriculum design and classroom research. *Science Education, 6*, 373–396.

Roth, W., & Van Eijck, M. (2010). Fullness of life as minimal unit: Science, technology, engineering, and mathematics (STEM) learning across the life span. *Science Education, 94*(6), 1027–1048.

Rounds, J. (2004). Strategies for the curiosity-driven museum visitor. *Curator, 47*, 389–412.

Scientific American. (2011). Energy and sustainability: Safecast. http://www.scientificamerican.com/citizen-science/project.cfm?id=safecast-open-information. Downloaded October 3, 2011.

St. John, M., & Perry, D. (1993). A framework for evaluation and research: Science, infrastructure and relationships. In S. Bicknell & G. Farmelo (Eds.), *Museum visitor studies in the 90s* (pp. 59–66). London: Science Museum.

Tran, L. (2006). Teaching science in museums: The pedagogy and goals of museum educators. *Science Education, 91*(2), 1–21.

Tran, L. (2008). The work of science museum educators. *Museum Management and Curatorship, 23*(2), 135–153.

U.S. Bureau of the Census. (2010). *Statistical abstracts of the United States, 2010*. Washington, DC: Government Printing Office.

Vygotsky, L. S. (1981). The development of higher forms of attention in childhood. In J. V. Wertsch (Ed.), *The concept of activity in Soviet psychology*. Armonk: Sharpe.

Wellington, J. (1998). Practical work in science: Time for re-appraisal. In J. Wellington (Ed.), *Practical work in school science: Which way now?* (pp. 3–15). New York: Routledge.

Wenger, E., McDermott, R., & Snyder, W. M. (2002). *Cultivating communities of practice*. Boston: Harvard Business School Press.

Wertsch, J. V. (1985). *Vygotsky and the social formation of mind*. Cambridge, MA: Harvard University Press.

Wertsch, J. (1998). *Mind as action*. New York: Oxford University Press.

Ignoring Half the Sky: A Feminist Critique of Science Education's Knowledge Society

Anita Hussénius, Kristina Andersson, Annica Gullberg, and Kathryn Scantlebury

Introduction

A Chinese proverb observes that women 'hold up half the sky', yet often in science education, we have ignored the knowledge generated by feminist researchers about how females engage and participate in science. Further, science education has often failed to consider the implications from feminist critiques of science on science education. This chapter will provide a feminist perspective on who generates knowledge in science education and what knowledge is acceptable as 'scientific' by the field. Second, we will discuss the culture of science education and discuss whether science educators value the knowledge produced by gender and feminist researchers. In particular, we will examine the integration (or lack thereof) of gender issues into the dominant areas in science education research, such as teachers' pedagogical content knowledge, the development of students' science knowledge through inquiry, the role of conceptual change, and teachers' preparation and professional development programmes. Third, we will provide examples of how gender theory and feminist perspectives in science education could generate new knowledge about gender and science education.

A. Hussénius • K. Andersson
Centre for Gender Research, Uppsala University, Uppsala, Sweden

A. Gullberg
University of Gävle, Gävle, Sweden

K. Scantlebury (✉)
Department of Chemistry and Biochemistry, University of Delaware, Newark, DE, USA

Center for Secondary Teacher Education, College of Arts and Sciences, University of Delaware, Newark, DE, USA
e-mail: kscantle@udel.edu

N. Mansour and R. Wegerif (eds.), *Science Education for Diversity: Theory and Practice*,   301
Cultural Studies of Science Education 8, DOI 10.1007/978-94-007-4563-6_14,
© Springer Science+Business Media Dordrecht 2013

Gender Perspective in Research

Gender research with its reflective and critical approach and with its strong roots in the philosophy of science has much to contribute to science education research. To be able to provide a feminist perspective on the generation and valuation of knowledge within science education research, we need to clarify what we mean with feminist research and how it differs from gender research. In a report from the Swedish Research Council, Anne Hammarström (2005) defines gender perspective in medicine as to critically study the prevailing paradigm and to understand and to apply gender theories in practice. For some researchers, feminist research is another name for the 'all'-gender research. For others, it has a specific meaning that is applicable to certain parts of the field and is characterised by including a political strive and movement for social change. We use Hammarström's critical approach to distinguish three different subareas within science education research with a gender perspective: research addressing gender, gender research and feminist research. These subareas are related to each other, with different criteria to be fulfilled (see Table 1). On the first level is research that addresses gender and uses it (or sex) as analytical categories. On the next level, gender research emanates from a gender theoretical framework that may involve power analysis. Finally, feminist research practises theories about the gender order to highlight power relations on different levels in different settings with an intention to change these imbalanced power relations. In other words, a feminist researcher cannot accept women's subordination. Taking a feminist stand within science education also involves critiquing science's positivistic view of rationality, objectivity and truth (Brickhouse 2001; Fox Keller and Longino 1996). Moreover, a feminist position holds that 'the epistemologies, metaphysics, ethics, and politics of the dominant forms of science are androcentric and mutually supportive' (Harding 1986, p. 9). Therefore, such an approach would influence how the researcher framed questions, viewed the object/subject, chose the research methods, and selected the techniques to analyse the data and communicate findings.

Table 1 Subareas for gender perspective in research

Research addressing gender	Research about sex or gender as analytical categories with:
	(a) No gender theoretical background and/or
	(b) No gender analysis of power
Gender research	Research projects that use:
	(a) A gender theoretical framework and/or
	(b) A gender perspective to analyse power
	(c) Critical review with a gender perspective of existing research
Feminist research	Research projects that use:
	(a) A gender theoretical framework and/or
	(b) A gender perspective to analyse power
	(c) Critical review with a gender perspective of existing research with the aim to change power imbalances

Gendered Knowledge and Knowledge About Gender in Science Education

When science education researchers have studied gender as a category, they conceptualised gender as an individual trait, rather than from a social context (Nyström 2007). Such studies fall under the subarea *addressing gender* shown in Table 1. Few studies on gender and science education adopt a critical stance, and many studies are restricted to comparing female and male students on variables such as students' achievement, participation, engagement and attitudes towards science (Brotman and Moore 2008; Scantlebury 2010). We mean that a feminist science criticism offers a theoretical platform, an alternative way of looking at the sciences and science education, as a starting point to challenge the hypotheses and prevailing research practices in the field. Therefore, we want to highlight the importance of integration of gender perspectives in science education research and that a limited and incomplete knowledge might be produced if such perspectives are ignored. To illustrate this we provide two examples from other disciplines: one from medicine and the perception of heart diseases and the second from students' skills in mathematics and the perception of their accomplishments. Ignoring gender in producing knowledge and ignoring knowledge about gender is, in fact, ignoring half the sky.

Heart Disease Symptoms

Until the mid-1980s, men were the primary subjects for research on cardiovascular diseases, and doctors extrapolated the results of these studies to all humans. But women admitted to the hospital with acute heart problems expressed symptoms that did not align with those of classic (male) angina patients. Misdiagnosed, that is, physicians did not recognise women's symptoms as heart disease, several women died in complications from heart attacks since they did not get the necessary treatment (Hammarström 2005). When these misdiagnoses were highlighted, the interpretation was that the generalisation of symptoms on heart problems was not valid for all humans. The conclusion was that biological-caused symptom differences might exist, and more knowledge about women's spectra of symptoms had to be examined. Recently, the symptoms for heart attacks as well as risk factors derived by research on men have been shown also to be applicable on women. Researchers that integrated knowledge of gender studied the interaction between the physicians and their patients (Swahn 2008). They found that women and men communicate their pain in different ways, expressing their feelings differently and using stronger or weaker words. Female patients are not taken as seriously as men by the hospital staff and run a risk of not getting sufficient treatment. The implication that follows is that knowledge about heart diseases in itself is not enough to make an accurate diagnosis but has to be accompanied by a knowledge of gender.

Students Skills in Mathematics

In the 1980s researchers explained boys' higher achievement in mathematics was because boys' brains were more logical compared to girls. The proportion of female mathematicians at a high level is still very small, and as recently as 2005, preeminent academics have stated that biological rather than cultural factors explained the differences. For example in his 2005 talk at the NBER conference, the former Harvard University president Lawrence Summers claimed that intellectual capacity varies more among men than women which explained why there were more men as 'great' mathematical minds than women (Summers 2005). This theory using biology to explain gender differences in mathematical aptitude was originally launched in the late 1800s by the British sex researcher Havelock Ellis (1897 in Hyde and Mertz 2009). Yet a recent study has invalidated this theory. A large American study analysed children's test scores from 41 countries within PISA project[1] (Hyde and Mertz 2009). In several countries, there was a similar variation of mathematical ability among girls and boys, and no difference between the two sexes in the most mathematically talented group. Hyde and Mertz concluded that the lack of a universal gender pattern negated a biological explanation for the different participation of women and men and that sociocultural factors influenced the extent to which girls/women choose mathematics as a career. The researchers also studied the sex distribution of the participants in the International Mathematics Olympiad. According to the gender equality index developed by the World Economic Forum, there are a higher proportion of women on teams from countries with higher index scores. Until recently, the reasoning that biological differences explained the pattern of more male mathematicians was incorrect and gendered.

In both examples given above, taking a knowledge of gender into account highlighted a bias in the research. Furthermore, that bias limits our knowledge and those limitations can have serious consequences and implications. Our concern is whether science education research is *gendered* and what consequences follow upon a lack of *knowledge about gender* in science education research. Traditionally, science education research has focussed on students' understanding of subject matter and is dominated by individualised perspectives of learning. Recently, the field has expanded to include issues of epistemology, affective factors (such as students' science attitudes, perceptions of science and scientists) and sociocultural studies, but these are marginal topics within the discipline (Roth 2010). In the next section, we examine if gender is included in science education research and, if so, how it is integrated.

---

[1]Programme for International Student Assessment.

Examining Gender in Science Education Research

To what extent are gender issues integrated into the dominant areas in science education research? Chang et al. (2010) conducted an automatic content analysis with four science education journals[2] to determine the topics, trends and dominant authors in the field. As shown in Table 2, they identified nine main areas, namely, *conceptual change and concept mapping, nature of science and socio-scientific issues, professional development, conceptual change and analogy, instructional practice, scientific concept, reasoning skills and problem-solving, attitudes and gender* and *design-based and urban education.*

Fifty-six percent of science education research has focussed on knowledge-building issues (topics 1, 4 and 6). These topics incorporate some aspect of studying learners' acquiring science concepts and/or how learners conceptualise science and/or how to change learners' alternative science concepts. The dominance of conceptual change research in science education was noted and acknowledged by the coeditors of *Cultural Studies of Science Education* (a journal not included in the analysis), Kenneth Tobin and Wolf Michael Roth, and they attempted to address this imbalance by focussing the second forum[3] sponsored by the journal on *Cultural studies and conceptions/conceptual change: reuniting psychological and sociological perspectives* (Roth 2010). Authors and forum presenters published their perspectives in a book with the same title. In that publication, Scantlebury and Martin (2010) noted that as a field conceptual change within science education research had ignored the knowledge generated with a gender perspective. Their review of the conceptual change literature found one article, which incorporated

Table 2 Nine main topics within science education (Chang et al. 2010)

Main topics	Numbers of articles $n = 1{,}401$	Frequency %
1. Conceptual change and concept mapping	553	39.5
2. Nature of science and socio-scientific issues	191	13.6
3. Professional development	149	10.6
4. Conceptual change and analogy	147	10.5
5. Instructional practice	90	6.4
6. Scientific concepts	88	6.3
7. Reasoning skill and problem-solving	80	5.7
8. Attitude and gender	64	4.6
9. Design-based and urban education	39	2.8

---

[2]The four journals were *International Journal of Science Education, Journal of Research in Science Teaching, Research in Science Education* and *Science Education.*

[3]Cultural Studies of Science Education has organised a small conference focussed on sociocultural issues impacting science education for journal editors, reviewers and contributors and researchers in this area of science education.

gender into the research methodology that moved beyond the practice of dividing the data set into female and male subjects and identifying gender differences, that is, focussing on gender as an individual trait. Bunce and Gabel's (2002) article reported that different pedagogical approaches could improve students' conceptual understanding of science and introducing students to different levels of representation in chemistry improved girls' understanding and achievement. This study falls into the category *research addressing gender* (see Table 1). In general, conceptual change research has ignored a gender perspective in research, in anyone of the three subareas identified and reported in Table 1.

In Chang et al.'s (2010) analysis, some articles had overlapping items from several main topics, e.g. topic 1 and 2 are overlapping to some extent. But the topic 'attitudes and gender' was independent and did not occur within other topics. Only 64 (5 %) of the 1,401 articles considered 'attitudes and gender', and among them only 18 included gender as a keyword. However, Chang et al. (2010) excluded articles from *Cultural Studies of Science Education, Journal of Science Teacher Education* or science education research published in other journals. In order to ascertain if gender is ignored in science education research and by using Chang's work as a foundation, we conducted ERIC literature searches on the different phrases and topics that dominate the field and when feasible conducted separate searches. We searched using each 'main topic' listed in Table 1 and then added the descriptor '*science*', repeated the search and added another descriptor, namely, '*gender*'. We replaced *gender* with '*feminist*' as a descriptor and repeated the search. As the search identified few studies, we used a broader descriptor – '*equity*' – and repeated the search. Finally, '*gender theory or feminist critique*' was included in the search with the other descriptors. In this way, we documented the extent to which gender issues are integrated into science education research, and '*gender theory or feminist critique*' identified research that used theories of gender and/or feminist critique. These search results are shown in Table 3. Besides Chang et al.'s (2010) main topics, we included pedagogical content knowledge (PCK). In the time, since Shulman (1986) introduced and defined PCK (which included taking gender into account), it has become a well-established 'concept' and another large research domain within science education.

Table 3 shows the number of science education articles by main topic and then the search descriptors of science, gender, feminist, gender/feminist/equity and gender theory/feminist critique in the ERIC database. To describe the procedure, pedagogical content knowledge (PCK) is taken as an example. An initial search on PCK generated 1,805 publications, with 1,142 in peer-reviewed journal articles. Adding science reduced the number to 583 publications, of which 414 were peer reviewed. Including *gender* as a search descriptor reduces the number of articles to 18. We replaced *gender* with *equity* as a descriptor and identified two peer-reviewed articles. Exchanging *gender* with *feminist* generated one reference (a symposium at an international conference). To eliminate the double-counting articles, e.g. if the words *gender* and *equity* appeared in the same article, we searched using '*gender or feminist or equity*'. That search resulted in eight peer-reviewed articles from a total of 20. All eight articles used gender (i.e. sex) only as a

Table 3 Number of articles in main topics of science education research (From ERIC search, June 2011)

Main topics	Articles in main topic[a]	Search descriptors added to main topic					%[b]
		Science[a]	Gender[a]	Feminist[a]	Gender/feminist/equity[a]	Gender theory/feminist critique[a]	
Pedagogical content knowledge	1,805 (1,142)	583 (414)	18 (6)	1 conf.	20 (8)	1 conf.[c]	1.9
Conceptual change	1,011 (639)	780 (503)	31 (14)	1 (0)	36 (15)	0	3
Nature of science	1,112 (617)		34 (18)	6 (4)	48 (24)	1 (1)	3.9
Socio-scientific issues	46 (36)	45 (35)	2 (0)	0	2 (0)	0	0
Professional development	27,247 (8,284)	3,128 (1,248)	86 (25)	5 (1 conf.[c])	145 (32)	2 (1)	2.6
Instructional practice	595 (199)	97(36)	1 conf.[c]	0	1 conf.[c]	0	0
Scientific concepts	7,630 (4,506)	6,999 (4,158)	83 (47)	6 (2 conf.[c])	91 (49)	0	1.2
Reasoning skills or problem-solving	387,774 (9,902)	8,152 (2,586)	199 (56)	6 (0)	237 (59)	1 (0)	2.3
Design-based or urban education	5,459 (907)	550 (142)	52 (7)	1 (1)	61 (12)	0	8.4

[a]Peer-reviewed articles in parenthesis

[b]Percentage of peer-reviewed articles with 'gender or feminist or equity' of total peer-reviewed articles within in the main topic including 'science'

[c]conf. = paper presented at a conference

category to analyse the results. Gender or feminist theories or gender research is not conducted in research studies of PCK. Although included in the original description of PCK (Shulman 1986), gender research is missing in this area 25 years later.

The investigation showed that very few peer-reviewed articles within science education research's main topics considered aspects of gender, feminism or equity. The exception is the area 'design-based and urban education' where 8.4 % of the articles contained at least one of the descriptors. In our ERIC search, we added the descriptor, *gender theory or feminist critique*, to identify gender and feminist research according to the definitions of the subareas used in Table 1. Overall, five articles aligned with these definitions, with a majority of the science education research using gender as a keyword is categorised as '*research addressing gender*'. In summary, science education research that includes a gender perspective is almost negligible – it is just a small glimpse of the sky.

In the following sections, we use gender theory to explain the lack of gender research in science education. Hammarström (2005) has described three different ways to apply gender theories to evaluate medical studies, and we have used her criteria to analyse science education research. All the three fall under the subarea *gender research* in Table 1. The first way is to conduct research, interpret data and explain results with gender theories. The second way is to develop gender concepts, models or theories within the research and finally to use gender theories to analyse how sex and gender is dealt with within the research field. This could be about, e.g. how underlying assumptions may influence or hinder the scientific scrutiny.

A Feminist Theoretical Framework for Improving the Field of Science Education Research

A gender perspective can enrich and improve the field of science education research. In this section, we will present two specific gender theories for analysing science education research. In our interpretation, gender is more complex than just adding social/cultural influences 'onto' the biological sexes. It is not just to sum 'social' and 'sex' like a simple mathematic formula $1 + 1 = 2$ (Hirdman 1990). The gender concept means you cannot understand 1 and 1 as really existing. From the very beginning with the development of the fetus and through the whole life, the human being is in intimate interplay between genetics and ambient factors. Like Hirdman puts it, there only exists a 2. Moreover, gender is constituted on different levels in society, namely, the *structural,* the *symbolic* and the *individual* level (Harding 1986; Hirdman 1990; Rubin 1975). The *structural* level is how the society is organised regarding gender, e.g. the division of labour. In most cultures, women and men have different tasks and roles. Women in one culture may have the same tasks as men in another culture, but within the same culture, labour often is divided by gender. What is consistent across cultures is that a higher status is attributed to men's, compared to women's, labour. Swedish historian Hirdman (1990) described this pattern from two perspectives: first, the separation of the two sexes and, second, the superior status of the male standard. The formation of gender consolidates differences between the sexes, and the female gender is always subordinate the male one, independent of status, class, time and space. One expression of this gender order is how the wages and professional status associated with careers decline when women enter work areas previously dominated by males. For example, in academe, there are a higher percentage of men in all disciplines at the more prestigious universities and also more men than women in higher-status disciplines. The faculties of Ivy League institutions in the United States and the world's top 100 universities employ more men than women at each level – lecturers and assistant, associate and full professors. There is also a lower participation of women in the natural sciences (e.g. physics) and engineering compared to men. In 2009, the percentage of female professors in natural sciences at the five biggest universities in Sweden ranged between 8.6 and 22, and the prediction is that during the coming years the proportion of women professors will decrease (Swedish Statistics). Thus, the structural gender of the natural sciences and engineering domains remains as masculine (Harding 1986).

According to Harding (1986), the *symbolic* level of gender is the archetypical idea of Woman and Man from different cultures as it is expressed in language, for example, by retaining dichotomies where the oppositional pairs are assigned a feminine and masculine meaning (e.g. emotion-rationality, subjectivity-objectivity, nature-culture). This symbolic meaning of gender generates conceptions of feminine and masculine and extends into appropriate practices for women and men, that is, how they should act, dress and engage in the appropriate work. On a symbolic level, physics is viewed as rational, difficult, and hard, with disembodied knowledge (Fox Keller 1992). From a feminist perspective, physics is a masculine gender practice both structurally and symbolically. This view may create tensions for primary school teachers where nurturing and caring, a feminine symbolic gender, are important features that permeates their school culture (Carrington 2002; Ford et al. 2008; Gannerud and Rönnerman 2003).

The *individual* level is individuals' socialisation into a gender identity where the structural and the symbolic level have a great impact on this process. But this process can also be a dialectic, that is, individual gender can impact and transform the structural and symbolic gender. For example, in recent Western history, football (soccer) was a totally male domain and an impossible activity for girls and women. On a structural level, there were no teams for women and no sports clubs offered football training for girls. A reason for that was the general image (on the symbolic level) that femininity and football playing were incompatible. Moreover, there were biological claims that athletic activity negatively affected female reproductive organs (Johannisson 1994). Nevertheless, individual girls challenged this norm because of their interest to play football at a competitive level. The 'breaking of new land' by these girls together with women's movement influenced and changed the gender of football.

The understanding of femininity and masculinity within the three levels changes from one culture to another and over time, but within one culture these three forms of levels of gender are related to each other (Harding 1986). The theory is also sensitive to other social categories such as ethnicity, religion, socio-economic status, sexuality and age. For example, all men are not superior to all women. In the next section, we use gender theory to analyse science education research.

Using Gender Theory for Analysing Science Education Research Field

A feminist perspective on science education provides researchers the tools for highlighting power and the hierarchies generated through power differentials. Those hierarchies influence knowledge production within science education and the knowledge produced about gender, science and science education. The result from our ERIC search shows on a structural level what can be seen as *research that matters*. The 'concept' area is still the dominant one with a huge amount of articles. One interpretation of this area's dominance is that 'concept research' has a high symbolic value within the science education research field and therefore is a high-status area.

The ways in which science curricula are planned and what content is chosen in school are historically situated. Science education has its base in the natural science disciplines, and although evolved into different fields of research and education, there is still a close connection between the two. Not surprisingly, science knowledge and those who produce that knowledge have had and still have a strong influence on the content of 'school science' and on the curricula for K-12 students and their teachers. Scientists decide what science is of most importance to include in the science curriculum for those who learn and also for those who teach science.[4] But this transformation of scientific knowledge to a school context is problematic. Concepts, laws and theories are commonly presented as rigid truths, compared to the way they are communicated and debated within the scientific community itself and thereby convey a stereotyped positivistic view of science. Moreover, while the curriculum in school addresses the scientific phenomena and concepts explicitly, it also mediates an implicit message of a hierarchy of science practices and who can access and participate in that practice (Andrée 2007; Harding 1986; Lemke 1990; Roberts and Östman 1998).

A feminist critique can provide new and important areas of research within science education. To illustrate this we give an example of how individual researchers can use their knowledge of gender on a structural and symbolic level to contribute to research design. Schmader and Johns (2003) showed how different expectations because of individuals' sex can generate performance differences. They investigated

---

[4]For example, in the United States, the 'committee of ten' decided that students would first study biology, then chemistry and lastly physics. Scientists in the nineteenth century introduced this 'layer cake' approach to the curriculum which remains dominant in the twenty-first century (DeBoer 1991).

how 'stereotypical threat' could affect women's and men's achievement on a mathematical test. The researchers assigned the same task to two groups and told the control group that their task was a working memory test. The subjects in the other group received a 'stereotype threat' when they were told that different mathematical performances might stem from underlying gender differences. In the control group, there were no significant differences between the sexes on the math task. However, in the 'stereotypical threat' group, the women performed 40 % lower than the control group. But the threat did not affect the men's performance. This is an example of how the symbolic gender, i.e. in this case that logical thinking has a masculine connotation, influenced women and results in a physiological reaction, a brain stress. Just telling the 'stereotype threat' group that gender difference *might* have importance for performance and without mentioning how send the implicit message of symbolic gender mediating that women are subordinated men. The experimental design indicates that the underlying hypothesis emanated from a theory of the gender order and the results from the empirical data expose and verify this theory. In the next section, we will give other examples from completed and ongoing feminist science education research projects.

Some Examples of Feminist Research in Science Education

In general, the research focussing on science teachers' professional development has not taken a critical perspective on the design, implementation and enactment of these programmes. Martin et al. (2006) found that left unchallenged, teachers enrolled in advanced degree programmes (typically white males) developed into target students. As target students, these teachers disrespected their peers and joined and supported cliques that contributed to a hostile and negative classroom climate for others to the extent that one teacher noted it was like 'being in middle school'. Research with high school African American girls using cogenerative dialogues provided them voice to share with teachers their perspectives on changing the classroom culture to increase their engagement in science. Increasing the girls' engagement leads to their improved achievement in science.

During a 5-year, action research project, one of the authors (Andersson) followed a group of preschool teachers and studied the process of their work on science and gender (Andersson 2010, 2011). She found that it was difficult for the teachers to think beyond their stereotypical assumptions about girls and boys, although they had engaged on examining education from an equity perspective since the 1990s. They also expressed strong negative experiences of school science and had feelings of stupidity and low self-esteem with regard to science. The project offered teachers the opportunity to build their science confidence, to learn more about themselves and to reflect on their internalised views of gender and of science. The study foregrounded other competencies than subject matter knowledge that are of importance for elementary and preschool teachers to engage children with meaningful science activities (Andersson and Gullberg 2011). For example, teachers of young children need the skill to reformulate children's questions, making them possible to explore, investigate and discuss (Jelly 2001). They do not need to 'answer' children's questions about scientific phenomena, by providing the correct scientific explanations. However, in using a feminist/gender perspective on the research design and in analysing the data, we turned the focus from the teachers' lack of subject matter knowledge to question what is science or what can it be for preschool students. Moreover, we moved away from an individual level where the teachers were 'the problem that should be fixed' to a structural level where the teachers are part of a value system with an imbalanced power order. In this value system within science education, elementary teachers together with students are in the bottom of a hierarchy where science subjects are in the top. The amount of research that is engaging within the area of scientific concepts and conceptual change (see Tables 2 and 3) demasks this hierarchy. It is taken for granted that the concepts explored really are necessary for all teachers and students, and as a consequence, these concepts stand firm and are to a very little degree questioned and problematised. Furthermore, when teachers' science subject matter knowledge is studied, the choice of participants is often those at the bottom of the hierarchy with little or no power, that is, elementary teachers and students (Abell 2007).

Through these new insights gained by the action research project referred to above (Andersson 2011), we recently have began a new study to explore the integration of gender knowledge into science teacher education. In this ongoing research project, we are studying if an increased awareness of gender issues in science and in science teaching among K-6 student teachers influences their identities as teachers and their teaching of science. The project explores the process of pre-service teachers' ongoing negotiations of gender, science and science teaching. Engaging pre-service teachers in critical reflections on science, teaching and gender can provide them with the opportunity to examine their alienation from science and the tools to reflect upon their own and their future pupils' participation in science (Barton 1997).

We have introduced critical perspectives on gender as related to the nature of science, the culture of science and a feminist critique of the sciences as part of new, teaching sequences for the pre-service teachers' first and second semester of science courses. The new parts are integrated into ordinary science subject and science education courses and consist of lectures, seminars, compulsory written tasks, gender theory readings, etc. In the beginning of the first science semester and after the science year, the students write short essays focussed on personal experiences related to issues of gender and science. We are interested in exploring the process of the student teachers' ongoing negotiations of gender, science and science teaching. In order to capture this complex process, we collect data continuously using a variety of methods such as essays, semi-structured group interviews and documentation of the student teachers' participation in various teaching activities. Group discussions about 'cases' highlighting different issues of gender and the teaching of science (Andersson et al. 2009) have been used to intervene as well as a data collection. The aim with the described activities is to develop tools for reflection which may increase the student teachers' possibility for articulating thoughts about science and gender through verbal and written forms.

Although the analysing phase of this project has just started, we venture to assert that the symbolic image of science is important and has an impact on students' self-image. In a group discussion on how the pre-service teachers related to science, one of the female participants admitted that she found science easy in school:

> ...but that was nothing that you would admit in school. Instead you complained about how difficult it was, even thought it was not!

Thus, girls pretended to find science difficult to fit into the norm of being girls. Thereby they reconstruct themselves as girls; they adjust to an expected girl identity, instead of opposing to and criticising the image of science. The ability in science is masculine gendered on a symbolic level and affects the individual's thoughts and actions. Nevertheless, a majority of the project's close to 100 female pre-service teachers have expressed a negative experience of science education during their schooling. They have commented on the objective and static culture of science. That is, science knowledge in science stands firm, and you cannot influence it. According to these pre-service teachers, a typical answer from a science teacher when asked to explain something you do not understand is: 'It is just like this', and the impression you get as a student is that science is something 'you do not have to understand, just memorize'.

This view of science is conflicting with the stereotypic view of femininity where relations and relationships are central. By using a feminist theory, it is possible to understand female students' lack of interest in science as a problem for which science itself is responsible, rather than individual deficiencies of the students.

Conclusions: Considering the Whole Sky

Gender permeates whole societies and is interwoven with science as a subject and interacts with teachers, students and researchers in a complex manner. Gender knowledge raises other research questions and adds a different theoretical frame-work for analysing and interpreting data. Gender remains the 'missing paradigm' in science education (Schulman 1986). While feminist researchers continue to develop knowledge within the field, we suggest science education researchers need to utilise feminist/gender theoretical frameworks to extend and expand upon knowledge and thus begin to consider the whole sky. We will also call upon the use of gender theoretical framework for analysing the science education research field. For example, within pedagogical content knowledge, two areas are dominant: 'students' difficulties' and 'instructional strategies'. The research on 'students' difficulties' focusses on how and why students struggle to learn a topic at an individual level. But the research does not consider that there may be factors at a symbolic or structural level that could contribute to 'student difficulties'. In math, the stereotypic threat, a symbolic factor, hindered the female participants' performance, not the individual's lack of math knowledge (Schmader and Johns 2003). If researchers continue to focus on individual's difficulties, the problems of females' participation and achievement in science will remain unsolved. By shifting the focus to a symbolic level, researchers can analyse why science has developed a culture that is alien to the students. And through problematising the image of science and analysing what knowledge is considered important and why, insights can emerge that will complement and extend the research area of 'student difficulties'. When researchers have identified the obstacles, it will be possible to find more effective instructional strategies and challenge the stereotypic image of science that may hinder students' development of their scientific knowledge. If science education researchers could expand their studies to include a gender theoretical framework and use a gender perspective to analyse the power dynamics, then by doing so, we could begin to see a complete sky.

References

Abell, S. K. (2007). Research on science teacher knowledge. In S. K. Abell, & N. G. Lederman (Eds.), *Handbook of research in science education* (pp. 1105–1149). Mahwah, NJ: Lawrence Erlbaum.

Andersson, K. (2010). "It's funny that we don't see the similarities when that's what we're aiming for" – Visualizing and challenging teachers' stereotypes of gender and science. *Research in Science Education.* doi:10.1007/s11165-010-9200-7.

Andersson, K., Hussénius, A., & Gustafsson, C. (2009). Gender theory as a tool for analysing science teaching. *Teaching and Teacher Education, 25*, 336–343.

Andersson, K., & Gullberg, A. (2011). What is science in preschool and what do teachers have to know to empower the children? *Culture Studies of Science Education.*

Andersson, K., & Gullberg, A. (2012). What is science in preschool and what do teachers have to know to empower the children? *Culture Studies of Science Education.* doi:10.1007/s11422-012-9439-6.

Andrée, M. (2007). *Den levda läroplanen. En studie av naturorienterade undervisningspraktiker i grundskolan* [The lived curriculum. A study of science classroom practices in lower secondary school]. Stockholms universitet, Studies in Educational Sciences, 97. Stockholm: HLS förlag.

Barton, A. C. (1997). Liberatory science education: Weaving connections between feminist theory and science education. *Curriculum Inquiry, 27*(2), 141–163.

Brickhouse, N. W. (2001). Embodying science: A feminist perspective on learning. *Journal of Research in Science Teaching, 38*(3), 282–295.

Brotman, J. S., & Moore, F. M. (2008). Girls and science: A review of four themes in the science education literature. *Journal of Research in Science Teaching, 45*(9), 971–1002.

Bunce, D., & Gabel, D. (2002). Differential effects on the achievement of males and females of teaching the particulate nature of chemistry. *Journal of Research in Science Teaching, 39*(10), 911–927.

Carrington, B. (2002). A quintessentially feminine domain? Student teachers' constructions of primary teaching as a career. *Educational Studies, 28*(3), 287–303.

Chang, D., Chang, D., & Tseng, K. (2010). Trends of science education research: An automatic content analysis. *Journal of Science Education and Technology, 19*, 315–331.

DeBoer, G. (1991). *A history of ideas in science education: Implications for practice.* New York: Teachers College Press.

Ford, D. J., Fifield, S., Qian, X., Allen, D., Donham, R., & Gwekwerere, Y. (2008). *Preservice K-8 teachers' developing pedagogical context knowledge within an integrated science and education continuum.* Paper presented at the National Association of Research in Science Teaching (NARST), Baltimore, MD.

Fox Keller, E. (1992). *Secrets of life. Secrets of death.* London: Routledge.

Fox Keller, E., & Longino, H. E. (1996). *Feminism and science.* Oxford: Oxford University Press.

Gannerud, E., & Rönnerman, K. (2003). *Lärande och omsorg i förskola och skola*: IPD-rapporter, nr 2003:03 [Learning and care in preschool and school]. Institutionen för pedagogik och didaktik, Göteborgs universitet.

Hammarström. (2005). *Genusperspektiv på medicinen – två decenniers utveckling av medvetenheten om kön och genus inom medicinsk forskning och praktik* [Gender perspective in medicine – Two decades of gender awareness development in medical research and practice]. Swedish National Agency for Higher Education.

Harding, S. (1986). *The science question in feminism.* Ithaca: Cornell University Press.

Hirdman, Y. (1990). *Genussystemet. In SOU 1990:44, Demokrati och makt i Sverige* [Swedish Government Official Report, SOU 1990:44 Democracy and Power in Sweden]. Stockholm.

Hyde, J. S., & Mertz, J. E. (2009). Gender, culture, and mathematics performance. *Proceedings of the National Academy of Sciences of the U S A, 106*(22), 8801–8807.

Jelly, S. (2001). To teach the children to ask questions – And to answer them. In W. Harlen (Ed.), *Primary science: Taking the plunge* (pp. 64–76). Portsmouth: Heinemann.

Johannisson, K. (1994). *Den mörka kontinenten: kvinnan, medicinen och fin-de-siecle* [The dark continent: The woman, the medicine and the fin-de-siecle]. Stockholm: Norstedt.

Lemke, J. L. (1990). *Talking science: Language, learning and values.* Norwood: Ablex Publishing Company.

Martin, S., Milne, C. E., & Scantlebury, K. (2006). Eyerollers, jokers, risk-takers and turn sharks: Target students in a professional science education program. *Journal of Research in Science Teaching, 43*(8), 819–851.

Nyström, E. (2007). *Talking and taking positions. An encounter between action research and the gendered and racialised discourses of school science.* Doktorsavhandlingar i pedagogiskt arbete, nr 16. Umeå universitet.

Roth, W. M. (Ed.). (2010). *Re/structuring science education: Reuniting sociological and psychological perspectives.* Dordrecht: Springer.

Rubin, G. (1975). The traffic in women: Notes on a 'political economy' of sex. In R. Reiter (Ed.), *Towards an anthropology of women* (pp. 157–210). London: Monthly Review Press.

Roberts, D. A., & Östman, L. (1998). *Problems of meaning in science curriculum.* New York: Teachers College Press.

Scantlebury, K. (2010). Still part of the conversation: Gender issues in science education. In B. Fraser, C. McRobbie, & K. Tobin (Eds.), *Second international handbook of science education.* Boston: Kluwer Academic Publishers.

Scantlebury, K., & Martin, S. (2010). How does she know? Re-visioning conceptual change from feminist perspectives. In W. M. Roth (Ed.), *Re/structuring science education: Reuniting sociological and psychological perspectives* (pp. 173–186). Rotterdam: Springer.

Shulman, L. S. (1986). Those who understand: Knowledge growth in teaching. *Educational Researcher, 15*(2), 4–14.

Schmader, T., & Johns, M. (2003). Converging evidence that stereotype threat reduces working memory capacity. *Journal of Personality and Social Psychology, 85*(3), 440–452.

Summers, L. (2005). *Remarks at NBER conference on diversifying the science & engineering workforce.* Retrieved August 18, 2011, from http://www.harvard.edu/president/speeches/summers_2005/nber.php

Swahn, E. (2008). *Genusperspektiv vid kranskärlssjukdom: praktisk handledning för öppenvården* [Gender perspectives in coronary heart disease: Practical guide for non-institutional care]. Södertälje: AstraZeneca.

Religion in Science Education

Michael J. Reiss

Does Religion Have a Place in Science Education?

To the bemusement of many science educators, in school and elsewhere, issues to do with religion seem increasingly to be of importance in school science lessons and some other educational settings. Before discussing how science educators might deal with religion, the first issue to be addressed is the possibility that they shouldn't.

The argument that science and school science lessons should not deal with religion, in my experience, relies on the assumption that the question can be addressed by epistemological reasoning. Granted this assumption, which, again in my experience, is generally unquestioned, the argument generally proceeds along one of two lines: either religion and science have different epistemologies so it is simply inappropriate or invalid to deal with religious matters in the science classroom or religion itself is epistemologically invalid and so the question almost doesn't arise.

First of all, let us suppose that we grant that the question is an epistemological one by which I mean that we accept that science and school science should restrict themselves to knowledge resulting from science. An initial objection to this supposition is that it could be argued that science and school science should therefore not avail themselves of mathematical reasoning on the grounds that mathematical knowledge is arrived at in ways that are wholly distinct from scientific knowledge. As I have argued elsewhere:

> I believe that the internal angles of any flat triangle add to 180° but the truth of this statement is arrived at differently in mathematics (i.e. through logical proof – cf. Euclid's *Elements*, Book 1, Proposition 32) than it would be if it were a scientific statement along the lines 'All vertebrates have four limbs', to test which one would look at large numbers

M.J. Reiss (✉)
Institute of Education, University of London, London, UK
e-mail: m.reiss@ioe.ac.uk

N. Mansour and R. Wegerif (eds.), *Science Education for Diversity: Theory and Practice*,
Cultural Studies of Science Education 8, DOI 10.1007/978-94-007-4563-6_15,
© Springer Science+Business Media Dordrecht 2013

of vertebrates. In mathematics, it doesn't help (except when teaching pupils about the truth of the proposition) to corral large numbers of triangles and then carefully measure and sum their internal angles. (Reiss 2013)

However, the epistemological argument is used incorrectly if it is used to maintain that mathematical reasoning should not be used in science or school science. One might as well argue that logical reasoning and the English (or French, Swahili or whatever) language should not be used in science or school science. The point is that even though it is the case that certain areas of science, notably theoretical physics, are basically applied mathematics, it nevertheless is the case that they are either tested or, at least in principle, are capable of being tested by comparison with the material world (even if the equipment to test certain theories is almost unaffordable) – that is precisely why they are called *applied* rather than *pure* mathematics. So mathematics and the English (or French, Swahili or whatever) language are used merely as tools in science, much as logic is. Science does not itself make contributions to the disciplines of mathematics, languages or logic.

Much the same argument applies to the question as to whether ethics has a place in science. It seems to be useful to make a clear distinction between the two disciplines. Science is about attempting to explain the observable features of the material (whether natural or manufactured) world. Ethics is fundamentally about attempting to discern or decide what it is that is morally right or wrong for moral agents to do in the world. Of course, deciding this often takes account of facts that scientists have helped establish about the world. For example, issues about the moral acceptability of abortion are *affected* by such matters as the age at which individuals are capable of suffering and the consequences for the mother-to-be in terms of health and well-being of either having or not having the baby. But issues about the moral acceptability of abortion are not *determined* by such matters. There are issues to do with autonomy, justice and rights that simply cannot be reduced to science; ethics is a distinct branch of knowledge.

Epistemologically, therefore, there is little of ethics that can be included within science (just as there is nothing of science that can be included within mathematics). The discipline of aesthetics is similar in that part of what it is that causes us to decide that something, for instance, a natural landscape or a work of art, is of high quality has a scientific basis but aesthetics can almost certainly not be reduced entirely to science.

However, the question of whether religion has a place in science education is not the same as the question of whether it has a place in science. It is perfectly possible to conclude that religion has no place in science but that it does in science education. The reason for this is simply that science education is a broader field of study than is science. Just as we might conclude that ethics has a role to play in science education (Jones et al. 2010), even if it doesn't in science, we need to examine whether religion has a role to play in science education.

Science education is fundamentally about introducing people to the knowledge that science has accumulated and to an understanding of how this knowledge has been and is being produced. The best reason therefore for including issues of religion in the science classroom would be if, by so doing, learners gain a better understanding of science. Precisely the same argument would seem to hold for why we might include some history in science education, namely, that it helps learners better to learn science.

In the case of history, we can envisage a number of ways in which it might help learning about science. Consider, for example, the periodic table. Telling students a bit about Lavoisier's early classification of 33 'elements', Newlands' law of octaves and Mendeleev's 1869 table can help in the teaching of chemistry in a number of ways. Some learners find it motivating; others simply appreciate a bit more how difficult it was to arrive at the periodic table; others, if well taught, come to understand that the questions these early chemists struggled to answer can more easily be answered with today's knowledge of atomic nuclei and electron shells.

So under what circumstances might one wish matters of religion to be included within the teaching of science? I shall examine two obvious possibilities: when teaching about the nature of science and when teaching about evolution.

Teaching About the Nature of Science

The importance attached to 'the nature of science' in school science education has grown in recent years (Lederman 2007), despite certain detractors. The term 'nature of science' is, not surprisingly, understood in a number of ways but at its heart is knowledge about how, and to a lesser extent why, science is undertaken. So the nature of science includes issues about the fields of scientific enquiry and the methods used in that enquiry.

A key point about the fields of scientific enquiry is that these have shifted over time. In large measure this is simply because of developments in instrumentation. We can now study events that happen at very low temperatures, at distances, at speeds and at magnifications that simply were not possible a few decades ago. What is still unclear is the extent to which certain matters currently outside of mainstream science will one day fall within the compass of science. Take dreams, for example. It may be that these will remain too subjective for science but it may be that developments in the recording of brain activity will mean that we can obtain a sufficiently objective record of dreams for them to be amenable to rigorous scientific study.

But the scope of science has also shifted for reasons that are more to do with theorisation than with technical advances. Consider beauty. Aesthetics for a long time fell out with science. But there is now, within psychology and evolutionary biology, a growing scientific study of beauty and desire (e.g. Buss 2003). Indeed, a number of the social sciences are being nibbled away at by the natural sciences and if you believe some scientists, almost the only valid knowledge is scientific knowledge (Atkins 2011).

Despite such movements in the fields of scientific enquiry and in the actual methods employed by scientists, the overarching methods of science (what a social scientist might term its methodology) have shifted far less, certainly for several hundreds of years, arguably for longer than that.

As is well known, Robert Merton characterised science as open-minded, universalist, disinterested and communal (Merton 1973). For Merton, science is a group activity: even though certain scientists work on their own, science, within its various subdisciplines, is largely about bring together into a single account the contributions of many different scientists to produce an overall coherent model of one aspect of reality. In this sense, science is (or should be) impersonal. Allied to the notion of science being open-minded, disinterested and impersonal is the notion of scientific objectivity. The data collected and perused by scientists must be objective in the sense that they should be independent of those doing the collecting. This is the main reason why the data obtained by psychotherapists are not really scientific: they depend too much on the relationship between the therapist and the client. The data obtained by cognitive behavioural therapists, on the other hand, are more scientific.

Karl Popper emphasised the falsifiability of scientific theories (Popper 1934/1972): unless you can imagine collecting data that would allow you to refute a theory, the theory isn't scientific. The same applies to scientific hypotheses. So, iconically, the hypothesis 'all swans are white' is scientific because we can imagine finding a bird that is manifestly a swan (in terms of its anatomy, physiology and behaviour) but is not white. Indeed, this is precisely what happened when early White explorers returned from Australia with tales of black swans.

Popper's ideas easily give rise to a view of science in which knowledge accumulates over time as new theories are proposed and new data collected to distinguish between conflicting theories. Much school experimentation in science is Popperian: we see a rainbow and hypothesise that white light is split up into light of different colours as it is refracted through a transparent medium (water droplets); we test this by attempting to refract white light through a glass prism; we find the same colours of the rainbow are produced and our hypothesis is confirmed. Until some new evidence causes it to be falsified, we accept it (Reiss 2008).

Thomas Kuhn made a number of seminal contributions, but he is most remembered nowadays by his argument that while the Popperian account of science holds well during periods of *normal science* when a single paradigm holds sway, such as the Ptolemaic model of the structure of the solar system (in which the Earth is at the centre) or the Newtonian understanding of motion and gravity, it breaks down when a scientific *crisis* occurs (Kuhn 1970). At the time of such a crisis, a scientific revolution happens during which a new paradigm, such as the Copernican model of the structure of the solar system or Einstein's theory of relativity, begins to replace (initially to coexist with) the previously accepted paradigm. The central point is that the change of allegiance from scientists believing in one paradigm to their believing in another cannot, Kuhn argues, be fully explained by the Popperian account of falsifiability.

A development of Kuhn's work was provided by Lakatos (1978) who argued that scientists work within research programmes. A research programme consists of a set of core beliefs surrounded by layers of less central beliefs. Scientists are willing to accept changes to these more peripheral beliefs so long as the core beliefs can be defended. So, in biology, we might see in contemporary genetics a core belief in the notion that development proceeds via a set of interactions between the actions of genes and the influences of the environment. At one point, it was thought that the passage from DNA to RNA was unidirectional. Now we know (reverse transcriptase, etc.) that this is not always the case. The core belief (that development proceeds via a set of interactions between the actions of genes and the influences of the environment) remains unchanged, but the less central belief (that the passage from DNA to RNA is unidirectional) is abandoned.

The above account of the nature of science portrays science as what John Ziman (2000) has termed 'academic science'. Ziman argues that such a portrayal was reasonably valid between about 1850 and 1950 in European and American universities but that since then we have entered a phase largely characterised by 'post-academic science'. Post-academic science is increasingly transdisciplinary and utilitarian, with a requirement to produce value for money. It is more influenced by politics, it is more industrialised and it is more bureaucratic. The effect of these changes is to make the boundaries around the domain of science a bit fuzzier. Of course, if one accepts the contributions of the social study of science (e.g. Yearley 2005), one finds that these boundaries become fuzzier still. My argument in this chapter does not *rely* on such a reading of science though someone who is persuaded by the 'Strong Programme' within the sociology of scientific knowledge (i.e. the notion that even valid scientific theories are amenable to sociological investigation of their truth claims) is much more likely to accept the worth of science educators considering the importance of religion as one of many factors that influence the way science is practised and scientific knowledge produced.

I am very aware that to many science educators even raising the possibility that religion might be considered within science raises suspicions that this is an attempt to find a way of getting religion into the science classroom for religious rather than scientific reasons. This is not my intention. In terms of the nature of science, considering religion is useful simply for helping learners better understand why certain things come under the purview of science and others don't.

Consider, first, the scriptures as a source of authority. To the great majority of religious believers, the scriptures of their religion (the Tanakh, the Christian Bible, the Qur'an, the Vedas, including the Upanishads, the Guru Granth Sahib, the various collections in Buddhism, etc.) have an especial authority by very virtue of being a scripture. This is completely different from the authority of science. Newton's *Principia* and Darwin's *On the Origin of Species* are wonderful books, but they do not have any permanence other than that which derives from their success in explaining observable phenomena of the material world. Indeed, as is well known, Darwin knew almost nothing of the mechanism of inheritance despite the whole of his argument relying on inheritance, so parts of *The Origin* were completely out of date over a 100 years ago.

Then consider the possibility of miracles where we use miracle not in its everyday sense (and the sense in which it is sometimes used in scripture), namely, 'remarkable', 'completely unexpected' or 'wonderful', but in its narrower meaning of 'contrary to the laws of nature'. Scientists can react to this latter notion of miracles in one of two ways: either miracles are impossible (because they are contrary to the laws of nature) or they are outside of science (because they are contrary to the laws of nature).

I hold that it can be a useful exercise with some students for science educators to get students to consider whether such topics as astrology, ghosts, paranormal phenomena and miracles fall within the scope of science or not. The aim, I would again emphasise, is not to smuggle such topics into science but to get students more rigorously to think about what science is and how it proceeds.

Teaching About Evolution

The Scientific Consensus Concerning Evolution

As with any large area of science, there are parts of what we might term 'front-line' evolution that are unclear, where scientists still actively work attempting to discern what is going on or has gone on in nature. But much of evolution is not like that. Evolution is a well-established body of knowledge that has built up over 150 years as a result of the activities of many thousands of scientists. The following are examples of statements about evolution that lack scientific controversy:

- All of today's life on Earth is the result of modification by descent from the simplest ancestors over a period of several thousand million years.
- Natural selection is a major driving force behind evolution.
- Evolution relies on the inheritance of genetic information that helps its possessor to be more likely to survive and reproduce.
- Most inheritance is vertical (from parents) though some is horizontal (e.g. as a result of viral infection carrying genetic material from one species to another).
- The evolutionary forces that gave rise to humans do not differ in kind from those that gave rise to any other species (Reiss 2013).

For those, such as I, who accept such statements and the theory of evolution, there is much about the theory of evolution that is intellectually attractive. For a start, a single theory provides a way of explaining a tremendous range of observations; for example, why it is that there are no rabbits in the Precambrian, why there are many superficial parallels between marsupial and placental mammals, why monogamy is more common in birds than in fish and why sterility (e.g. in termites, bees, ants, wasps and naked mole rats) is more likely to arise in certain circumstances than in others. Indeed, I have argued elsewhere that evolutionary biology can help with some theological questions, including the problem of suffering (Reiss 2000a).

Rejecting Evolution

The theory of evolution is not a single proposition that a person either wholly accepts or wholly rejects. At one pole are materialists who, eschewing any sort of critical realist distinction between the empirical, the actual and the real (Bhaskar 1978), maintain that there is no possibility of anything transcendent lying behind what we see of evolution in the results of the historical record (fossils, geographical distributions, comparative anatomy and molecular biology) and today's natural environments and laboratories. At the other pole are the advocates of creationism as inspired by a literal reading of certain scriptures. But in between lie many others including those who hold that evolutionary history can be providential as human history is.

In addition, there are a whole set of nonreligious reasons why someone may actively reject aspects of the theory of evolution. After all, it may seem to defy common sense to suppose that life in all its complexity has evolved from non-life. And then there is the tremendous diversity of life we see around us. To many it hardly seems reasonable to presume that giant pandas, birds of paradise, spiders, orchids, flesh-eating bacteria and the editors of this book all share a common ancestor – yet that is what mainstream evolutionary theory holds.

It is, though, for religious reasons that many people reject evolution. Creationism exists in a number of different forms but something like 50 % of adults in Turkey, 40 % in the USA and 15 % in the UK reject the theory of evolution and believe that the Earth came into existence as described by a literal (i.e. fundamentalist) reading of the early parts of the Bible or the Quran and that the most that evolution has done is to change species into closely related species (Miller et al. 2006; Lawes 2009). Christian fundamentalists generally hold that the Earth is nothing like as old as evolutionary biologists and geologists conclude – as young as 10,000 years or so for Young Earth creationists. For Muslims, the age of the Earth is much less of an issue.

Allied to creationism is the theory of intelligent design. While many of those who advocate intelligent design have been involved in the creationism movement, to the extent that the US courts have argued that the country's First Amendment separation of religion and the state precludes its teaching in public schools (Moore 2007), intelligent design can claim to be a theory that simply critiques aspects of evolutionary biology rather than advocating or requiring religious faith. Those who promote intelligent design typically come from a conservative faith-based position (though there are atheists who accept intelligent design). However, in their arguments against evolution, they typically make no reference to the scriptures or a deity but argue that the intricacy of what we see in the natural world, including at a subcellular level, provides strong evidence for the existence of an intelligence behind this (e.g. Meyer 2009). An undirected process, such as natural selection, is held to be incapable of explaining all such intricacy.

Evolution in School Science

Few countries have produced explicit guidance as to how schools might deal with the issues of creationism or intelligent design in the science classroom. One country that has is England (Reiss 2011). In the summer of 2007, after months of behind-the-scenes meetings and discussions, the then DCSF (Department of Children, Schools and Families) Guidance on Creationism and Intelligent Design received ministerial approval and was published (DCSF 2007). The Guidance points out that the use of the word 'theory' in science (as in 'the theory of evolution') can mislead those not familiar with science as a subject discipline because it is different from the everyday meaning, when it is used to mean little more than an idea. In science the word indicates that there is a substantial amount of supporting evidence, underpinned by principles and explanations accepted by the international scientific community.

The DCSF Guidance goes on to state 'Creationism and intelligent design are sometimes claimed to be scientific theories. This is not the case as they have no underpinning scientific principles, or explanations, and are not accepted by the science community as a whole' (DCSF 2007) and then goes on to say:

> Creationism and intelligent design are not part of the science National Curriculum programmes of study and should not be taught as science. However, there is a real difference between teaching 'x' and teaching *about* 'x'. Any questions about creationism and intelligent design which arise in science lessons, for example as a result of media coverage, could provide the opportunity to explain or explore why they are not considered to be scientific theories and, in the right context, why evolution is considered to be a scientific theory. (DCSF 2007)

This seems to me a key point and one that is independent of country, whether or not a country permits the teaching of religion (as in the UK) or does not (as in France, Turkey and the USA). Many scientists, and some science educators, fear that consideration of creationism or intelligent design in a science classroom legitimises them. For example, the excellent book *Science, Evolution, and Creationism* published by the US National Academy of Sciences and Institute of Medicine asserts, 'The ideas offered by intelligent design creationists are not the products of scientific reasoning. Discussing these ideas in science classes would not be appropriate given their lack of scientific support' (National Academy of Sciences and Institute of Medicine 2008, p. 52).

As I have argued (Reiss 2008), I agree with the first sentence of this quotation but disagree with the second. Just because something lacks scientific support doesn't seem to me a sufficient reason to omit it from a science lesson. Nancy Brickhouse and Will Letts (1998) have argued that one of the central problems in science education is that science is often taught 'dogmatically'. With particular reference to creationism, they write:

> Should student beliefs about creationism be addressed in the science curriculum? Is the dictum stated in the California's *Science Frameworks* (California Department of Education, 1990) that any student who brings up the matter of creationism is to be referred to a family member of member of the clergy a reasonable policy? We think not. Although we do not

believe that what people call 'creationist science' is good science (nor do scientists), to place a gag order on teachers about the subject entirely seems counterproductive. Particularly in parts of the country where there are significant numbers of conservative religious people, ignoring students' views about creationism because they do not quality as good science is insensitive at best. (Brickhouse and Letts 1998, p. 227)

It seems to me that school biology and earth science lessons should present students with the scientific consensus about evolution and that parents should not have the right to withdraw their children from such lessons. Part of the purpose of school science lessons is to introduce students to the main conclusions of science – and the theory of evolution is one of science's main conclusions. At the same time, science teachers should be respectful of any students who do not accept the theory of evolution for religious (or any other) reasons. Indeed, nothing pedagogically is to be gained by denigrating or ridiculing students who do not accept the theory of evolution.

My advice for science teachers is not to get into theological discussions, for example, about the interpretation of scripture. Stick to the science and if you are fortunate enough to have one or more students who are articulate and able to present any of the various creationist arguments against the scientific evidence for evolution (e.g. that the theory of evolution contradicts the second law of thermodynamics, that radioactive dating techniques make unwarranted assumptions about the constancy of decay rates, that evolution from inorganic precursors is impossible in the same way that modern science disproved theories of spontaneous generation), use their contributions to get the rest of the group to think rigorously and critically about such arguments and the standard accounts of the evidence for evolution.

My own experience of teaching the theory of evolution for some 30 years to school students, undergraduate biologists, trainee science teachers, members of the general public and others is that people who do not accept the theory of evolution for religious reasons are most unlikely to change their views as a result of one or two lessons on the topic, and others have concluded similarly (e.g. Long 2011). However, that is no reason not to teach the theory of evolution to such people. One can gain a better understanding of something without necessarily accepting it. Furthermore, some studies suggest that careful and respectful teaching about evolution can indeed make students considerably more likely to accept at least some aspects of the theory of evolution (Winslow et al. 2011).

Evolution in Science Museums

Education about evolution does not only take place in schools. It takes place through books, magazines, TV, the Internet, radio and science museums. Science museums have long had exhibits about evolution. Tony Bennett (2004) provides an historical analysis to look at how science museums have presented evolution.

Using a Foucauldian framework of governmentality, he attempts to discern the modes of power that lie behind the manifestations of particular forms of knowledge. Bennett concludes that:

> In their assembly of objects in newly historicised relations of continuity and difference, evolutionary museums not only made new pasts visible; they also enrolled those pasts by mobilising objects – skulls, skeletons, pots, shards, fossils, stuffed birds and animals – for distinctive social and civic purposes. (Bennett 2004, p. 189)

In one sense, this is hardly surprising – museums have to make selections about what to display and how to curate such displays, and these are clearly cultural decisions whether one is referring evolution or anything else. However, visitor to science museums can easily presume that they are being presented with objective fact.

Monique Scott too has produced a book about evolution in museums (Scott 2007) though her work, unlike Bennett's, has more to do with the present than with history. Using questionnaires and interviews, Scott gathered the views of nearly 500 visitors at the Natural History Museum in London, the Horniman Museum in London, the National Museum of Kenya in Nairobi and the American Museum of Natural History in New York. Perhaps her key finding is that many of the visitors interpreted the human evolution exhibitions as providing a linear narrative of progress from African prehistory to a European present. As she puts it:

> Despite the distinctive characters of each of the four museums considered here and the specific cultural differences among their audiences, it is clear that museums and their visitors traffic in common anthropological logic – namely the color-coded yardstick of evolutionary progress. In fact, visitors equipped with a weighty set of popular images – imagery derived from such things as *Condé Nast Traveler* magazines, *Planet of the Apes* films, and *National Geographic* images – occupy the nexus between the evolutionary folklore circulating outside the museum and that which has been generated within it. This collection of images often urges Western museum visitors to negotiate between the "people who stayed behind" and their own fully evolved selves (defined often by such culturally coded "evolutionary leaps" as clean-shaven-ness and white skin). (Scott 2007, p. 148)

So how might one hope that science museums would treat religion when putting together exhibitions about evolution? Museums have a number of advantages over classroom teachers; for one thing, they have much more time in which to prepare their teaching. So we might hope that a science museum, while not giving the impression that the occurrence of evolution is scientifically controversial today (it isn't), might convey something of the history of the theory of evolution. This would include the fact that evolution was once scientifically controversial and that religious believers have varied greatly as to how they have reacted to the theory of evolution. On the one hand, we have today's creationists; on the other, we have Charles Kingsley, the Anglican divine and friend of Charles Darwin who read a prepublication copy of *On the Origin of Species* and wrote to Darwin:

> I have gradually learnt to see that it is just as noble a conception of Deity, to believe that he created primal forms capable of self development into all forms needful pro tempore & pro loco, as to believe that He required a fresh act of intervention to supply the lacunas w^h. he himself had made. (Kingsley 1859)

Of course, there are an increasing number of creationist museums (e.g. http://creationmuseum.org/) and zoos (e.g. www.noahsarkzoofarm.co.uk/). Perhaps somewhat optimistically, I would ask those running such creationist places of learning to make one concession to evolution. I do not expect them to promote evolution but it is reasonable to ask them to make it clear that the scientific consensus is that the theory of evolution and not creationism is the best available explanation for the history and diversity of life. Of course, it is perfectly acceptable for those running creationist institutions to critique evolution and to try to persuade those visiting such institutions that the standard evolutionary account is wrong. But just as science teachers with no religious faith should respect students who have creationist views, so creationists should not misrepresent creationism as being in the scientific mainstream. It is not.

Conclusions

Science education for diversity has long striven to take account of issues to do with gender, socio-economic class, ethnicity and disability. However, it has traditionally made rather less effort to consider issues to do with religious faith (Reiss 2000b). In a well-known mapping of the possible ways in which the relationship between science and religion might be understood, Barbour (1990) suggested four: conflict, independence, dialogue and integration. As is evident from the above, there is a tension whether the relationship is understood epistemologically or ontologically. I am happy to identify as someone who, while holding that science and religion are ontologically integrated, believes that epistemologically it makes considerable sense to treat them as independent. Of course, others will see the relationship between science and religion differently. Science education needs to take account not only of student diversity but also of teacher diversity. We should strive for curricula, for pedagogies and for assessment regimes that are respectful of science, of learners and of teachers.

References

Atkins, P. (2011). *On being: A scientist's exploration of the great questions of existence.* Oxford: Oxford University Press.

Barbour, I. G. (1990). *Religion in an age of science: The Gifford Lectures 1989–1991* (Vol. 1). London: SCM.

Bennett, T. (2004). *Pasts beyond memory: Evolution, museums, colonialism.* London: Routledge.

Bhaskar, R. (1978). *A realist theory of science.* Sussex: Harvester Press.

Brickhouse, N. W., & Letts, W. J., IV. (1998). The problem of dogmatism in science education. In J. T. Sears & J. C. Carper (Eds.), *Curriculum, religion, and public education: Conversations for an enlarging public square* (pp. 221–230). New York: Teachers College, Columbia University.

Buss, D. M. (2003). *The evolution of desire: Strategies of human mating, revised ed.* New York: Basic Books.

California Department of Education (1990). *Science framework for California public schools.* Sacramento, CA.

DCSF. (2007). *Guidance on creationism and intelligent design.* http://webarchive.nationalarchives.gov.uk/20071204131026/, http://www.teachernet.gov.uk/docbank/index.cfm?id=11890. Accessed 22 Jan 2012.

Jones, A., McKim, A., & Reiss, M. (Eds.). (2010). *Ethics in the science and technology classroom: A new approach to teaching and learning.* Rotterdam: Sense.

Kingsley, C. (1859, November 18). *Letter to Charles Darwin.* http://www.darwinproject.ac.uk/entry-2534. Accessed 29 Jan 2012.

Kuhn, T. S. (1970). *The structure of scientific revolutions* (2nd ed.). Chicago: University of Chicago Press.

Lakatos, I. (1978). *The methodology of scientific research programmes.* Cambridge: Cambridge University Press.

Lawes, C. (2009). *Faith and Darwin: Harmony, conflict, or confusion?* London: Theos.

Lederman, N. G. (2007). Nature of science: Past, present, and future. In S. K. Abell & H. G. Lederman (Eds.), *Handbook of research on science education* (pp. 831–879). Mahwah: Lawrence Erlbaum.

Long, D. E. (2011). *Evolution and religion in American education: An ethnography.* Dordrecht: Springer.

Merton, R. K. (1973). *The sociology of science: Theoretical and empirical investigations.* Chicago: University of Chicago Press.

Meyer, S. C. (2009). *Signature in the cell: DNA and the evidence for Intelligent Design.* New York: HarperCollins.

Miller, J. D., Scott, E. C., & Okamoto, S. (2006). Public acceptance of evolution. *Science, 313*, 765–766.

Moore, R. (2007). The history of the creationism/evolution controversy and likely future developments. In L. Jones & M. J. Reiss (Eds.), *Teaching about scientific origins: Taking account of creationism* (pp. 11–29). New York: Peter Lang.

National Academy of Sciences and Institute of Medicine. (2008). *Science, evolution, and creationism.* Washington, DC: National Academies Press.

Popper, K. R. (1934/1972). *The logic of scientific discovery.* London: Hutchinson.

Reiss, M. J. (2000a). On suffering and meaning: An evolutionary perspective. *Modern Believing, 41*(2), 39–46.

Reiss, M. (2000b). Teaching science in a multicultural, multi-faith society. In J. Sears & P. Sorensen (Eds.), *Issues in the teaching of science* (pp. 16–22). London: RoutledgeFalmer.

Reiss, M. J. (2008). Should science educators deal with the science/religion issue? *Studies in Science Education, 44*, 157–186.

Reiss, M. J. (2011). How should creationism and intelligent design be dealt with in the classroom? *Journal of Philosophy of Education, 45*, 399–415.

Reiss, M. J. (2013). Beliefs and the value of evidence. In J. K. Gilbert & S. M. Stocklmayer (Eds.), *Communication and engagement with science and technology: Issues and dilemmas* (pp. 148–161). New York: Routledge.

Scott, M. (2007). *Rethinking evolution in the museum: Envisioning African origins.* London: Routledge.

Winslow, M. W., Staver, J. R., & Scharmann, L. C. (2011). Evolution and personal religious belief: Christian university biology-related majors' search for reconciliation. *Journal of Research in Science Teaching, 48*, 1026–1049.

Yearley, S. (2005). *Making sense of science: Understanding the social study of science.* London: Sage.

Ziman, J. (2000). *Real science: What it is and what it means.* Cambridge: Cambridge University Press.

Students' Perceptions of Apparent Contradictions Between Science and Religion: Creation Is Only the Beginning

Berry Billingsley

Introduction

Within a typical class of secondary school pupils, there are likely to be pupils who hold beliefs which they associate with their religion. In my research (first at PhD level and now via a more substantial project which will be described shortly), I am interested to know how and whether pupils' religious beliefs interact with the teaching they receive in science. When I began teaching science in secondary school with some knowledge of religion and particularly Christianity, I presumed that in the minds of my pupils, such interactions would be taking place. Newtonian mechanics seemed to describe a universe where one thing follows another in ways that can be determined; religion, it seemed to me, described a world in which nature is at the command of a Creator. It seemed possible that if pupils reflected on what they had learnt in my physics lessons, some would perceive these descriptions as conflicting. Secondly, would pupils be concerned, I wondered, to know whether religious ideas about how the world works should be subject to the same process of testing as scientific ideas, particularly when science and religion are addressing a common topic, such as how the universe began? Fortunately, perhaps, in my own practice as a science teacher, these questions were largely hypothetical as questions of this type rarely arose. This observation became one of the points of interest in my research. Why were pupils not inclined to ask these kinds of questions and how, as a science teacher, should I respond if at all to their silence?

In this chapter I will present three ways to conceptualise the interactions that can take place between science and religion. The first conceptualisation looks at the epistemological relationships between science and religion; for the second, I examine children's ideas about the epistemological relationships between science and

B. Billingsley (✉)
Institute of Education, University of Reading, Reading, UK
e-mail: b,billingsley@reading.ac.uk

N. Mansour and R. Wegerif (eds.), *Science Education for Diversity: Theory and Practice*,   329
Cultural Studies of Science Education 8, DOI 10.1007/978-94-007-4563-6_16,
© Springer Science+Business Media Dordrecht 2013

religion; for the third, I draw on comments made by pupils about their perceptions of what happens in their classrooms to discuss how the relationships between science and religion might be spoken about in secondary school classrooms. My aim in this chapter is to show, using selected illustrations, that there are cognitive and social barriers that seem in practice to prevent pupils from exploring the questions that are described in scholarship.

How Do Science and Religion Relate: From an Epistemic Perspective

Descriptions of the relationships between science and religion vary and there are some people who say that there is no relationship to consider as science and religion are concerned with different matters entirely. It is the case, however, that in public forums science and religion are not always discussed in ways that suggest they are distinct. At the very least, some religions are perceived to make claims that are about the physical world, not just spiritual and moral matters, while some scientists are vocal about the supernatural world. Dawkins (2006, p. 57), for example, advises readers that 'I am not attacking any particular version of God or gods. I am attacking God, all gods, anything and everything supernatural, wherever and whenever they have been or will be invented'.

Barbour (2000) and Brooke (1991) have each provided historical reviews to the present day of the stances taken in academic and popular literature about the relationships between science and religion. Both authors identify that what makes the theme particularly complex is that there is no universal agreement on the natures of science and religion which means that, when seeking to understand the relationships between them, many different but internally consistent perspectives can be given.

As Brooke (1991) notes:

'There is no such thing as the relationship between science and religion. It is what different individuals and communities have made of it in a plethora of different contexts'. (p. 321)

In his review, Brooke groups the approaches seen in the literature into three themes. The warfare theme or conflict model is the first. Here, says Brooke, writers claim that characteristically and fundamentally science and religion are opposed. This, however, says Brooke, offers 'a historical reconstruction that is only concerned with extreme positions' (p. 35).

The second theme in Brooke's scheme is that science and religion are complementary on the basis that they ask and address different kinds of questions. Applied to the context of origins, science and religion are said to be complementary or independent because religion is concerned not with scientific but with teleological explanations for our existence, thus 'why' we are here, rather than 'how' (Bauser and Poole 2002). The accounts presented in the Judaeo-Islamic-Christian story of Creation are argued to be figurative not literal (Berry 1996). Alexander (2008, p. 44), writing about the Galileo affair, argues that 'The moral of the tale is that we should be resistant to the idea that biblical passages can be removed from their original contexts to score scientific points'.

Brooke's third model of the relationship refers to even warmer interactions between the two fields. Brooke says:

> 'Contrary to the first – the conflict model – it is asserted that certain religious beliefs may be conducive to scientific activity. And contrary to the second – the separationist position – it is argued that interaction between religion and science, far from being detrimental, can work to the advantage of both'. (p. 4)

In the scholarly literature, then, the relationship between science and religion is often presented as something of an epistemic conundrum. To make sense of the arguments offered in this debate, pupils need a conceptual framework that sees 'different' as potentially 'independent'. Only then can pupils understand the reasoning which underpins the view that science and religion are not necessarily incompatible.

Young People's Thinking

There is a body of research which indicates that the view that science and religion are opposed is widespread among young people (Astley and Francis 2010; Billingsley et al. 2012; Francis et al. 1990; Fulljames 1996; Fulljames et al. 1991; Fulljames and Stolberg 2000; Taber et al. 2011a, b).

Of particular interest here is a study by Roth and Alexander (1997) which looked at high school students' thinking about controversial issues such as abortion, euthanasia and the origins of humankind. The participants were 23 boys in a Canadian boarding school. Three boys were selected to take part in in-depth interviews because their views were felt to be representative of the different positions held by the cohort.

Brent saw science and religion as exclusive ways to understand reality and rejected his view of science. He is cited as saying, for example, 'From my perspective there are no similarities between science and religion at all . . . Science completely goes against what God has created, the so called power person who created everything; and people who are in the sciences are saying that 'you know, it wasn't him at all.' To me that is a direct insult' (Roth and Alexander 1997, p. 142).

For Brent, the worldviews presented by science and religion were so discordant that they were perceived to be exclusive. This notion that scientific explanations replace religious ones seems to stem from a perception that science and religion have apparently dichotomous ways to describe how things happen in the world; thus,

> 'Science it is said operates as a worldview that regards natural phenomena as the produce of impersonal forces. By contrast, religious and magical systems involve personalised gods, spirits or demons'. (Brooke 1991, p. 17)

In contrast, a second pupil, Todd, said that science and religion are both valid. We are told that Todd's view rested on a sense that science explained the workings of nature but did not explain the spiritual connection he felt with the beauty of nature. In Todd's view, scientists who reject religion do so because 'people in science who are atheist have not had a good experience with religion or either they became so rational that they ignore the emotional and the spiritual side' (Roth and Alexander 1997, p. 129).

The case of a third pupil, Ian, highlights a different way of responding to what are perceived by a pupil to be two conflicting worldviews. We are told that for Ian, 'Institutional science and religion were incompatible and he kept the domains clearly separate' (p. 134). This strategy of conscious compartmentalisation between conflicting schema has been noted by Jegede (1997) in his study of the responses to science teaching by pupils with African cultural backgrounds. It was also a strategy that was found in a study I carried out in Australia of undergraduate perceptions of the relationships between science and religion (Billingsley 2004). For that study, 40 undergraduates were selected on a convenience basis from the refractory area of a city university to each take part in a 45-min interview about science and religion. One student, for example, said, 'If I'm thinking about religion, I take a religious kind of view, but if I'm thinking in a science way, I take the science view' (p. 283).

The aim of the Australian study was to map the different approaches that students took when they were asked to consider science-religion dilemmas. These dilemmas were instances where some people say that science and religion conflict. The dilemmas looked at the beginning of the universe and the beginning of life, whether God can change what happens in response to human prayer and whether supernatural miracles are acceptable within a scientific worldview. In addition to the strategy of conscious compartmentalisation of science and religion, a number of other approaches were noted. Three students in the sample of 40 undergraduates explained that until the interview, they had had not cause to compare the beliefs they associated with science and those they associated with religion. These comments, each by a different student, illustrate the finding:

'I haven't thought about them together like this before. I've always thought of religion and science and they're totally different'. (p. 214)

'Either something greater made us or evolution made us from mud. It's too hard and you can't bring them together. I've not thought about this before. Now I think they are linked. I used to think they were separate'. (p. 214)

'I guess because I haven't done a lot of science studies, I haven't been forced to question everything I believe... and I've been able to say, yeah, there's no contradiction still for me'. (p. 230)

One way to contextualise these findings is to refer to sociocultural theory, in which it is said that the brain preserves a sense of the social and cultural settings in which learning takes place (Edwards 2009; Fawns and Sadler 1996; Wegerif et al. 1999).

The study also looked at how, where contradictions were perceived to exist between science and religion, students decided what to believe. The following comments, each by a different student, illustrate the varying criteria that seemed to apply:

'I am looking for an answer that catches my interest. I'm an imaginative kind of person; I'm not someone who cares if it's wrong or right. I like reading science fiction and I like the imagination of it'.

'I am seeking a view that makes me happy'. (p. 251)

'I am looking for an answer that is consistent with scientific evidence but still leaves room for religious faith'. (p. 252)

'I've seen a lot of evidence for the scientific version and not much for the religious viewpoint. I'm an evidence girl'. (p. 283)

In the following example, the student's rationale is that a universe without meaning would be too 'depressing' to be true:

'I still tend to ignore the evolution side of it and go more on the religious side of it because the evolution side doesn't give you any answers about why man is here anyway. It may be possible we're not here for any reason at all and we're just some kind of creation that came by itself, but how depressing is that. We go to school and we go to work then we retire and then we die. There's got to be more to it than that. So that's why I go for the God side of it, not the science side of it'. (p. 252)

These studies highlight that in a number of ways, young people's thinking about the relationships between science and religion can differ from the models that are commonly described in scholarly literature. I argued previously that Brooke (1991) argues for the importance of noticing that individuals are drawing on different interpretations of science and religion when they form a view of the relationship. Here, I am highlighting that some young people have not begun to make connections between their scientific and religious beliefs, while some others hold the ideas they associate with each of science and religion consciously apart. Among those in this study who had examined their perceptions of what science and religion say, some drew on objective reasoning, while some others described their emotional well-being as their overriding concern. In the next section, I move from looking at ways of thinking about science and religion to consider how the relationships between science and religion might be discussed in classrooms.

Moving into a School Setting

For this section I will refer to a study recently undertaken in England as part of the LASAR (Learning about Science and Religion) Project (Taber et al. 2011a, b). The LASAR Project is a collaborative research project by the Institute of Education at University of Reading and the Faraday Institute for Science and Religion, St. Edmund's College, Cambridge, funded for 3 years by the John Templeton Foundation. The project was motivated by a concern that there is a strong public perception, reinforced on a more than occasional basis by the popular media, that science and religion are in some sense opposed. We were concerned that school pupils may come to see science as, to an extent, an atheistic activity, a perspective which could deter students who hold religious faith from considering science as a suitable basis of future study and career.

The LASAR Project sought to find out more about what secondary age students do think about science and religion and the factors which they feel influence their views.

In England, science and Religious Education (RE) are compulsory curriculum subjects for pupils up to the age of 16, although parents can choose to withdraw their children from RE lessons. In the study described here, semi-structured interviews with pupils and teachers were organised in four secondary schools in different national regions, identified using an educational directory (Tierney et al. 2005).

The selection of three pupils in each school was made by a teacher appointed to the task by the head teacher. During transcription, the schools and pupils were renamed. The details of those discussed here are as follows:

- Alisha is a pupil at Abbey School, a Church school in the centre of a small city.
- Brenda is a pupil at Borough School, a large comprehensive in the suburb of a major city.
- Chas, Christine and Colin are pupils at Ceeside School, a smaller comprehensive school in a coastal town.
- David and Dean are pupils at Dalesford Grammar, a state-maintained grammar school in a rural town.

One of the aims in gathering the data was to look at whether when it comes to this theme (science and religion), the compartmentalisation of teaching is an added challenge for pupils who are already struggling to see how the two domains relate. None of the pupils in this sample of 12 had experienced collaborative teaching. Among the 12, some said that they had never looked at the relationships between science and religion in their lessons, while others said that the theme had been addressed.

Chas, for example, said, 'we've never done like science in religion ... we don't do science and religion, we don't bond them together, we have two different lessons'.

Alisha said that her experience of lessons is that

'they've never put religion and science together' and that 'I think like the science teachers ... do try and like avoid them [discussions on the theme] a bit because what- if they do like answer ... people could be against it because of their religion could be different'.

One of the pupils who reported a session that had explored the relationship was Colin. He said, 'We did have a very detailed discussion in science about what we believed in, about religion, and science, and comparing them together – it was really interesting'.

The significance of an opportunity to explore the relationship within lessons seemed to be in evidence in this comment by Christine, who became aware of the potential for science and religion to be linked only when she discussed her thinking about the relationship to that point:

> 'I've met religion and I do science but I've never had 'em both together, like. I never knew it could link in such a way'.

There was also evidence in the interviews that unless a science teacher introduces the theme, some pupils will perceive that questions about science and religion are not appropriate ones to raise in science lessons. Brenda told us:

> 'We don't really talk about RS in science, I don't think the teacher really brings it up, and no-one ever asks about it, so there's no need for her to bring it up. And the same with RS, no-one really asks the science questions because you'd really more ask your science teacher about that instead of asking your RE teacher'.

In the study of English pupils' thinking, David also presented a view which suggests that questions are only asked in what pupils deem to be the appropriate place. He said:

> 'We don't ask science teachers questions any more at the moment, because we don't think that they'd answer them. We wouldn't have thought (pause) – oh they won't answer that because it's not on their topic'.

The notion that pupils know and keep the rules of the classroom fits into a wider theme called 'the grammar of schooling' (Tyack and Tobin 1994). Bernstein (1971, 2000) devised the concept of *frame* to describe the strength of the boundary, understood by pupils and teacher, between what may be taught and learned in a subject and what may not. For well-defined subjects like the sciences, argues Bernstein, these boundaries are sharp.

The absence of questions about the relationships between science and religion in science lessons is not necessary because pupils see scientific and religious explanations as complementary and/or unproblematic but may for some be more to do with the 'grammar of schooling' which persuades pupils to keep 'inappropriate' questions to themselves. For example, Brenda told us 'I guess I've just been taught about God and everything, so I guess that's what I've been taught to believe, so I just believe it'. Did this mean that during science lessons, questions or concerns relating to religious ideas did not occur? This did not appear the case. Brenda said:

> 'Like the Big Bang theory – I don't believe in that one. I think some theories – most theories are true but like some of them I think are just made up, because they can't find any other explanation so they kind of think to try and explain it'.

And then, when Brenda is asked how she thinks the universe began, she explains:

> 'I think it was the way it says in the Bible'.

The explanation that Brenda gives for rejecting the Big Bang is interesting. It is a source of frustration to many scientists that the Big Bang theory is shrugged off by a significant percentage of people on the basis that it is 'just a theory' (Hogan 1998). In this pupil's thinking, the decision to reject the Big Bang theory seems tied into what is arguably the student's view that any scientific explanation comes as an alternative to Divine Creation. A number of researchers have looked at how pupils who have worldviews that are in a sense incongruent with a scientific view respond to the teaching they receive in science lessons (Aikenhead 1996; Aikenhead and Jegede 1999; Atwater 1996; Jegede 1997; Jegede and Okebukola 1991). Their studies have highlighted that the goal for a significant proportion of pupils is to succeed at school without 'expending excessive time or effort' (Loughran and Derry 1997, p. 935). As such, what is committed to memory in science lessons can for some pupils be a 'ritual' that is undertaken only for the purpose of passing examinations (Atwater 1996). With this in mind, it is interesting to note that Fysh and Lucas (1998) interviewed secondary school pupils in Australia and reported that while many adolescents felt that contradictions exist between science and religion, only nine of the 44 students interviewed in the study said there were more than 'rare clashes' between science and religious content in the classroom.

One of the challenges for individuals when they consider what science and religion say about a given topic is whether all claims should be subject to the same methods of testing (Barbour 2000). One student, Alisha, saw science and religion as competing on the basis that they seem to have different methods to verify their claims.

> 'I think in a way science and RE is kind of like rivals because they do – I think they would contradict each other because in RE they say you're meant to believe this because of God's word [whereas] in science they say well no, there must be a reason, so it's kind of hard, it's kind of like sides to sort of choose'.

Alisha's perception suggests that she sees science and religion as ways to address a common set of questions but that their criteria for justifying those answers are unmatched. Dean, when he gave his perspective on why questions about religion are not usually discussed in science lessons, also referred to what he perceived as the contrasting natures of science and religion:

> 'Science is – they want to tell you facts – they want to get you to learn equations, sort of thing. They d-don't want to talk about things that (pause) can change from individual to individual'.

The choice of words by these pupils seems to suggest that in their minds, science is associated with the science classroom and religion with the RE classroom. It is perhaps interesting to notice that in Alisha's comment, science and RE are described as 'rivals' which perhaps reflects a perception for this pupil that each classroom is the proponent of one view and opposed to the other.

Recommendations

In this chapter I have sought to highlight through a selection of illustrations the importance of noticing the social constraints that operate in classrooms when studying pupils' developing thinking about the relationships between science and religion. The comments presented here are not intended to be representative of the range of views held by young people, but are instead intended to reflect aspects of pupils' thinking that can be hidden from view in the classroom. Further, although the examples here are a small sample of the range of views that is likely to exist, it is interesting to notice some of the criteria used by these pupils to decide which questions to ask. Previously, I cited David as saying that he felt that teachers would dismiss questions that are 'not on their topic'. I also drew attention to Alisha's perception that science teachers would not welcome discussions about science and religion because 'people could be against it because of their religion could be different'. What these comments seem to suggest is that when making judgements about which questions to raise and which to keep silent, some pupils are drawing on their personal inferences of what their teachers want to hear. It seems to follow that if a teacher wants to encourage pupils to voice questions about the relationships between science and religion, then this may be achieved by telling pupils that such questions are welcome.

These recommendations build on a premise that it is important to ensure pupils have access via their school education to a range of perceptions of how science and religion relate. This is something that I would argue for on the basis that there is likely to be a proportion of pupils who do not have access outside the school setting to the view that science and religion are not necessarily incompatible. For these pupils, the notion of science as necessarily an atheistic perspective seems a distinct possibility. It also seems reasonable to suggest that where this perception exists, there is a strong chance that it will rest unchallenged. In so saying I am reminded of a conversation I experienced long ago when I was a student hoping for holiday work. I telephoned one potential employer to enquire if my letter had arrived safely. The employer's reply was 'yes and I thought my silence said it all'.

References

Aikenhead, G. (1996). Science education: Border crossing into the subculture of science. *Studies in Science Education, 27*(1), 1–52.

Aikenhead, G., & Jegede, O. (1999). Cross-cultural science education: A cognitive explanation of a cultural phenomenon. *International Journal of Research in Science Teaching, 36*(3), 269–287.

Alexander, D. (2008). *Creation or evolution: Do we have to choose?* Oxford: Monarch.

Astley, J., & Francis, L. J. (2010). Promoting positive attitudes towards science and religion among sixth-form pupils: Dealing with scientism and creationism. *British Journal of Religious Education, 32*(3), 189–200.

Atwater, M. M. (1996). Social constructivism: Infusion into the multicultural science education research agenda. *Journal of Research in Science Teaching, 33*, 821–837.

Barbour, I. (2000). *When science meets religion: Enemies, strangers or partners?* San Francisco: HarperCollins.

Bauser, J., & Poole, M. (2002). Science education and religious education: Possible links? *School Science Review, 85*(311), 117–124.

Bernstein, B. (1971). On the classification and framing of educational knowledge. In M. Young (Ed.), *Knowledge and control* (pp. 47–69). London: Collier-Macmillan.

Bernstein, B. (2000). *Pedagogy, symbolic control and identity.* Oxford: Rowman & Littlefield Publishers, Inc.

Berry, R. J. (1996). *God and the biologist: Faith at the frontiers of science.* Leicester: Apollos.

Billingsley, B. (2004). *Ways of approaching the apparent contradictions between science and religion*. Hobart: University of Tasmania.

Billingsley, B., Taber, K. S., Riga, F., & Newdick, H. (2012). Secondary school students' epistemic insight into the relationships between science and religion – A preliminary enquiry. *Research in Science Education*, published online. http://link.springer.com/article/10.1007%2Fs11165-012-9317-y#

Brooke, J. (1991). *Science and religion: Some historical perspectives*. Cambridge: Cambridge University Press.

Dawkins, R. (2006). *The God delusion*. London: Bantam Press.

Edwards, J. (2009). *Socio-constructivist and socio-cultural lenses on collaborative peer talk in a secondary mathematics classroom*. Paper presented at The British Society for Research into Learning Mathematics.

Fawns, R., & Sadler, J. (1996). Managing student learning in classrooms: Reframing classroom research. *Research in Science Education, 26*(2), 205–219.

Francis, L., Gibson, H., & Fulljames, P. (1990). Attitude towards Christianity, Creationism, Scientism and Interest in Science among 11–15 year olds. *British Journal of Religious Education, 13*(1), 4–17.

Fulljames, P. (1996). Science, creation and Christianity: A further look. In L. Francis, W. Kay, & W. Campbell (Eds.), *Research in religious education* (pp. 257–266). Leominster: Gracewing.

Fulljames, P., & Stolberg, T. (2000). Consonance, assimilation or correlation?: Science and religion courses in higher education. *Science & Christian Belief, 12*(1), 35–46.

Fulljames, P., Gibson, H., & Francis, L. (1991). Creationism, Scientism, Christianity and Science: A study in adolescent attitudes. *British Educational Research Journal, 17*(2), 171–190.

Fysh, R., & Lucas, K. B. (1998). Science and religion: Acknowledging student beliefs. *Australian Science Teachers Journal, 44*(2), 60–68.

Hogan, C. J. (1998). *The little book of the big bang: A cosmic primer*. New York: Springer.

Jegede, O. (1997). School science and the development of scientific culture: A review of contemporary science education in Africa. *International Journal of Science Education, 19*(1), 1–20.

Jegede, O., & Okebukola, P. A. (1991). The effect of instruction on socio-cultural beliefs hindering the learning of science. *Journal of Research in Science Teaching, 28*, 275–285.

Loughran, J., & Derry, N. (1997). Researching teaching for understanding: The students' perspective. *International Journal of Science Education, 19*, 925–938.

Roth, W. M., & Alexander, T. (1997). The interaction of students' scientific and religious discourses: Two case studies. *International Journal of Science Education, 19*(2), 125–146.

Taber, K. S., Billingsley, B., Riga, F., & Newdick, H. (2011a). Secondary students' responses to perceptions of the relationship between science and religion: Stances identified from an interview study. *Science Education, 95*(6), 1000–1025.

Taber, K. S., Billingsley, B., Riga, F., & Newdick, H. (2011b). To what extent do pupils perceive science to be inconsistent with religious faith? An exploratory survey of 13–14 year-old English pupils. *Science Education International, 22*(2), 99–118.

Tierney, J., Sinkie, E., & Gregory, J. (Eds.). (2005). *Education yearbook 2005/2206*. Harlow: Pearson Education.

Tyack, D., & Tobin, W. (1994). The 'grammar' of schooling: Why has it been so hard to change? *American Educational Research Journal, 31*(3), 453–479.

Wegerif, R., Mercer, N., & Dawes, L. (1999). From social interaction to individual reasoning: An empirical investigation of a possible socio-cultural model of cognitive development. *Learning and Instruction, 9*(6), 493–516.

Gender and Science in the Arab States: Current Status and Future Prospects

Saouma BouJaoude and Ghada Gholam

There is currently a pressing need to reform science education systems in the Arab states because of the perceived relationship between science and technology and competitiveness, wealth creation, and quality of life. This is happening at a time when the performance of a number of these states on international comparison studies in science and math such as TIMSS and PISA is weak. A myriad of reports by UNESCO (2008a), United Nations Development Program, Regional Bureau for Arab States (UNDP/RBAS) (2003), and the World Bank (2008, 2011) indicate that education in general and science education more specifically in Arab states are in a state of crisis. These reports suggest that two major problems have characterized Arab science education: access to and quality of science education. However, a closer analysis of the status of science education in Arab states shows that these two problems are multifaceted and that gender is a central factor to consider because of the apparent inequality between men and women, a situation that leads to women not having equal opportunities for success in science- and technology-related careers. Problems of quality are of the same nature and magnitude for males and females. In addition, it is doubtful whether access is truly equivalent for males and females in terms of equality in the learning process, educational outcomes, and external results and equal opportunities in employment and salaries. To address these issues, this chapter analyzes gender and science education in Arab states from a sociocultural perspective. Factors associated with this perspective have been shown to influence student achievement in general and girls' achievement more specifically, negatively or positively depending on the classroom and cultural contexts in which these girls live. This chapter starts by presenting the essential elements of the social-cultural

S. BouJaoude (✉)
Department of Education and Science and Math Education Center, American University
of Beirut, Beirut, Lebanon
e-mail: boujaoud@aub.edu.lb

G. Gholam
UNESCO Cairo Office, Cairo, Egypt

N. Mansour and R. Wegerif (eds.), *Science Education for Diversity: Theory and Practice*,
Cultural Studies of Science Education 8, DOI 10.1007/978-94-007-4563-6_17,
© Springer Science+Business Media Dordrecht 2013

perspective which is used to frame the questions and discussions. This is followed by an overview of the status of science education in the Arab state with a focus on the status of women and the sociocultural factors that constrain their ability to go beyond a certain stage in development and role in society. Finally, the chapter explicates the complex relationships between gender and science education by analyzing existing literature on the topic with the aim of identifying specific questions worthy of future investigation in Arab states.

Sociocultural Perspectives and Science Education

Teaching students about redox reactions, controlled experimentation, and acid–base reactions outside their context is a relatively inefficient and inappropriate way of learning (Lemke 2001). Teaching these and other concepts outside their social, economic, and even political contexts is not in congruence with modern views of scientific literacy and the natures of science. Lemke considers science as a human activity that is not to be viewed in separation from politics, society, and culture. To account for context of teaching and learning, the sociocultural perspectives in science education emerged as important research areas which should be taken into account while designing curricula and teaching concepts and developing views about students' understandings. According to Lemke (2001), a sociocultural perspective entails viewing "science, science education and research in science education as human social activities conducted within institutional and cultural frameworks" (p. 296) suggesting the need to consider context in which science is being taught and developed. Lemke carries this argument further by asserting that student learning should also be viewed in light of the context in which it is happening, accounting for student attitudes, motivations, interests and the like. Such a notion is also asserted by Robbins (2005) when he suggests that student thinking is rooted within a sociocultural context and is, in particular, influenced by other individuals with whom students interact.

Lemke further suggests that, from a sociocultural perspective, ignoring the various identities and attitudes of students is not healthy for their academic as well as social development. He claims that "we could succeed better at science literacy for all if we supported the much wider range of uses for science learning that fit with the lives and identities of a much larger fraction of the population" (p. 308). Similarly, Carter (2007) claims that teaching science in a fragmented and highly abstract manner is clearly irrelevant to students' contemporary lives. She emphasizes "the need for science education to develop culturally sensitive and sociocultural perspectives beyond the normative canonical knowledge and skills that have traditionally dominated its agenda" (p. 172). For Carter, it is essential to recognize all types of people in all contexts and with all their knowledge generating endeavors in this multicultural dynamic world in which we all live.

Examining the sociocultural factors that enhance or impede student learning in science revealed that various factors coalesce to determine students' overall achievement. Consequently, it is neither the cognitive abilities nor gender by themselves that enhance or impede success in science. Classroom culture and students' stereotypes greatly influence participation as well as achievement in science classes. Brand et al. (2006) affirmed that students in their study thought that only "smart" people can be high achievers in science and math. One student even expressed his belief that only white people can succeed in these subject areas. Such negative stereotypes, generated within the cultural context to which the student belongs, impede achievement in science. Another study by Cowie (2005) that investigated students' positions regarding classroom assessment revealed that these students related their performance in science to the relations between the teacher, students, and the subject of study. Students reported a more in-depth understanding and appreciation of science and attributed their success to teachers' encouragement and feedback in particular when the feedback came from teachers they respected. These findings suggest that students view learning as a social rather than an individual activity. Thus, and as claimed by Robbins (2005), adopting a sociocultural approach to research about students' understanding is of practical academic importance. This approach may shed light on student thinking which is "complex and fluid, and is constituted by many interpersonal and contextual factors" (p. 168).

To disentangle the complexity of classroom life, Von Secker and Lissitz (1999) conducted a study with tenth grade students in various schools during which they measured their higher-order thinking skills along with their understanding of various concepts using questions from biology, physics, chemistry, and earth science. Results of this study revealed that students' socioeconomic status (SES), class, and gender present a threat to achievement. Students from a low SES, females, and minorities were found to be at risk of failure. This is due to the high positive correlation established between being a female coming from a low socioeconomic background and low achievement in science. The investigators claimed that critical thinking exerts an indirect effect on achievement due to the interaction between student gender and minority status. This finding implies that, on average, females and minorities are at a higher risk of low achievement when teachers are encouraged to adopt instructional practices that emphasize critical thinking.

As indicated above, gender, class, race, language, and culture influence students' achievement negatively or positively depending on classroom context. Lemke (2001) affirmed that none of these elements has an objective definition and that "all represent misleading and harmful oversimplifications of the complexity of human similarities and differences" (p. 303). According to Lemke all these elements have their origins in politics rather than science and as a result using them in research necessitates investigating their histories beforehand. Lemke further claims that researchers are not aware of the origins of these elements while doing research due to the insufficient training they typically have in the areas from which sociocultural perspectives originate. The sociocultural areas of research that were most prominent in the past decade are those related to gender equity issues, classroom discourse, language, and minority. These will be detailed in the following sections.

Gender Equity

Spelke (2005) suggests that there is no innate ability among males toward science and math and that on the contrary, both males and females have equal cognitive abilities. Spelke also asserts that differences in achievement and cognitive profiles of males and females basically stem from "differing strategy choices" (p. 956). In this respect both should be provided with equal access to education. The United States Agency for International Development (USAID) asserts that "Gender equity entails an equal opportunity for both males and females to be granted their human rights and to participate in and benefit from economic, social, cultural and political development" (2008, p. 5). USAID carries these arguments further by affirming that despite the fact that educating boys and girls is of equal importance in developing their capabilities and increasing their opportunities, educating girls in specific is of a particular importance and leads to additional socioeconomic gains. According to this report, "the benefits include increased economic productivity, higher family incomes, delayed marriages, reduced fertility rates, and improved health and survival rates for infants and children" (p. 1).

Addressing issues of inequity requires the implementation of focused interventions that target specific identified needs. These interventions should be based on gender analysis and should encourage learning, result in systemic modifications, and transform the power dynamics between the sexes. More importantly, these interventions should be culturally sensitive and focused on a careful sociocultural analysis of the needs of boys and girls and not assume that science is a culture-free enterprise. In addition, USAID cautions against interventions in which the attention is mainly focused on insuring girls' access to education while disregarding the quality of that education because such practices put girls at a disadvantage. Furthermore, to have a lasting impact, interventions should insure equality in access to education, equality in the learning process, equality of educational outcomes, and equality of external results; these are detailed as follows:

Equality of access. Equal opportunities for boys and girls to gain admission to basic education in its various forms whether formal or informal.

Equality in the learning process. Equal opportunities to learn as well as equal attention and treatment in class. According to the USAID report this equality necessitates using the same curriculum and exposing both boys and girls to teaching methods and materials that are free of gender stereotypes and the like.

Equality of educational outcomes. Opportunities for achievement should be equal for both boys and girls. What is more important is that achievement should be based on a person's own abilities and skills and in no way be affected by gender.

Equality of external results. This implies equality in an individual's chance to gain access to various career opportunities as well as the right to have fair earnings based on qualifications.

Classroom Discourse, Language, and Minority Status

Concerning classroom discourse and language, Lemke (2001) investigated interactions in science classrooms in which he used the social and functional linguistics theory to analyze students' and teachers' utterances. This theory regards the use of language as a "socially and culturally contextualized meaning-making, in which language plays the part of a system of resources for meaningful verbal action" (p. 304). From his work emerged various recommendations including providing students with the necessary opportunities to talk and use scientific language to communicate with their teachers and with each other during classroom teaching. Lemke asserts that if differences are taken seriously, then curricula and teaching methods should be designed by considering students' class, gender, language and intellectual abilities. Thus, a sociocultural perspective recommends adopting science teaching approaches that are responsive to the different needs of students in a heterogeneous classroom. Moreover, Brand et al. (2006) asserted that negative stereotypes and students' lack of minority role models impeded achievement in science and math.

In the pages that follow, we describe the status of science education in Arab states with a focus on the role that sociocultural factors play in enhancing or hindering the success of girls in science education in Arab states. The focus will be on gender-related issues in education because of the primacy of these issues in Arab states and scarcity of research on other factors such as class and language in the educational literature in these states. Therefore, we first describe the status of science education in Arab states with special emphasis on access and quality issues in educations especially as they influence girls in science. This discussion addresses quality as demonstrated in science curricula, student learning in science (as evidenced in international comparisons such as TIMSS and PISA), results of public examinations, and assessment practices in science education. This is followed by a description of the status of Arab women in science fields and careers and the role of women in knowledge production in science, technology, and science education. Finally, we discuss (a) attempts to improve access to quality education for girls, (b) the sociocultural factors that constrain the ability of women to go beyond a certain stage in development and role in society, and (c) future directions in research that aims to understand the current situation in depth and propose real and feasible solutions to the problems associated with gender and science education.

Status of Science Education in Arab States

Advancements in science and technology are important educational goals in various Arab states (Dagher and BouJaoude 2011). Attaining these goals requires establishing reform projects aiming to develop educational systems that include

Table 1 Adult and youth literacy in a number of Arab states

	Adult literacy rates (%)		Youth literacy rates (%)	
	1990	2000–2004	1990	2000–2004
Algeria	52.9	68.9	77.3	89.9
Bahrain	82.1	88.5	95.6	98.6
Egypt	47.1	55.6	61.3	73.2
Iraq	35.7	40.0	41.0	–
Jordan	81.5	90.9	96.7	99.4
Kuwait	76.7	82.9	87.5	93.1
Lebanon	80.3	87.0	92.1	–
Libya	68.1	81.7	91.0	97.0
Mauritania	34.8	41.2	45.8	49.6
Morocco	38.7	50.7	55.3	69.5
Oman	54.7	74.4	85.6	98.5
Qatar	77.0	84.2	90.3	94.8
Saudi Arabia	66.9	77.9	85.4	93.5
Sudan	45.8	59.9	65.0	79.1
Syria	64.8	82.9	79.9	95.2
Tunisia	59.1	73.2	84.1	94.3
United Arab Emirates	71.0	77.3	84.7	91.4
Yemen	32.7	49.0	50.0	67.9

This table is adapted from Hammoud (2005)

updated curricula and quality instructional materials. Such a development should be associated with teacher development programs that aim to prepare teachers for challenges inherent in new reforms. Below we describe the status of science education in Arab states with a focus on access, quality, and knowledge production in science education. Quality is discussed from many facets including curriculum, assessment, and student learning. Finally, we analyze science education in Arab states from a sociocultural perspective.

Access to Education

According to Dagher and BouJaoude (2011), Arab states are not quite different from other developing countries in terms of access to and the quality of science education and the production of science and technology. As shown in Table 1, adult illiteracy rates are relatively high in a number of states such as Algeria, Egypt, Iraq, Mauritania, Morocco, Sudan, and Yemen. Also, youth illiteracy rates are relatively high in Egypt, Mauritania, Morocco, and Yemen.

Efforts to improve access to education have resulted in an increase in student enrollment at all educational levels in the past decades and a decrease in illiteracy rates among the population in general and among females more specifically. However, the illiteracy rates are still relatively high and this poses serious implications to the attainment of scientific and technological literacy for all (United Nations Development Program, Regional Bureau for Arab States [UNDP/RBAS] 2002, 2003; World Bank 2008).

According to the reports mentioned above, Arab states have achieved considerable strides in formal schooling for girls over the last 50 years, having accepted education as a basic human right and placed significant focus on enrollment. Compulsory public education laws enforced in most of the region's states have secured equal access to schooling and participation for girls. When compared to their counterparts in West Africa or South Asia, girls in Arab states are more likely to be enrolled in school. Yet recent evidence points out that the rapid growth in girls' school enrollment has slowed down or has even suffered a setback. Nearly one in four girls of primary school age in the Arab states is not in school. Finally, enrollment rates for girls in secondary and tertiary schooling continue to decline. Female illiteracy in the region is compounded by high dropout rates and number of girls who never enrolled in school, creating a staggering female illiteracy rate of 50 % on average for Arab women.

Quality of Science Education

Problems with science education quality are evident from the use of outdated curricula that do not focus on preparing future citizens who are capable of decision making in the twenty-first century (BouJaoude 2010). Moreover, this quality is caused by the adoption of instructional methods that emphasize theoretical science content and neglect inquiry teaching and learning and science as a way of knowing (Dagher and BouJaoude 2011), even though many science curricula and standards include explicit goals focused on inquiry and the nature of science (Dagher 2009).

Science Curricula. Dagher and BouJaoude (2011) assert, based on a research review of studies conducted in different Arab states, that curricula and teaching methods neglect students' backgrounds, interests, and motivations, fail to stimulate their creativity and imagination, and do not develop their problem-solving skills. Similarly, these studies reported that many standards and curricula in the Arab states are adopted from foreign ones without regard to the culture in which they are implemented thus affecting their quality and the ability of teachers to integrate science in everyday life. In addition, such curricula fail to integrate the use of technology in the teaching of science. On the other hand, and despite the inclusion of the nature of science (NOS) among the goals of science curricula and standards, detailed objectives and instructional activities produced and used are devoid of any mention of NOS in these curricula (BouJaoude 2002; Dagher 2009).

Student Learning in Science. In the absence of comparative data on achievement of Arab students, results of international comparisons in science and math such as TIMSS and PISA can be useful to gauge the quality of learning of students in the countries that participated in such comparisons. The number of Arab states participating in TIMSS has increased from 2 in 1999 to 12 in 2007. Results

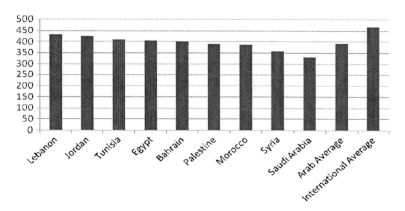

Fig. 1 Average math scores of Arab grade 8 students in TIMSS 2003

of TIMSS and PISA show that students in the Arab states scored lower than the international average on TIMSS in the years 2003 and 2007 with very few exceptions. Similarly, students who participated in PISA in 2006 and 2009[1] scored lower than the international average. These results suggest possible problems in the quality of science education at the precollege level. These problems require careful analysis of the results of international comparisons to identify factors contributing to the lower performance of students and propose possible solutions. Below is a detailed description of the results of TIMSS that includes a comparison of the performance of males and females.

TIMSS 1999. When analyzing results of TIMSS 1999 for items that exhibit gender-related differential item functioning (DIF) in math in Jordan, results revealed that all the DIF items on measurement favored male students, while most of the DIF items in algebraic and data analysis favored female students. Most of the DIF items that negatively impacted females were unfamiliar items that required some risk-taking such as estimation, expectation, or approximation. Most of the DIF items that favored females were familiar items which have one specific correct answer (Innabi and Dodeen 2006). Moreover, results of Jordanian fourth and eighth graders revealed that despite the lack of gender differences in mean achievement scores, there were significant gender differences favoring males when problem-solving skills were considered (Innabi and Dodeen 2006).

TIMSS 2003. TIMSS 2003 results (Fig. 1) indicated that Arab eighth graders scored 393 in math on average, placing them well below the international average of 467. Only a small percentage (less than 1 %) of Arab students reached the advanced international benchmark defined by TIMSS, while 45 % of the students did not reach the low international benchmark category (UNDP 2002). The gender differences in math between Arab eighth graders were negligible. At the country

---

[1]Qatar and Jordan participated in PISA in 2006 and 2009; Tunisia participated in 2003, 2006, and 2009, while Dubai participated in 2009.

level, girls outperformed boys in some countries, and boys scored higher grades in others (Lebanon, Tunisia, and Morocco), while similar achievements were attained in Egypt, Syria, Palestine, and Saudi Arabia. Among the entire pool of students participating in TIMSS, differences in achievement between boys and girls were negligible in about one third of the countries. In the remaining countries, girls had higher achievement than boys, especially in math (UNDP 2002).

Arab eighth graders' average score in science was 419, which was also below the international performance average of 474. Jordan is the only Arab country that scored above the international average by one point. As for gender differences in science, Arab girls outperformed boys. At the country level, girls had significantly higher average achievement than boys in Bahrain, Jordan, Palestine, and Saudi Arabia. Boys obtained higher average achievement in Morocco and Egypt, whereas no significant differences were found in Lebanon and Syria (UNDP 2002).

With regards to the fourth grade, only three countries participated in TIMSS 2003, namely, Tunisia, Morocco, and Yemen. Results in math were even below those of the eighth graders: students' average achievement was 321 as compared to the international average achievement of 495. A staggering 76 % of Arab fourth graders did not reach the low benchmark defined by TIMSS (UNDP 2002). Gender differences between Arab fourth graders were insignificant. Among the entire pool of students participating, differences in achievement between boys and girls were negligible in approximately half of the countries in both math and science. In the remaining countries, girls had higher achievement than boys (UNDP 2002).

PISA Results

At the outset, PISA assessment focused on reading and math. Since PISA focuses on real-world applications and out-of-school learning that seem to be gendered in nature, slightly larger gender difference were found in PISA than TIMSS because such kind of knowledge is more gender specific and accessible to boys through activities such as playing football and videogames and exploring their neighborhoods (Else-Quest et al. 2010). By 2009, PISA evolved into an internationally standardized assessment of reading, math, and science literacy for 15-year-old students, which includes a combination of multiple choice and open-ended questions (National Center for Education Statistics 2009). Qatar was the first Arab country to participate in PISA in 2006. In 2009, three Arab countries participated, namely, Qatar, Jordan, and the UAE. In PISA 2009, female students scored higher on average than male students in the combined reading literacy scale in all 65 participating countries and other education systems (National Center for Education Statistics 2009).

Results of Internal Examinations (Public Exams). Boys outperform girls on public examinations in some countries. In Sudan, for example, in each year from 1980/1981 to 1989/1990, boys achieved higher grades than girls in the primary school leaving and intermediate examinations, despite the fact that there were more boys taking the examinations than girls (Greaney and Kellaghan 1995). Research on reasons for girls' lower participation and achievement in examinations pointed to a number of factors, including cultural and religious beliefs regarding women's traditional roles, girls' obligation to carry out household chores, conflicting role expectations for girls and adolescents, and quality of schooling (Greaney and Kellaghan 1995). Recently, however, there has been increasing evidence that when girls are provided with access to education, they outperform boys in most academic subjects (Koushki et al. 1999; Queen's University 2007). According to UNICEF (2005) data for the Arab states shows that girls outperform boys in almost every academic area in the past decade. Moreover, when Arab girls are enrolled in primary school, they usually achieve higher than boys and have lower repetition rates than boys.

Assessment Practices in Science Education. Assessment practices in science education in many Arab states seem to be focused on recall and lower level cognitive questions. A study of end of secondary school public exams in Egypt, Jordan, Lebanon, Morocco, and Tunisia in addition to Iran revealed that most of these public exams focus on recall of traditional content (Valverde 2005). Only assessment goals in Lebanon and Morocco had performance expectations that included interpretation of data from investigations. Most other countries, except Lebanon, placed their performance expectations on understanding simple information. These practices still take place even though ministry of education documents in many states specify learning objectives associated with the development of knowledge, skills, and attitudes in science, the nature of science, and science technology and society (e.g., Jordanian Ministry of Education 2003).

Similarly, in the Sultanate of Oman, the evaluation guidelines categorize skills into five broad areas: initiating and planning, collecting and presenting evidence, analyzing and interpreting data, communicating and working in teams, and writing reports (Ambusaidi and Elzain 2008). These and other similar guidelines are not used appropriately and teachers in many Arab states tend to prepare students to succeed on exams following specific algorithmic criteria without any regard to the broader curriculum goals and objectives.

In summary, it is evident from results of international exams in math and science as well as public exams in specific countries that there is a trend of girls achieving as well as or better than boys in both subjects. This trend is evident also in higher education. What is unfortunate, however, is that despite these changes "Arab women remain poorly prepared to participate effectively and fruitfully in public life by acquiring knowledge through education" (UNDP 2006).

Arab Women in Science Fields and Careers

According to UNESCO (2010), school systems and curricula in Arab states generally reinforce gender bias against girls. Female students are mostly tracked into arts and humanities rather than science streams at the secondary school level. In vocational programs, females are more likely to be placed into fields like nursing, home economics, or simple bookkeeping, as opposed to the more technical fields.

The same phenomenon of relegating females to nontechnical positions persists in higher education as shown in Table 2 which presents percentages of females in the Arab states distributed according to field of study in tertiary education. The figures indicate that the field of study with the highest percentage is education, followed by humanities and social sciences. While a fair percentage of girls are studying science, this percentage is substantially lower in the fields of engineering, manufacturing, and construction, as well as, agriculture and services. The reasons for the above trends are complex. However, very little research has attempted to understand the complexity of students' views about science and investigate women's career choices within this context, an area of research that has been tackled in Western countries. For example, Haste (2004) found that students in the United Kingdom do not seem to see science as a unitary entity but rather as different "sciences" to which they relate differently. The results of her research show that individuals between the ages of 11 and 21 were found to belong to four distinct groups: "greens" interested in environmental issues but with a specific agenda, "techno-investors" enthusiastic about the potential of science, the "science oriented" keen on science as a way of thinking, and the "alienated from science" who were mostly young and female. These findings are echoed in the results of the ROSE project in Europe (Sjøberg and Schreiner 2005).

As a consequence of the above, we use research findings from other context to conjecture about these reasons in Arab states. Research indicates that some women purposely choose not to pursue careers in science and technology because they believe they will feel "cultural discomfort." Moreover, many women perceive that entering what is commonly viewed as male terrain will have a personal and social cost (University of Wisconsin 2008). When venturing into the fields of science and engineering, women find themselves in the midst of systems and performance criteria strictly designed by men for men (Loughborough University 2000). Similarly, a recent UNESCO (2011) report examined factors behind girls' reluctance to take up science and technology subjects in school and their lack of interest in pursuing careers in these fields. These factors included societal pressures placed on girls to conform to stereotypical roles and status of women and a school environment and management which can affect girls' choices and their academic performance. This report recommends revamping career guidance programs to provide needed support to women in order for them to confront the phenomenon of female underrepresentation in science and technology careers.

Surprisingly, the same factors discussed above appeared in a study conducted in the United Kingdom which concluded that the main reasons that hinder young women's advancement in science, engineering, and technology include stereotypical attitudes of girls, boys, teachers, media, and the society at large. This study also revealed the unexpected result that even those women who choose to study science at the university often end up pursuing careers in fields totally unrelated to their field of study (Loughborough University 2000). The same study showed that fear of math remains to be a factor prohibiting young women from studying physics and chemistry, even though serious efforts had been exerted to make math more accessible to girls.

S. BouJaoude and G. Gholam

Table 2 Percentages of females in the Arab states distributed according to field of study in tertiary education

Country	Education (%)	Humanities and arts (%)	Social science, business, and law (%)	Science (%)	Engineering, manufacturing, and construction (%)	Agriculture (%)	Health and welfare (%)	Services (%)	Not known or unspecified (%)
Algeria	69	75	59	61a	31	47	60	29	45
Bahrain	51	83	70	75	21	–	85	69	72
Djibouti	–	48	47	22	21	–	–	49	–
Jordan	84	63	39	51	29	54	48	53	60
Lebanon	94	67	52	53	24	54	68	53	60
Mauritania	17	24	26	21	–	–	–	–	25
Morocco	38	52	50	41	29	38	67	48	–
Oman	63	69	43	56	23	74	66	–	48
Palestinian Authority	70	66	40	46	30	18	57	31	40
Qatar	85	85	65	68	25	–	76	–	40
Saudi Arabia	73	73	53	59	2	23	44	–	24
the United Arab Emirates	92	76	55	55	29	74	80	30	70

Source: UNESCO (2010)

Even though there are serious gender-related problems in the Arab states, there are still many Arab women who have excelled in the sciences. For example, the annual L'Oreal/UNESCO Awards for Women in Science grants 5 women $100,000 each. In the period from 1998 to 2010, 5 out of the 13 recipients of this award for the Africa and Arab states region came from Arab countries. They are Egyptian immunologist Rashika El Ridi (2010), Egyptian physicist Karimat El Sayed (2004), Tunisian physicist Zohra Ben Lakhdar (2005), Habiba Bouhamed (2007), and Lihadh Al-Ghazali (2008) from the UAE (UNESCO 2010a, b). Other achievements of Arab women in science are highlighted in the Arab Human Development Report (2001).

Knowledge Production in Science, Technology, and Science Education

According to BouJaoude (2006), science and technology input indicators in the Arab states are lagging behind those of the advanced and leading developing countries. In the period 1996–2000, Arab states devoted about 0.2 % of their gross domestic product to research and development compared to industrial advanced countries like Sweden, which devoted about 3.7 % of gross domestic product to research and development during the same period (Nour 2003). While a number of Arab states such as Egypt and Saudi Arabia spend more than other Arab states, they still fall short of the amounts spent by developed and a number of developing countries. It is worth noting, however, that the expenditure of many Arab states on education is almost the same as advanced countries. Additionally, the number of scientists and engineers in research and development is low in Arab states compared to both advanced and leading developing countries like Singapore and Korea. Moreover, the majority of science and technology researchers are employed by public and university sectors, while the percentage share of private sector is very marginal. Additionally, Arab states lag behind in the percentage of students enrolled in scientific fields.

It is clear from the above that science and technology research in Arab states is not flourishing. This situation, combined with the fact that women are underrepresented in some areas such as engineering, manufacturing, and construction and agriculture (Table 2), points to the fact that women do not play a major role in knowledge production in science and technology.

When considering science education research, according to BouJaoude and Dagher (2009), there is a healthy level of science education research activity in some Arab states such as Jordan, Lebanon, Morocco, and the Sultanate of Oman. BouJaoude and Dagher have identified the following trends from a review of research studies in the four countries: (1) dominance of quantitative research methods, (2) limited access to published science education research studies published in Arabic journals and limited publication of science education research in international journals, (3) lack of attention to science learning in informal contexts and the public's understanding of science, and (4) the formulaic nature of research studies possibly because most of them were completed to satisfy promotion decisions with colleges and universities.

Attempts to Improve Access to Education for Girls

In the following pages, we describe projects and programs to address issues related to access and quality science education for women in the Arab regions. These attempts are mainly implemented by governments or private institutions in collaboration with international organizations such as UNESCO and national and regional organizations for women in science and technology. However, these projects and programs do not seem to have been evaluated and thus it is not possible to report actual impact results. We present them because they offer a promise to improve the status of science education for females in Arab states.

UNESCO's Medium-Term Strategy (2002–2007) focused on eliminating gender disparities in primary and secondary education as a means of achieving gender equality and female empowerment. A Science Career Guidance and Counseling Training Module was developed by UNESCO's Section for Science and Technology Education in response to women's underrepresentation in the field of science and technology in most developing countries, particularly in Africa (UNESCO 2008b). The module targets policy makers, teacher trainers, education and career advisors, teachers, and inspectors. Its objective is to help reduce gender disparities in science and technology and provide women with a path toward having a career in science. Specific objectives of the module include (a) promoting a positive image of women in scientific and technological careers; (b) sensitizing parents, teachers, educators, school administrative staff, curriculum developers, and trainers to counter gender stereotypes with regard to science careers; (c) improving access of girls to scientific and technological education by providing clear ideas of career opportunities; and providing teachers with the necessary career guidance tools to meet the needs of female learners seeking careers in science and technology.

One of the early regional efforts to tackle the issue of women in science in the Arab countries was the Abu Dhabi Declaration, which was adopted by the World Conference on Science 1999 in the associated meeting entitled "The Interaction of Arab Women with Science and Technology." With regards to education, the Declaration underlined the need to make science and technology more attractive for Arab girls by establishing science clubs in schools and universities and encouraging girls to join them and assume leadership positions in these clubs. Moreover, the Declaration emphasized the necessity to increase the participation of Arab women specialized in science and technology (S&T) in research and development of new technologies and creating new job opportunities for women in these fields. The Declaration tackled legal issues of Arab women in S&T, such as the need for passing new laws or amending existing ones to include incentives for the private sector to employ women in S&T fields, guaranteeing equality between men and women in

S&T wages in terms of wages and career growth and providing women in S&T with fair reward systems and retirement plans (UNESCO 1999).

In line with the recommendations of the Abu Dhabi Declaration, the Saudi Science Club established a division for women to support preuniversity science students. Moreover, the Arab Science and Technology Foundation (ASTF) in the United Arab Emirates recently established a committee specifically for women members (Islam 2007). Similarly, the King Khaled Charitable Foundation in Saudi Arabia endowed one million Saudi riyals annually to support postgraduate research by Saudi women, while Al-Nahda Society offers young Saudi women scientist scholarship for graduate and postgraduate study abroad (Islam 2007). Additionally, the Joint Supervision Program (JSP) established by the King Abdul Aziz University (KAAU) in Saudi Arabia helped local women to enroll in UK universities while working and being supervised by the Saudi staff at KAAU. The program offers women an opportunity to have international academic experiences and obtaining a PhD from a UK university while remaining with their families in Saudi Arabia, thus taking into account the cultural context of Saudi Arabia while giving women the opportunity to pursue higher education. A total of 34 women obtained their PhDs through the JSP, 68 % in science (Islam 2007).

A number of national and regional organizations for women in science and technology were established to strengthen the participation of women in these fields. They include the Arab Network for Women in Science and Technology (ANWST), which was established in Bahrain to compile and disseminate achievements by Arab women in science and technology and promote the active participation of women in science and technology careers (UNESCO 2004). ANWST attempts to help in providing access to careers in science and technology for Arab women and addressing the gender imbalance in these fields. The network also collaborates with Arab and international organizations to offer scientific and technological training for women (UNESCO 2004).

At the international level, the Academy for Educational Development (AED) developed an innovative tool which can be used in the Arab region to garner the power of social media toward increasing girls' interest in science and technology. The program, entitled *Science: It's a Girl Thing*, offers user-friendly web-based resources for conducting science activities and experiments at home. What makes the tool interactive and dynamic is that updates, videos, and links are regularly posted on certain web sites and social media sites, thereby making science activities tailored specifically for girls available outside the school setting (AED 2009).

Sociocultural Factors that Constrain the Ability of Women to go Beyond a Certain Stage in Development and Role in Society

It is evident from the description of the status of science education in Arab states that access to education has improved significantly for both males and females at all educational levels. It is also evident that when girls are given the opportunity to attend school their academic performance is superior to that of boys. However, the problem still exists in the lack of equity in the job market and in the type of specialization that girls "decide" to pursue in universities. This problem is persistent despite the fact that many regional and international organizations have developed programs focused on improving the quality of science education for girls and enhancing the opportunities of women to pursue careers in science.

It could be conjectured that cultural pressures in the Arab states are still significantly influencing the career choices of women even though more education and employment opportunities are becoming available. What is intriguing and consequently open for investigation is the fact that even though girls are achieving higher, their numbers are not increasing significantly in science- and technology-related fields. Reasons for this state of affairs are discussed below.

Several factors influence the fact that girls are participating less and show less motivation to take part in science careers. Many of these factors are associated with cultural and societal influences. The attitude of teachers, parents, classmates, and business people as well as the level of confidence of girls in their science skills might be determining the often observed gender gap in science education. Teachers sometimes, consciously or unconsciously, encourage girls to pursue nonscientific options in higher education because of the persisting belief that some careers are "feminine" and other are "masculine" (UNICEF 2003), typically associating masculinity with "hard" science- and technology-related careers. Parents typically have the same orientations and beliefs about stereotypical roles and careers of their female children: they succumb to cultural and societal norms that determine what is appropriate or inappropriate for girls to do. Using the same logic, business people promote the same type of thinking as parents and teachers. Unfortunately, girls' self-confidence in their abilities to pursue what society identifies as "masculine" careers becomes very low. It takes a very courageous women or parent to take a "road less travelled" by others. It thus is clear that to achieve gender parity in science and technology education (STE), it is important not only to motivate girls themselves but also to address the surrounding sociocultural and economic factors as well.

Another possible factor that might influence the participation of girls in science is that science educators in schools and universities have not accepted the social-cultural perspective, including the notion that science is a human activity that is not to be viewed in isolation from politics, society, and culture (Lemke 2001). Thus science might still be taught as a "culture-free" subject thus ignoring the possibility that there might be women's ways of knowing and other cultural, political, and cultural factors that influence career choices made by women.

Conclusion

It is evident from the above that science in Arab states is still male-dominated, especially in science- and technology-related careers, to a large extent even with the increasing access to education for females, the fact that females are achieving higher than males in academic science, and the significant numbers of programs to encourage females to pursue science-related specializations and careers. One of the possible reasons for the current situation could be the lack of attention to sociocultural factors that influence the choices that females make. This situation requires concerted efforts to understand the situation in depth and propose real and feasible solutions to the problem. Results of questions like the following are essential to move forward in the future:

1. To what extent is a social-cultural perspective being considered when designing curricula and instructional materials?
2. Are there any attempts to establish a gender-responsive school management system and if so what are the characteristics of such a system?
3. Are there any attempts to establish gender-responsive social and physical environments in schools and universities and if so what are the characteristics of such environments?
4. Do teachers encourage girls to opt for science subjects? If yes, what specific approaches have been used to do so? If not, what are the reasons for not doing so?
5. What specific activities do teachers organize to promote science learning for girls and for boys? If such activities are organized, what are their characteristics and results? If no such activities are organized, why not?
6. What strategies and techniques do teachers use to ensure that girls and boys participate equally in science subjects including laboratory hands-on activities?
7. What strategies and techniques do teachers use to help students, especially girls, overcome their fears, inhibitions, and lack of confidence in science subjects and careers?
8. How does the performance of girls and boys in science subjects compare in national examinations and what are the trends over the years?

The above questions are not meant to provide an exhaustive list of research questions whose answers would provide a road map to address the issue of gender bias. They are provided to emphasize the fact that gender inequity in science education and science-related careers has not been taken as seriously as it should and has not been understood well in Arab states. This lack of understanding has resulted from using findings of research conducted in contexts that are not akin to local contexts to develop programs that are meant to solve a problem that is culturally and socially bound. What is needed is research that results in home-grown programs based on culturally and socially sensitive locally produced research findings.

Acknowledgments We wish to acknowledge the support of Ms. Christine Asaad and Ms. Dina Selim during the writing of this chapter.

References

AED. (2009). *AED uses social media to show that science is 'a girl thing'*. Retrieved from http://www.aed.org/News/Releases/science_girl_thing.cfm

Ambusaidi, A., & Elzain, M. (2008). The science curriculum in Omani schools: Past, present and future. In R. Coll & N. Taylor (Eds.), *Science education in context* (pp. 85–97). Rotterdam: Sense Publishers.

BouJaoude, S. (2002). Balance of scientific literacy themes in science curricula: The case of Lebanon. *International Journal of Science Education, 24*, 139–156.

BouJaoude, S. (2006). *Bridging the gap between scientists and science educators in the Arab Region*. Report presented at the Expert Group Meeting on "Bridging the Gap between Scientists and Science Educators", organized and sponsored by the UNESCO Office, Cairo, Egypt, from January 29 – February 1, 2006. (An executive summary of this article appeared in a UNESCO Cairo Office document with the same title.)

BouJaoude, S., & Dagher, Z. (2009). Introduction: Science education in Arab states. In S. BouJaoude & Z. Dagher (Eds.), *The world of science education: Arab states* (pp. 1–8). Rotterdam. The Netherlands: Sense Publishers.

BouJaoude. (2010). *Competencies and educational structures needed to prepare secondary students for the 21st century*. Paper presented at a symposium organized by the Ministry of Education in the United Arab Emirates in cooperation with the Arab Bureau of Education for the Gulf States, February 24, 2010.

Brand, B. R., Glasson, G. E., & Green, A. M. (2006). Socio-cultural factors influencing students' learning in science and mathematics: An analysis of the perspectives of African American students. *School Science and Mathematics, 106*(5), 228–236.

Carter, L. (2007). Socio-cultural influences on science education: Innovation for contemporary times. *Science Education, 92*, 165–181.

Cowie, B. (2005). Student commentary on classroom assessment in science: A sociocultural interpretation. *International Journal of Science Education, 27*(2), 199–214.

Dagher, Z. (2009). Epistemology of science in curriculum standards of four Arab countries. In S. BouJaoude & Z. Dagher (Eds.), *The world of science education: Arab states* (pp. 41–60). Rotterdam: Sense Publishers.

Dagher, Z., & BouJaoude, S. (2011). Science education in Arab states: Bright future or status quo? *Studies in Science Education, 47*, 73–101.

Else-Quest, N., Hyde, J., & Linn, M. (2010). Cross-national patterns of gender differences in mathematics: A meta-analysis. *Psychological Bulletin, 136*(1), 103–127.

Greaney, V., & Kellaghan, T. (1995, March). *Equity issues in public examinations in developing countries* (World Bank Technical Paper Number 272). World Bank Publications, Retrieved from http://www.u4.no/document/showdoc.cfm?id=81

Hammoud, H. (2005). *Illiteracy in the Arab world: Background paper prepared for the Education for All Global Monitoring Report 2006*. Retrieved from http://unesdoc.unesco.org/images/0014/001462/146282e.pdf. The data were compiled from EFA Global Monitoring Report 2005

Haste, H. (2004). *Science in my future: A study of values and beliefs in relation to science and technology amongst 11–21 year olds*. London: Nestle Social Research Programme.

Innabi, H., & Dodeen, H. (2006). Content analysis of gender-related differential item functioning TIMSS items in mathematics in Jordan. *School Science and Mathematics, 106*(8), 328–337.

Islam, S. (2007). *Women in science: The regional perspective*. Paper presented at the International Conference on Women Leaders in Science, Technology and Engineering, Kuwait, January 2007. Retrieved from http://www.authorstream.com/Presentation/Mertice-20041-0201kuwait-presentation-si-Please-Note-Objectives-Status-Women-Arab-Region-University-Education-Scie-as-Entertainment-ppt-powerpoint/

Jordanian Ministry of Education. (2003). *General framework: Curriculum and assessment*. Amman: Author.

Koushki, P. A., Al-Sanad, H. A., & Larkin, A. M. (1999, January). Women engineers in Kuwait: Perception of gender bias. *Journal of Engineering Education, 88*, 93–97.

Lemke, J. L. (2001). Articulating communities: Socio-cultural perspectives on science education. *Journal of Research in Science Teaching, 38*(3), 296–316.

Loughborough University – United Kingdom. (2000). *Barriers to tertiary education. opportunity.* Retrieved from http://www.lboro.ac.uk/orgs/opp2000/chap2c.htm

National Center for Education Statistics. (2009). *Program for international student assessment (PISA).* Retrieved from http://nces.ed.gov/surveys/pisa/faq.asp

Nour, S. (2003, December 16–18). *Science and Technology (S & T) Development Indicators in the Arab Region: A comparative study of Arab Gulf and Mediterranean countries.* Paper Submitted for the ERF 10th Annual Conference, Morocco.

Queen's University. (2007, March 13). *Palestinian girls living in war zones outperform boys academically.* Retrieved from http://www.sciencedaily.com/releases/2007/03/070312073650. htm

Robbins, J. (2005). 'Brown Paper Packages'? A sociocultural perspective on young children's ideas in science. *Research in Science Education, 35*, 151–172.

Sjøberg, S., & Schreiner, C. (2005). How do learners in different cultures relate to science and technology? *Asia-Pacific Forum on Science Learning and Teaching, 6*(2), 1–17.

Spelke, E. S. (2005). Sex differences in intrinsic aptitude for mathematics and science? A critical review. *American Psychologist, 60*(9), 950–958.

UNDP. (2002). *Arab human development report 2002.* New York, NY: Author.

UNDP. (2006). *Towards the rise of women in the Arab World, Arab human development report 2005.* New York: Author.

UNDP/RBAS. (2002). *Creating opportunities for future generations. Arab human development report 2002.* New York: Author.

UNDP/RBAS. (2003). *Building a knowledge society. Arab human development report 2003.* New York: Author.

UNESCO. (1999). *Women, science and technology towards a new development.* Retrieved from http://unesdoc.unesco.org/images/0011/001181/118131e.pdf

UNESCO. (2004). *Arab network for women in science and technology (ANWST).* Retrieved from http://www.unesco.org/new/en/natural-sciences/science-technology/sti-policy/global-focus/gender-issues/arab-networks

UNESCO. (2008a). *EFA global monitoring report 2008: Education for all by 2015 will we make it?* Retrieved from http://unesdoc.unesco.org/images/0015/001547/154743e.pdf

UNESCO. (2008b). *Improving science education in the Arab States: Lessons learned from education practices in four developed countries.* Retrieved from http://www.unesco.org/new/fileadmin/MULTIMEDIA/FIELD/Cairo/pdf/Education/Improving_Science_Education.pdf

UNESCO. (2010a). *UNESCO world science report 2010: The current status of science around the world.* Retrieved from http://unesdoc.unesco.org/images/0018/001899/189958e.pdf

UNESCO. (2010b). Education for all global monitoring report: Reaching the marginalized. Retrieved from http://unesdoc.unesco.org/images/0018/001866/186606E.pdf

UNESCO. (2011). *EFA Global Monitoring Report: The hidden crisis: Armed conflict and education.* Retrieved from http://unesdoc.unesco.org/images/0019/001907/190743e.pdf

UNICEF. (2003). *The state of the world's children 2004.* New York: UNICEF.

UNICEF. (2005). *Progress for children: A report card on gender parity and primary education.* Retrieved from http://www.unicef.org/progressforchildren/2005n2/PFC05n2en.pdf

University of Wisconsin – Milwaukee. (2008, September 8). Tracking the reasons many girls avoid science and math. *ScienceDaily.* Retrieved July 4, 2011, from http://www.sciencedaily.com/releases/2008/09/080905153807.htm

USAID. (2008). *Education from a gender equality perspective.* Retrieved from http://www.ungei.org/resources/files/Education_from_a_Gender_Equality_Perspective.pdf

Valverde, G. (2005). Curriculum Policy seen through high-stakes examinations: Mathematics and biology in a selection of school leaving examinations from the Middle East and North Africa. *Peabody Journal of Education, 80*, 29–55.

Von Secker, C. E., & Lissitz, R. W. (1999). Estimating the impact of instructional practices on student achievement in science. *School Science and Mathematics, 36*(10), 1110–1126.

World Bank. (2008). *The road not traveled: Education reform in the Middle East and North Africa.* Retrieved from http://siteresources.worldbank.org/INTMENA/Resources/EDU_Flagship_Full_ENG.pdf

World Bank. (2011). *Learning for all: Investing in people's knowledge and skills to promote development.* Retrieved from http://siteresources.worldbank.org/EDUCATION/Resources/ESSU/Education_Strategy_4_12_2011.pdf

Author Biographies

Kristina Andersson has a Ph.D. in science education. She teaches at the teacher education program at the University of Gävle and is also a guest researcher at the Centre for Gender Research at Uppsala University in Sweden.

Berry Billingsley is a Lecturer in science education at the Institute of Education at the University of Reading, UK. Her Ph.D. looked at tertiary level students' thinking about the apparent contradictions between science and religion in the Australian context. She is now the Principal Investigator of a 3-year project in England looking at secondary school students' thinking on this area, working with Keith S. Taber, Fran Riga and Helen Newdick. The LASAR project (Learning about Science and Religion) was set up under the auspices of the Faraday Institute for Science and Religion, Cambridge.

Saouma BouJaoude graduated from the University of Cincinnati, Cincinnati, Ohio, USA in 1988 with a doctorate in Curriculum and Instruction with emphasis on science education. From 1988 to 1993 he was Assistant Professor of science education at the Department of Science Teaching, Syracuse University, Syracuse, New York, USA. In 1993 he joined the American University of Beirut (AUB). Between 2003 and 2009 he was the Chairman of the Department of Education and Professor of Science Education at AUB. Presently he is the Director of the Center for Teaching and Learning and of the Science and Math Education Center at AUB. Dr. BouJaoude has published numerous research articles in international journals such as the *Journal of Research in Science Teaching*, *Science Education*, *International Journal of Science Education*, *Journal of Science Teacher Education*, *The Science Teacher*, and *School Science Review*, among others. In addition, he has written chapters in edited books in English and Arabic and has been an active presenter at local, regional and international education and science education conferences. Dr. BouJaoude presently serves on the editorial boards of the *Journal of Science Teacher Education* and *International Journal of Science and Mathematics Education*, *Eurasia Journal of Mathematics, Science, and Technology Education*, is a consulting editor for *International Review of Education*, a contributing

N. Mansour and R. Wegerif (eds.), *Science Education for Diversity: Theory and Practice*, 359
Cultural Studies of Science Education 8, DOI 10.1007/978-94-007-4563-6,
© Springer Science+Business Media Dordrecht 2013

international editor for *Science Education*, and a reviewer for *Journal of Research in Science Teaching*. Dr. BouJaoude has been involved in educational project in Dubai, Jordan, Egypt, Oman, Qatar, Saudi Arabia, in addition to Lebanon.

MeiLing Chow has been a secondary school teacher for 6 years, specializing in science and mathematics education for middle school. Her teaching is guided by a commitment to education for sustainability and by her Buddhist ethics. She is currently completing a master of philosophy degree at the Science and Mathematics Education Centre, Curtin University. Email: meiling_003@hotmail.com. Address: Science and Mathematics Education Centre, Curtin University, GPO Box U1987, Bentley, Western Australia 6845.

Lynn D. Dierking is Sea Grant Professor in Free-Choice Science, Technology, Engineering and Mathematics (STEM) Learning, College of Science, and Interim Associate Dean for Research, College of Education, Oregon State University. Her research expertise involves lifelong STEM learning, particularly free-choice, out-of-school time learning (in after-school, home- and community-based organizations), with a focus on youth, families, and communities, particularly those underrepresented in STEM. She is currently working on two research projects: an NSF-funded investigation of the long-term impact of gender-focused free-choice learning experiences on girls' interest, engagement, and involvement in science and a Noyce Foundation-funded 4-year longitudinal study of the STEM learning ecologies of 10-year-olds, in school and out of school. Lynn has published extensively and serves on the Editorial Boards of *Journal of Research in Science Teaching* and the *Journal of Museum Management and Curatorship*. She received a 2010 John Cotton Dana Award for Leadership from the American Association of Museums.

Michiel van Eijck (M.Sc. Biology 1993, Ph.D. Science Education 2006, University of Amsterdam) is Assistant Professor of Science Education at the Eindhoven University of Technology. His research centers on the re-production of scientific knowledge in educational systems worldwide. He published in major academic journals and is on the editorial board of several others. His latest books are *Cultural Studies and Environmentalism: The Confluence of Ecojustice, Place-Based (Science) Education, and Indigenous Knowledge Systems* (with Deborah Tippins, Michael Mueller, and Jennifer Adams) and *Authentic Science Revisited: In Praise of Diversity, Heterogeneity, Hybridity* (with Wolff-Michael Roth, Giuliano Reis, and Pei-Ling Hsu).

Sibel Erduran is Professor of Science Education at University of Bristol, UK. She has a Ph.D. from Vanderbilt University, M.Sc. from Cornell University, B.A. from Northwestern University, and has taught science and chemistry in a secondary school in northern Cyprus. She has published widely on the applications of nature of science in science education including argumentation in professional development of science teachers. She is the Science Studies Section Co-Editor for *Science Education Journal* and serves on the Editorial Boards of numerous other journals including the *International Journal of Science Education*. She is the recipient of NARST Best Paper Award and currently serves on the Executive Board of NARST.

Mariona Espinet is a Professor of science education at the Departament de Didàctica de la Matemàtica i de les Ciències Experimentals de la Universitat Autònoma de Barcelona, Spain. She began her career as a secondary science teacher and later on moved to the USA to undertake her doctoral work in science education. Her current research interests are in modeling and language in science education, environmental education and education for sustainability in school and communities, and science teacher education and development at all educational levels: infant, primary and secondary from sociocultural and sociolinguistic perspectives. Email: Mariona.Espinet@uab.cat. Address: Dep. Didàctica de la Matemàtica i de les Ciències Experimentals, Edifici G-5, Despatx 120, Campus de Bellaterra s/n, 08193 Cerdanyola del Vallès, Barcelona, Spain.

Ghada Gholam obtained her master's degree from the American University of Beirut in Lebanon in 1975 while a research assistant at the Science and Mathematics Education Center (SMEC) of the University. She joined UNESCO Regional Office for Education in the Arab States in Lebanon from 1980 to 1985. Between 1986 and 1988 she was appointed as expert manager for the Hariri Foundation Testing Project in Lebanon and developed the arithmetic proficiency tests for secondary students. In Sana'a, Yemen she joined the British Council as examination manager until 1990. Dr. Gholam graduated from King's College London, London University, UK, in 1997 with a doctorate in Mathematics Education. She joined UNESCO Cairo Office in 1998 as program specialist in education responsible for all education programs in Egypt, Sudan and Yemen, and as advisor for science education in the Arab Region. She contributed to the training and capacity building of a large number of science and mathematics teachers, as well as to the production of training material for trainers and teachers, in addition to workshops for the promotion of Science and Math education in the Arab Region through IT. Dr. Gholam has many publications in international journals and is the author of several book chapters dealing with various topics in education.

Annica Gullberg holds a Ph.D. in genetics. During the last 8 years her research interest has been in science education. Annica is a Senior Lecturer at the University of Gävle and a guest researcher at the Centre for Gender Research at Uppsala University.

Lindsay Hetherington is a Lecturer in science education at Graduate School of Education at University of Exeter. She has worked in teaching and teacher education for 10 years, beginning as a Teacher of Science and Chemistry in the UK state school system before moving to Exeter University. She teaches on the PGCE Secondary Science course and professional M.Ed. courses. Her research interests include using complexity theory to explore teaching, learning and curriculum, with a particular emphasis on science teaching and science teacher education.

Anita Hussénius is Associate Professor in Organic Chemistry at the University of Gävle and Director of the Centre for Gender Research at Uppsala University, Sweden. She also coordinates the research program *Nature/Culture Boundaries and Transgressive Encounters*, the aim of which is to study empirically and reflect theoretically on the ways in which knowledge about gender and gendered knowledge are produced in the intersection between the natural and cultural sciences. Her main research interest is about gender perspectives on science education and, more specifically, how an increased awareness of gender issues in science and in science teaching influences prospective teachers' identities as teachers and their teaching of science.

Martin Kracheel studied as a researcher with a strong focus on the dynamics of multimodal interactions on learning and development across educational, professional and everyday settings.

He currently works as a research assistant at the Technische Universität Kaiserslautern in a research project about bilingual language acquisition with Prof. Shanley Allen. Furthermore he works as a German teacher for immigrants for the NGO "Amitié Portugal Luxembourg".

Before he worked as a Research Associate in the CODI-SCILE-A (Competences for Organizing Discourse-In-Interaction & Science Learning: Analyzing knowledge building as activity of collaborative inquiring) research project.

He graduated from the trilingual (English, French, German) Master Program "Learning and Development in Multilingual and Multicultural Contexts", at the University of Luxembourg. He holds a Bachelor degree as a teacher for Political Sciences and French Philology from the Freie Universität Berlin.

Gerald H. Krockover received his B.A., M.A., and Ph.D. degrees from The University of Iowa. He is a former middle and high school science teacher and currently holds a joint appointment as an Emeritus Professor between the Department of Curriculum and Instruction in the College of Education and the Department of Earth and Atmospheric Sciences in the College of Science at Purdue University, West Lafayette Indiana, USA. Since joining the Purdue University faculty in 1970, Prof. Krockover has published 11 books, 124 refereed journal manuscripts, and has been awarded nearly $7 million in external funding. His specialty areas for his research include differentiated science instruction, informal learning, and equity issues in science education.

Xiufeng Liu is Professor of science education at the State University of New York–Buffalo. His research interests include science assessment, student conceptual progression from elementary to high school, and science education policies related to opportunities-to-learn. He publishes regularly in *Journal of Research in Science Teaching* and *Science Education*. He is also the author of three most recent books: *Essentials of Science Classroom Assessment* (Sage, 2009), *Linking Competence to Opportunities to Learn* (Springer, 2009), and *Using and Developing Measurement Instruments in Science Education: A Rasch Modeling Approach* (IAP, 2010).

Nasser Mansour is a Senior Lecturer in science education at the Graduate School of Education at Exeter University and works at Faculty of Education, Tanta University, Egypt. He is Fellow of the Higher Education Academy (HEA). Dr. Mansour graduated from University of Exeter, UK, in 2008. His Ph.D. looked at teachers' beliefs and practices about STS with emphasis on the interaction between cultural issues e.g. religious beliefs and the teaching of science. Dr. Mansour published in prestigious education journals such as *Science Education, International Journal of Science Education, Journal of Science Teacher Education, Cultural Studies of Science Education, Research in Science Education, Computer and Education* and *European Educational Research*. Dr. Mansour has been awarded the best paper Award at the European Educational Research Association conference in 2007 and The University of Exeter Merit Award in 2011. Dr. Mansour has been involved in international educational projects in Egypt, Saudi Arabia and UK. Dr. Mansour is a member of national organizations including ESERA, NARST, ASTE, BAAS and BERA and is currently associate editor of the journal *Thinking Skills and Creativity* and was President of the Junior Researcher JURE 2011 Pre-conference of European Association of Research on Learning and Instruction (EARLI) Conference held at Exeter. For further information see http://education.exeter.ac.uk/staff_details.php?user=nm259

Charles Max is a Professor in Educational Sciences, specializing in the learning sciences focusing on learning with educational media at the University of Luxembourg. Having a background in early learning and special needs education within multicultural contexts with a 10 years record as a practitioner in the field (teacher, inspector), he is a specialist in the area of socio-cultural theory, activity theory and developmental work research, focusing on competence development, the monitoring of professional growth and the implementation of change and development in diversified settings. He is co-heading the working group on science learning for the Luxembourgish pre-primary, primary and lower secondary curricula planning.

Alun Morgan taught Geography and Integrated Science in Secondary schools in England and Wales for 10 years before becoming Education for Sustainable Development Officer for Worcestershire County Council, a teacher advisory role. From 2002 he worked as Lecturer in Geography Education in the Institute of Education, London. He moved to London South Bank University in January 2009 to take up the post of Director of the Education for Sustainability Program. Currently he is a Research Fellow at the Graduate School of Education, University of Exeter where he works on the FP7 Science Education for Diversity project. His work focuses on the interface between geography and science education, environmental education and education for sustainability and global citizenship.

Loran Carleton Parker received her B.S. in atmospheric science, her M.A. in science education, and her Ph.D. in science education from Purdue University. She has formerly taught courses at Purdue University for pre-service elementary teachers, for informal learners, and has served as a research specialist for externally funded science education projects. She received her Ph.D. degree in 2009 and currently holds an appointment in the Discovery Learning Research Center at

Purdue University as an assessment specialist. Her dissertation was entitled, "The Use of Zoo Exhibits by Family Groups to Learn Science." She has published six refereed journal manuscripts and has made 11 refereed presentations at national meetings. Her specialty areas include atmospheric chemistry, the nature of science in science education, and informal science education.

Keith Postlethwaite began his career as a physics teacher in a large comprehensive secondary school in England. While there he completed a research project on ability grouping which led him into a research post in the University of Oxford. He subsequently moved to teaching and research posts in Reading and the University of the West of England. In 1999 he moved to the University of Exeter where he is now Associate Professor of Education. His teaching is focused on initial teacher education in physics, and on research methodology and methods. His research interests are in professional learning and in science education. He is particularly interested in viewing these fields through the lens of activity theory and in researching them through the use of mixed methods.

Michael J. Reiss is Pro-Director: Research and Development and Professor of Science Education at the Institute of Education, University of London, Chief Executive of Science Learning Centre London, Vice President and Honorary Fellow of the British Science Association, Honorary Visiting Professor at the Universities of Birmingham and York, Honorary Fellow of the College of Teachers, Docent at the University of Helsinki, Director of the Salters-Nuffield Advanced Biology Project, a member of the Farm Animal Welfare Committee and an Academician of the Academy of Social Sciences. For further information see www.reiss.tc.

Silvia Lizette Ramos-De Robles is a Lecturer of biology and environmental health sciences in the Department of Environmental Sciences at the University of Guadalajara. She began her career as an elementary school teacher and later on moved to pre-service and in-service teacher education. Her current research interests focuses on sociological aspects around the language development in the teaching and learning of science, especially in multilingual contexts. She has looked it through sociocultural and sociolinguistic perspectives in order to understand the scientific literacy practices in depth manner. Her research also focuses on socio-environmental aspects related with Environmental Health and Climate Change Education. Email: lramos@cucba.udg.mx Address: Las Agujas, Km 15.5 Carretera Guadalajara-Nogales. Zapopan, Jalisco, México. C. P. 45110

Wolff-Michael Roth is Research Professor and learning scientist at Griffith University. For the past 20 years, he has studied knowing and learning of mathematics and science across the course of the life span and in formal (kindergarten, school, university) and informal settings (workplace, activism). He has published over 40 books, 170+ chapters, and 350 peer-reviewed journal articles. His latest publications include *Passibility: At the Limits of the Constructivist Metaphor* (Springer, 2011), *Geometry as Objective Science in Elementary Classrooms: Mathematics in the Flesh* (Routledge, 2011), and *Language, Learning, Context: Talking the Talk* (Routledge, 2010).

Kathryn Scantlebury is a Professor in the Department of Chemistry and Bio-chemistry at the University of Delaware, Director of Secondary Education in the College of Arts and Sciences and Coordinator of Science Education. She taught high school chemistry, science and mathematics in Australia before completing her doctorate at Purdue University. Her research interests focus on gender issues in various aspects of science education, including urban education, preservice teacher education, teachers' professional development, and academic career paths in academe. She recently co-edited two books, *Re-visioning Science Education from Feminist Perspectives: Challenges, Choices and Careers* and *Coteaching in International Contexts: Research and Practice*. Scantlebury is the Research Director for the National Science Teachers Association and a Fellow of the American Association for the Advancement of Science.

Nigel Skinner, B.Sc. Ph.D. PGCE taught Science and Biology at state secondary schools in England for 10 years before moving into teacher education in 1990. He is currently a senior lecturer and PGCE secondary science course leader at the University of Exeter where he also carries out research into science teacher education and development using Cultural Historical Activity Theory as a theoretical framework. He works with beginning teachers as they enter the profession and also with experienced teachers from many different countries as they work towards master's degrees and doctorates.

Keith S. Taber worked as a science teacher in state comprehensive schools and in further education, earning his master's and doctoral degrees through part-time study, before joining the Faculty of Education at Cambridge, where he is a Senior Lecturer. He was the Royal Society of Chemistry's Teacher Fellow for 2000–2001. His research is focused on aspects of science learning, particularly related to conceptual understanding and integration. He has written widely on science education, and his books include *Progressing Science Education* (Springer, 2009), *Science Education for Gifted Learners* (as editor, Routledge, 2007), and *Chemical Misconceptions* (Royal Society of Chemistry, 2002). He is currently researching secondary students' perceptions of the relationship between science and religion.

Elisabeth (Lily) Taylor (nee Settelmaier) holds a Ph.D. from Curtin University where she currently specializes as a Lecturer in curriculum studies in the School of Education. Her research focuses on socially responsible science and sustainability education and uses auto/ethnographic methodologies. She is particularly interested in social and cultural aspects of secondary schooling. Elisabeth is an adjunct lecturer of Ibaraki University, Japan, and of Curtin's Science and Mathematics Education Centre. Email: Elisabeth.taylor@curtin.edu.au Address: School of Education, Curtin University, GPO Box U1987, Bentley, Western Australia 6845.

Peter Charles Taylor is Associate Professor of Transformative Education at the Science and Mathematics Education Centre, a graduate centre at Curtin University, Western Australia. One of his main research interests is education for biocultural diversity. Peter draws on a range of theoretical perspectives, including critical constructivism, reconceptualist curriculum theory, research as reflective/imaginative praxis, the cultural/linguistic nature of science and mathematics, postcolonial theorizing and integral philosophy. Email: p.taylor@curtin.edu.au Address: Science and Mathematics Education Centre, Curtin University, GPO Box U1987, Bentley, Western Australia 6845.

Norman Thomson is an Associate Professor in Mathematics and Science Education at the University of Georgia. He taught and has worked in East Africa for many years and his research interests include hominin evolution, climate change, and indigenous science knowledge. His endeavor is to explore – what do we teach in science and why?

Deborah J. Tippins is a Professor in the Department of Mathematics and Science Education at the University of Georgia where she specializes in science for K-8 learners. She draws on anthropological and sociocultural frameworks to study questions of social and ecojustice in science teaching and learning. Building on her experience as a Fulbright scholar in the Philippines, her current research focuses on citizen science, socio-scientific issues, and culturally relevant pedagogy. She is the co-author of six books, the most recent entitled *Cultural Studies and Environmentalism: The Confluence of Ecojustice, Place-Based (Science) Education and Indigenous Knowledge Systems*.

Rupert Wegerif is Professor of Education and Director of Research at the Graduate School of Education in the University of Exeter in the UK. He has researched and published widely on dialogic approaches to education usually with technology and sometimes in relation to teaching thinking. He is currently directing and working on several projects investigating (1) talk in classrooms, (2) tools to support computer supported collaborative learning and (3) dialogue across cultural differences in science education. He is co-editor of *Thinking Skills and Creativity*, an international journal with Elsevier.

Siu Ling Wong obtained her Ph.D. degree in physics from University of Oxford. She then worked as a research fellow in Clarendon Laboratory and a college lecturer at Exeter College, Oxford. She taught physics at Diocesan Girls' School after she moved back to Hong Kong. She is currently an Associate Professor in the Faculty of Education, The University of Hong Kong. Her key research interests include nature of science and teacher professional development. She serves on the editorial boards of *Science and Education* and *Canadian Journal of Science Education*, and an executive member of the East-Asian Association for Science Education.

Gudrun Ziegler is specializing in the research domains of plurilinguism, language acquisition and applied linguistics and directs the trilingual, 2-years research Master program "Learning and Development in Multilingual and Multicultural Contexts" at the University of Luxembourg (http://www.multi-learn.org).

Author Index

Subject Index

N. Mansour and R. Wegerif (eds.), *Science Education for Diversity: Theory and Practice,*   371
Cultural Studies of Science Education 8, DOI 10.1007/978-94-007-4563-6,
© Springer Science+Business Media Dordrecht 2013